Lecture Notes in Mathematics

1994

Editors:
J.-M. Morel, Cachan
F. Takens, Groningen
B. Teissier, Paris

W0090836

Árpád Baricz

Generalized Bessel Functions of the First Kind

 Springer

Árpád Baricz
Babes-Bolyai University
Department of Economics
Cluj-Napoca 400084
Romania
bariczocsi@yahoo.com

ISBN: 978-3-642-12229-3   e-ISBN: 978-3-642-12230-9
DOI: 10.1007/978-3-642-12230-9
Springer Heidelberg Dordrecht London New York

Lecture Notes in Mathematics ISSN print edition: 0075-8434
                        ISSN electronic edition: 1617-9692

Library of Congress Control Number: 2010926688

Mathematics Subject Classification (2000): 33C05, 33C10, 33C15, 33C75, 30C45, 26D05, 26D07, 39B62

Cover design: SPi Publisher Services

Printed on acid-free paper

springer.com

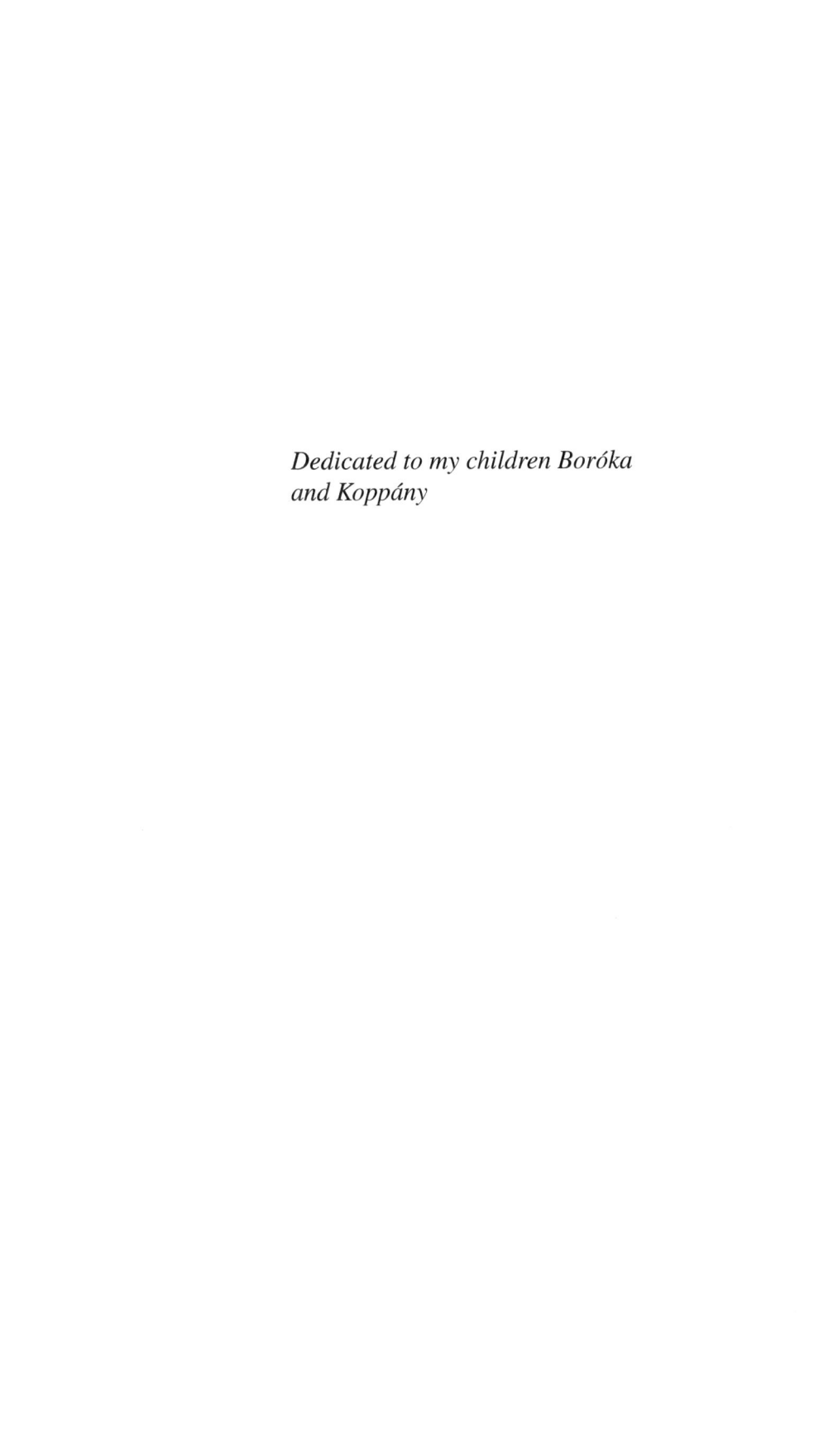

Dedicated to my children Boróka and Koppány

Preface

Bessel functions are indispensable in many branches of mathematics and applied mathematics. Thus, it is important to study their properties in many aspects. Recently, there has been a vivid interest on Bessel and hypergeometric functions from the point of view of geometric function theory and functional inequalities. Although many inequalities and geometric properties of Bessel and hypergeometric functions appear in works of many mathematicians, there is no unified treatment of the topic. I have written this monograph in order to partially fill this gap in the literature. The major part of this monograph is taken from my Ph.D. thesis [54] with the same title as this monograph and that is why most results are due to myself and my coauthors. The literature has grown very quickly during the past few years and everything could not have been covered. I have tried to follow closely the structure of my thesis.

Most of my papers used in this monograph were supported partially by the Institute of Mathematics, University of Debrecen, Hungary and some of these papers were completed during my visit to University of Debrecen. I take this opportunity to thank this institution for its excellent research facilities and to thank Péter T. Nagy for his constant encouragement during the course of my work. My research was also partially supported by the János Bolyai Research Scholarship of the Hungarian Academy of Sciences since September 2009. I would also like to thank this Institution for the financial support.

I would like to express my gratitude to Szilárd András, Wolfgang W. Breckner, Edward Neuman, Saminathan Ponnusamy and Matti Vuorinen for carefully reading the earlier versions of the manuscript and for their numerous constructive suggestions and helpful remarks. I am also indebted to the referees for their very constructive comments, and to the staff of Springer for their assistance.

Finally, I thank my family, children Boróka and Koppány and especially my wife Katalin, for support and love.

Cluj-Napoca
January 2010

Árpád Baricz

Contents

Notation

Following is a list of notation used in this book:

$\mathbb{N}, \mathbb{R}, \mathbb{C}$	Set of natural, real and complex numbers
$\mathbb{D}$	Open unit disk in the complex plane
$(a)_n$	Pochhammer (or Appell) symbol
$F(a,b,c,z)$	Gaussian hypergeometric function
$\mathscr{K}(x)$	Complete elliptic integral of the first kind
$\Phi(a,c,z)$	Kummer hypergeometric function
$_qF_r(a_1,\ldots,a_q,b_1,\ldots,b_r,z)$	Generalized hypergeometric function
J_p, I_p	Bessel and modified Bessel function of the first kind of order p
j_p, i_p	Spherical and modified spherical Bessel function of the first kind of order p
w_p	Generalized Bessel function of the first kind of order p
u_p	Generalized and normalized Bessel function of the first kind of order p
$\Gamma(z)$	Euler's gamma function
$B(p,q)$	Euler's beta function
$\log$	Natural logarithm function on $(0,\infty)$
Log	Principal branch of the logarithm function
γ	Euler-Mascheroni constant
$\mathscr{S}, \mathscr{S}^*, \mathscr{C}$	Class of univalent, starlike and convex functions
$\mathscr{S}^*(\alpha), \mathscr{C}(\alpha)$	Class of starlike and convex functions of order α
$\mathscr{H}^\mu$	Hardy space of analytic functions
$*$	Hadamard product $(f * g)$ of two power series
$\sim$	Asymptotic to $(f(x) \sim g(x)$ as $x \to a$ if $\lim_{x \to a} f(x)/g(x) = 1)$
$\mathscr{O}(r)$	Landau order symbol
$\sin, \cos, \sinh, \cosh$	Sine, cosine, hyperbolic sine, hyperbolic cosine

$\mathrm{Re}\,z, \mathrm{Im}\,z$ Real and imaginary part of the complex number z
$A(r,s), G(r,s), H(r,s)$ Arithmetic, geometric and harmonic mean of r and s
$A_2(r,s)$ Second-order power mean of r and s
$H_p(r,s)$ p-order power mean (or Hölder mean) of r and s
$j_{p,n}$ nth positive zero of the Bessel function J_p
$\mathrm{Si}(x), \mathrm{Shi}(x)$ Sine and hyperbolic sine integral

Survey

The aim of this brief survey is to give the reader a short overview of the topics discussed in this book.

Special functions. Special functions are some particular mathematical functions which have more or less established names and notations due to their importance in mathematical analysis, functional analysis, physics, or other applications. However, there is no general formal definition, but the list of mathematical functions contains functions which are commonly accepted as special. In particular, elementary functions, especially trigonometric functions are also considered as special functions. The theory of special functions has been developed essentially in the nineteenth century due to the contributions of C.F. Gauss, C.G.J. Jacobi, F. Klein, and many others. Because of their remarkable properties, special functions have been used for centuries. For example, since they have numerous applications in astronomy, trigonometric functions have been studied for over a thousand years. However, in the twentieth century the theory of special functions has been overshadowed by other fields such as real and functional analysis, topology, algebra and differential equations. All the same, in that times appeared G.N. Watson's book [227] entitled "A treatise on the theory of Bessel functions," which is a very important book in the theory of special functions, especially in the asymptotic expansions of Bessel functions. Nowadays this book is a classic and because of their remarkable properties special functions, like Bessel and hypergeometric functions are frequently used in probability, statistics, mathematical physics and engineering sciences. This is why the Hungarian mathematician Paul Turán, as Richard Askey said, considered "special functions" a misnomer: they should be called useful functions.

Special functions in geometric function theory. Special functions play an important role in geometric function theory. Maybe the most known application is the solution of the famous Bieberbach conjecture by L. de Branges. In the proceedings of the 1986 meeting to celebrate the proof of L. de Branges' theorem, L.V. Ahlfors [4] wrote:

> To my knowledge, de Branges may be the first who has tapped the rich reservoir of knowledge hidden in the volumes on special functions and sometimes relegated to a corner of the library, and applied it to the coefficient problem. In this case, what he used was original work by contemporary mathematicians, but it raises the question whether special functions, as a powerful tool, have not been unduly neglected.

The surprising use of generalized hypergeometric functions by L. de Branges has generated considerable interest, and the geometric properties of the generalized, Gauss and Kummer hypergeometric functions have been investigated by many authors in the last few decades. Although the geometric properties of these functions are interesting in their own right, they have shown to be useful in many other problems in geometric function theory. For example, recently the starlikeness of the normalized Gaussian hypergeometric functions was used by J.H. Choi et al. [82] in order to deduce sharp   norm estimates for the Alexander transforms of convex functions of order alpha. The investigation of the geometric properties of Bessel functions – discussed in details in this book – was motivated by the immense research about hypergeometric functions and some further researches are in progress.

Functional inequalities involving special functions. In 1991 M. Hazewinkel, as the editor of Kluwer Academic Publishers (Dordrecht/Boston/London) wrote:

> Inequalities are everywhere. Whole series of conferences are devoted to them. Indeed in my more despondent moments, when struggling with one or another problem, I sometimes have the feeling that mathematics (especially analysis) is all inequalities.

The functional inequalities which involve hypergeometric functions, especially elliptic integrals are very useful in quasiconformal analysis, and many of them have been deduced to solve some interesting problems in this topic. A large number of such inequalities has been collected by G.D. Anderson et al. in [19], which is the first book containing a systematic and detailed treatment of Gaussian hypergeometric functions from the point of view of inequalities. Some of these inequalities have been motivated naturally by the beautiful identities of elliptic integrals and nowadays the research of such inequalities is in active progress.

Since the Bessel and modified Bessel functions can be viewed as generalizations of sine, cosine, hyperbolic sine and hyperbolic cosine functions, it is natural to ask whether many identities and inequalities involving these trigonometric functions can be extended to Bessel functions. Although, this problem is interesting in its own right, recently it has been shown that such inequalities can be useful in engineering sciences, such as information and communication theory (see for example the papers of Á. Baricz and Y. Sun [56, 61]).

Topics discussed in this book. In this book our aim is to study the generalized Bessel functions of the first kind from the point of view of the complex analysis on the one hand, and on the other hand from the point of view of the classical analysis. The study of these functions from the point of view of the complex analysis is mostly related to univalence, starlikeness, convexity, close-to-convexity and subordination property, while the study from the point of view of the classical analysis is related to the extensions of some trigonometric inequalities, like Mitrinović, Mahajan, Redheffer, Cusa, Jordan, Grünbaum, Landen, van der Corput, Askey, to generalized Bessel functions of the first kind. Moreover, we deduce some chain of inequalities for Gaussian and Kummer hypergeometric functions, generalized Bessel functions, modified Bessel functions, general power series with positive coefficients.

Chapter 1
Introduction and Preliminary Results

Abstract We begin with a brief outline of Bessel functions, which will be needed in the next chapters. We recall here the Bessel, modified Bessel, spherical Bessel and modified spherical Bessel functions and define the generalized Bessel function. Some general properties of generalized Bessel functions are discussed in this chapter. These include: recursive formulas, differentiation formula, integral representations. We recall here also the Gaussian hypergeometric function with its basic properties which will have applications in Chap. 3. Finally, at the end of this chapter we list some classical inequalities which will be used in the sequel.

1.1 Overview

Let $(a)_0 = 1$ for $a \neq 0$ and let $(a)_n$ be the Pochhammer (or Appell) symbol defined by $(a)_n = a(a+1)\ldots(a+n-1)$ for all $n \geq 1$. The (Gaussian) hypergeometric function (series)

$$F(a,b,c,z) = {}_2F_1(a,b,c,z) = \sum_{n \geq 0} \frac{(a)_n (b)_n}{(c)_n} \frac{z^n}{n!}, \quad |z| < 1, \tag{1.1}$$

where $a, b, c \in \mathbb{C}$ and $c \neq 0, -1, -2, \ldots$, was studied by L. Euler about two centuries ago (for historical remarks see the papers of J. Dutka [91] and J. Kampé de Fériet [127]). C.F. Gauss's work on this series starting in 1812 stimulated great interest in this function. Among the first researchers of this subject were L. Fuchs, E.E. Kummer, B. Riemann, H.A. Schwarz and F. Klein.

By using (1.1) we can verify that actually $z \mapsto F(a,b,c,z)$ is a particular solution of the second-order hypergeometric differential equation

$$z(1-z)w''(z) + (c - (a+b+1)z)w'(z) - abw(z) = 0.$$

If the parameters a and b are negative integers or 0, then the solution of the above differential equation is a polynomial and hence analytic in the whole plane. For the other values of the parameters the series (1.1) converges absolutely for $|z| < 1$. It also

Á. Baricz, *Generalized Bessel Functions of the First Kind*,
Lecture Notes in Mathematics 1994, DOI 10.1007/978-3-642-12230-9_1,
© Springer-Verlag Berlin Heidelberg 2010

converges absolutely for $|z| = 1$ if $\text{Re}(a+b-c) < 0$; it converges conditionally for $|z| = 1, z \neq 1$ if $0 \leq \text{Re}(a+b-c) < 1$ and diverges for $|z| = 1$ if $\text{Re}(a+b-c) \geq 1$.

We next recall some known properties of the Gaussian hypergeometric function. Let us start with the behavior of the function $z \mapsto F(a,b,c,z)$ near 1.

1. If $\text{Re}(c-a-b) > 0, \text{Re}\,c > 0, \text{Re}(c-a) > 0$ and $\text{Re}(c-b) > 0$, then (see the book of E.D. Rainville [197, p. 49])

$$\lim_{z \to 1} F(a,b,c,z) = \frac{\Gamma(c)\Gamma(c-a-b)}{\Gamma(c-a)\Gamma(c-b)},$$

where Γ denotes Euler's gamma function defined by Euler's integral

$$\Gamma(z) = \int_0^\infty e^{-t} t^{z-1}\, dt \quad \text{for all } \text{Re}\,z > 0.$$

2. If $\text{Re}\,a > 0, \text{Re}\,b > 0, \text{Re}(a+b) > 0$ and if $\text{Re}(c-a-b) = 0$, then we have the formula (see the paper of R.J. Evans [97]) due to C.F. Gauss:

$$\lim_{z \to 1} \frac{F(a,b,a+b,z)}{\text{Log}(1-z)} = -\frac{\Gamma(a+b)}{\Gamma(a)\Gamma(b)}.$$

3. For $\text{Re}(c-a-b) < 0, \text{Re}\,a > 0, \text{Re}\,b > 0, \text{Re}\,c > 0$ the corresponding formula (see the book of E.T. Whittaker and G.N. Watson [228, p. 299]) is

$$\lim_{z \to 1}(1-z)^{a+b-c} F(a,b,c,z) = \frac{\Gamma(c)\Gamma(a+b-c)}{\Gamma(a)\Gamma(b)}.$$

Here and throughout in this book $\text{Log}\,z$ is the principal branch of the logarithm of z. Namely, if z is a nonzero complex number, then the principal branch of the logarithm of z is given by $\text{Log}\,z = \text{Log}\,|z| + i\theta$, where $\theta \in (-\pi, \pi]$ is the argument of z and $\text{Log}\,|z|$ denotes the usual natural logarithm of the positive real number $|z|$. Similarly, here and throughout in the sequel every many-valued function is taken with the principal branch. It is important to note that many special functions may be written in terms of the Gaussian hypergeometric function. Extensive lists of such particular cases are given almost in each classical book about special functions (see for example the books of M. Abramowitz and I.A. Stegun [1], G.E. Andrews et al. [26] and N.M. Temme [223]).

Following is a short list of particular cases for appropriate values of the arguments:

$$F(1,1,2,-z) = \frac{1}{z}\text{Log}(1+z),$$

$$F(a,b,b,z) = (1-z)^{-a},$$

$$F\left(\frac{1}{2},1,\frac{3}{2},z^2\right) = \frac{1}{2z}\text{Log}\left(\frac{1+z}{1-z}\right),$$

$$F\left(\frac{1}{2},\frac{1}{2},\frac{3}{2},z^2\right) = \frac{1}{z}\arcsin z,$$

$$F\left(\frac{1}{2},1,\frac{3}{2},-z^2\right) = \frac{1}{z}\arctan z.$$

We note that the complete elliptic integrals of the first and second kinds are also particular cases of the Gaussian hypergeometric function. More precisely, the Legendre complete elliptic integrals of the first and second kinds are the functions $z \mapsto \mathscr{E}(z)$ and $z \mapsto \mathscr{K}(z)$ defined for $|z| < 1$ by

$$\mathscr{E}(z) = \int_0^{\pi/2}\sqrt{1 - z^2\sin^2\theta}\,d\theta = \int_0^1\sqrt{\frac{1 - z^2t^2}{1 - t^2}}\,dt = \frac{\pi}{2}F\left(-\frac{1}{2},\frac{1}{2},1,z^2\right),$$

and

$$\mathscr{K}(z) = \int_0^{\pi/2}\frac{d\theta}{\sqrt{1 - z^2\sin^2\theta}} = \int_0^1\frac{dt}{\sqrt{(1 - t^2)(1 - z^2t^2)}} = \frac{\pi}{2}F\left(\frac{1}{2},\frac{1}{2},1,z^2\right),$$

respectively. Some other beautiful properties of Gaussian hypergeometric functions, like Ramanujan's differentiation formula and Ramanujan's refined asymptotic formula can be found in the first two sections of Chap. 3. For more information about these functions, for example Legendre and Elliott identities we refer to the papers of E.A. Karatsuba and M. Vuorinen [130], R. Balasubramanian et al. [32] and to references therein (see also the first section of Chap. 3).

The (Gaussian) generalized hypergeometric function (series) is defined as follows

$$_qF_r(a_1,\ldots,a_q,b_1,\ldots,b_r,z) = \sum_{n\geq 0}\frac{(a_1)_n\ldots(a_q)_n}{(b_1)_n\ldots(b_r)_n}\frac{z^n}{n!}, \qquad (1.2)$$

where $q \geq 1$ and $r \geq 1$ are integers, while $a_1,a_2,\ldots,a_q, b_1,b_2,\ldots,b_r$ are complex numbers such that $b_k \neq 0,-1,-2,\ldots$, for all $k \in \{1,2,\ldots,r\}$. We note that the series in (1.2) converges absolutely for $|z| < \infty$ if $q < r + 1$ and for $|z| < 1$ if $q = r + 1$. In particular for $q = 2$, $r = 1$, $a_1 = a$, $a_2 = b$ and $b_1 = c$ the series in (1.2) becomes the series in (1.1) and usually is denoted by $_2F_1(a,b,c,z)$ or simply by $F(a,b,c,z)$. Moreover, we note that a series $_{r+1}F_r(a_1,\ldots,a_{r+1},b_1,\ldots,b_r,z)$ is called balanced (or zero-balanced) if $a_1 + \ldots + a_{r+1} = b_1 + \ldots + b_r$.

The generalized hypergeometric function (series) plays an important role in function theory, especially in the solution by L. de Branges [77] of the famous Bieberbach conjecture: If $f(z) = z + a_2z^2 + a_3z^3 + \ldots$ is analytic and univalent in $|z| < 1$, then $|a_n| \leq n$ for all $n \in \{2,3,4,\ldots\}$. The surprising use of this class of functions has prompted renewed interest in function theory in the last few decades. This is evidenced by the almost 1530 papers listed just in the last ten years in Mathematical Reviews under the topic "hypergeometric function." For an extensive bibliography and history see the papers of G. Almkvist and B.C. Berndt [7], G.D. Anderson et al. [19], G.E. Andrews et al. [26].

In 1961 E.P. Merkes and W.T. Scott [151] investigated the starlikeness and univalence of Gaussian hypergeometric functions by using continued-fraction representations, while in 1984 B.C. Carlson and D.B. Shaffer [80] obtained a dense subset of certain classes of hypergeometric functions for each of the following classes: the convex functions of order α, the starlike functions of order α, and the prestarlike function of order α. In 1986 S. Ruscheweyh and V. Singh [201] obtained the exact order of starlikeness of $z \mapsto zF(a,b,c,\rho z)$ when $a > 0$, $c = a+1$, $-1 \leq \rho b \leq 1 + \rho a$ and estimated, using the continued fraction of C.F. Gauss, the order of starlikeness when $0 \leq a \leq b \leq c$. In 1987 S. Owa and H.M. Srivastava [168] investigated the geometric properties of generalized hypergeometric functions, defined by (1.2), using the well known Jack's lemma (see Lemma 2.4 below). In 1990 S.S. Miller and P.T. Mocanu [154] employed the method of differential subordinations [152, 153, 155] to investigate the local univalence, starlikeness and convexity of certain hypergeometric functions. H. Silverman [212] in 1993 investigated also the starlikeness and convexity of Gaussian hypergeometric functions, while in 1996 J.H. Choi et al. [84] presented some generalizations of the results of S.S. Miller and P.T. Mocanu, and gave some applications involving generalized hypergeometric functions associated with the Hardy space of analytic functions. In 1998 S. Ponnusamy and M. Vuorinen [176, 187, 188] presented also some generalizations of the results of S.S. Miller and P.T. Mocanu, and determined conditions of close-to-convexity of Gaussian and confluent (or Kummer) hypergeometric functions. Some important results on Gaussian and Kummer hypergeometric functions related to the papers of S. Ponnusamy and M. Vuorinen [176, 187, 188] were obtained in 1997 by A.P. Acharya [2]. In 2002 and 2007 R. Küstner [139, 140] by using among others the continued fraction of C.F. Gauss determined the order of convexity of hypergeometric functions $z \mapsto F(a,b,c,z)$ as well as the order of starlikeness of shifted hypergeometric functions $z \mapsto zF(a,b,c,z)$, for certain ranges of the real parameters a,b and c.

Further various other properties of the Gaussian hypergeometric functions, defined by (1.1), were examined under diverse conditions on the parameters a,b and c. We refer to papers by R. Balasubramanian et al. [37], Y.C. Kim and S. Ponnusamy [133], I.R. Nezhmetdinov and S. Ponnusamy [162], S. Ponnusamy [177–179], S. Ponnusamy and F. Rønning [181–184], S. Ponnusamy and S. Sabapathy [185], T.N. Shanmugam [209], H.M. Srivastava et al. [215], A. Swaminathan [217–221] and references therein for some important results. Hereafter there is an enormous research of G.D. Anderson et al. [11, 15, 17, 19, 21], R. Balasubramanian et al. [36], S. Ponnusamy and M. Vuorinen [186], S.L. Qiu and M. Vuorinen [194–196] about the Gaussian hypergeometric functions for real variable (see also the references therein).

When $r = q+1$, $a_i = b_i$ for each $i \in \{1, 2, \ldots, q\}$ and $b_r = \kappa$, then the generalized hypergeometric function $_qF_r(a_1, \ldots, a_q, b_1, \ldots, b_r, \cdot)$ is denoted by $_0F_1(\kappa, \cdot)$. In other words, we have

$$_0F_1(\kappa, z) = \sum_{n \geq 0} \frac{1}{(\kappa)_n} \frac{z^n}{n!}, \tag{1.3}$$

where $\kappa \in \mathbb{C}$ such that $\kappa \neq 0, -1, -2, \ldots$. This hypergeometric function is related to an obvious transform of Bessel functions, which was studied earlier, for example by R. Askey [28], E. Neuman [161] and V. Selinger [208].

In this monograph our aim is to continue the above mentioned investigations for this special case of generalized hypergeometric function (series), i.e. for the function (series) denoted by (1.3). A concept of generalized Bessel function is introduced, see Definition 1.1 on page 10, and the relations (1.15), (1.16). This generalization allows to handle the classical Bessel functions – defined by (1.5) – the modified Bessel functions – defined by (1.7) – the spherical Bessel functions – defined by (1.9) – the modified spherical Bessel functions – defined by (1.11) – and some other classes of special functions together by a simple manipulation of the parameters.

The book is divided into three chapters. The first chapter contains the definition of three classes of functions: the generalized Bessel functions of order p (Definition 1.1), the generalized Bessel functions of the first kind and order p (see (1.15)) and the generalized and normalized Bessel functions of the first kind and order p (see (1.20)). Here the basic properties of these functions such as power series representations, recursions (Lemma 1.1 on page 11 and Lemma 1.2 on page 14) and integral representations (Lemma 1.3 on page 16) are included. From these results we obtain as a special case the classical formulas for Bessel functions (see the book of G.N. Watson [227]). The original results from the second section of this chapter were published by S. András and Á. Baricz [24], Á. Baricz [52], Á. Baricz and E. Neuman [58]. Note that at the end of this chapter we list some classical inequalities like the Cauchy–Bunyakovsky–Schwarz inequality, Hölder-Rogers inequality, Jensen inequality, Hermite-Hadamard inequality and Chebyshev integral inequality. These inequalities will be used in the sequel.

In the first two sections of the second chapter the following geometric properties of the generalized and normalized Bessel functions are studied:

1. univalence (Theorems 2.1, 2.2, 2.3, 2.7 on pages 27, 29, 29, 38);
2. starlikeness (Theorems 2.8, 2.9, 2.11, 2.12, 2.13, 2.14 on pages 39, 41, 46, 48, 51);
3. convexity (Theorems 2.10, 2.11, 2.12, 2.13, 2.15 on pages 40, 46, 48, 53);
4. close-to-convexity (Theorems 2.4, 2.5 on pages 30, 33).

In proving the results of these sections several methods are used. One of the main tools is the method of differential subordination developed by S.S. Miller and P.T. Mocanu [152, 153], but there the Alexander transform [5] is also used together with L. Fejér's result [98], Jack's lemma [124] and the results of S. Owa and H.M. Srivastava [168], H. Silverman [210], W. Kaplan [129] and S. Ozaki [167]. In the last section of this chapter there are given some conditions on the parameters which guarantee that the generalized and normalized Bessel function belongs to a certain Hardy space (see Theorems 2.18, 2.19 on pages 59, 59). Moreover in this section we give (as another immediate application of convexity) a monotonicity property of generalized and normalized Bessel functions of complex order (see Theorem 2.22 and Corollary 2.14 on page 66). The original results from this chapter were published by S. András and Á. Baricz [24], and Á. Baricz [38, 40, 42, 43, 52].

The third chapter is devoted to the study of some inequalities involving:

1. generalized Bessel functions of the first kind (Theorems 3.1, 3.7, 3.13, 3.20, 3.21, 3.22, 3.27, 3.28, 3.30, 3.31, 3.33, 3.34, 3.37 on pages 80, 91, 100, 113, 115, 117, 129, 132, 137, 142, 144, 157);
2. modified Bessel functions of the first kind (Theorems 3.23, 3.24, 3.25, 3.26, 3.36, 3.39, 3.41 on pages 120, 121, 125, 127, 152, 162, 176);
3. Kummer hypergeometric functions (Theorems 3.8, 3.9 on page 93);
4. Gaussian hypergeometric functions (Theorems 3.2, 3.4, 3.5, 3.6, 3.15, 3.16, 3.17 on pages 82, 85, 86, 88, 105, 106);
5. general power series with positive coefficients (Theorems 3.3, 3.10, 3.14, 3.18 on pages 83, 95, 102, 108).

The first three sections of this chapter were motivated by two open questions proposed by G.D. Anderson et al. [17]. There a new technique is developed based on the study of the logarithmic convexity of some functions using a lemma proved by S. Ponnusamy and M. Vuorinen [186]. This approach enables one to sharpen some older results of S.L. Qiu and M. Vuorinen [195] and to discuss simply and solely some inequalities in a very general frame (see Theorems 3.3, 3.10, 3.14 on pages 83, 95, 102). In the fourth section we investigate the convexity of zero-balanced Gaussian hypergeometric functions and general power series with respect to Hölder means (see the paper of the author [48]). Here our main motivation is to extend the results obtained in the second section. In the fifth section some Askey type inequalities are proved (see Theorem 3.20 on page 113) and there some new lower and upper bounds are given for generalized Bessel functions (see Theorem 3.22 on page 117). In this section the advantage of the Gauss-Gegenbauer quadrature formula is exploited and the integral representations play an important role here (see the paper of Á. Baricz and E. Neuman [58]). In the sixth section some results concerning the intrinsic behavior of the modified Bessel functions (see for example Theorem 3.24 on page 121) are proved. These complete an earlier work of E. Neuman [159] and they were obtained by the author in a joint work with E. Neuman [59]. Moreover, at the end of this section we extend an inequality of J.G. van der Corput to modified Bessel functions.

In the seventh section of this chapter we extend some known classical inequalities (see the paper of Á. Baricz [49]), like Mahajan's inequality, Mitrinović's inequality, improvements of Jordan's inequality, Redheffer's inequality for generalized and normalized Bessel functions using an adequate integral representation and the monotone form of l'Hospital's rule. Moreover, in order to extend and improve many known results in the literature we establish some sharp Jordan and Kober type inequalities for Bessel and modified Bessel functions of the first kind by using the monotone form of l'Hospital's rule, and by using the classical Cauchy mean value theorem inductively we deduce new series expansions for the Bessel and modified Bessel functions (see Theorems 3.35, 3.36 and 3.37 on pages 145, 152 and 157).

Finally in the last section, by using mathematical induction and infinite product representations of the functions $\mathscr{J}_p : \mathbb{R} \to (-\infty, 1]$ and $\mathscr{I}_p : \mathbb{R} \to [1, \infty)$, defined by

$$\mathscr{J}_p(x) = 2^p \Gamma(p+1) x^{-p} J_p(x) \text{ and } \mathscr{I}_p(x) = 2^p \Gamma(p+1) x^{-p} I_p(x),$$

an extension of Redheffer's inequality for the function $\mathscr{J}_p$ and a Redheffer-type inequality for the function $\mathscr{I}_p$ are established. Moreover, by using some known results on the zeros of Bessel functions of the first kind sharp exponential Redheffer-type inequalities are established for $\mathscr{J}_p$ and $\mathscr{I}_p$. Here J_p and I_p, denotes the Bessel function and modified Bessel function, while Γ stands for the Euler gamma function. At the end of this section a lower bound for the Γ function is deduced, using Euler's infinite product formula. Our main motivation to write this section are the publications of C.P. Chen et al. [81], L. Zhu and J. Sun [243], which we wish to complement.

Although the methods and notions in this chapter are mostly from the classical analysis, the whole chapter is rooted on a series of recent articles of G.D. Anderson et al. [11, 17, 21], R. Balasubramanian et al. [36, 37], E. Neuman [159, 161], S. Ponnusamy and M. Vuorinen [186], S.L. Qiu and M. Vuorinen [194, 195]. The original results of this chapter were already published by Á. Baricz [39, 41, 45, 47–49], Á. Baricz and E. Neuman [58, 59], and Á. Baricz and S. Wu [62, 63].

Finally, we note that all the results of this book were taken also from the Ph.D. thesis [54] of the author, except the main results of Sects. 3.7.5 and 3.8.2, which were published recently by Á. Baricz and S. Wu [62, 63].

1.2 Generalized Bessel Functions of the First Kind

Bessel functions are named after the astronomer F.W. Bessel (1784–1846), however, D. Bernoulli is generally credited with being the first to introduce the concept of Bessel's functions in 1732. Bernoulli used the Bessel function of zero order as a solution to the problem of an oscillating chain suspended at one end. In 1764 L. Euler employed Bessel functions of integer order in an analysis of vibrations of a stretched membrane, an investigation which was further developed by L. Rayleigh in 1878, where he demonstrated that Bessel's functions are particular cases of Laplace's functions. Bessel, while receiving named credit for these functions, did not incorporate them into his work as an astronomer until 1817. The Bessel function was the result of Bessel's study of a problem of Kepler for determining the motion of three bodies moving under mutual gravitation. In 1824, he incorporated Bessel functions in a study of planetary perturbations where the Bessel functions appear as coefficients in a series expansion of the indirect perturbation of a planet, that is the motion of the Sun caused by the perturbing body. It was likely Lagrange's work on elliptical orbits that first suggested to Bessel to work on the Bessel functions. The notation $J_{z,n}$ was first used by P.A. Hansen [111] and subsequently by O.X. Schlömilch [207] and later modified to $J_n(2z)$ by G.N. Watson. Subsequent studies of Bessel functions included the works of G.B. Mathews in 1895, "A treatise on Bessel functions and their applications to physics" written in collaboration with A. Gray. It was the first major treatise on Bessel functions in English and covered topics such as applications of Bessel functions to electricity, hydrodynamics and diffraction. In 1922, G.N. Watson first published his comprehensive examination of

Bessel functions "A treatise on the theory of Bessel functions." Nowadays, this book is a classic and a wide range of problems concerning the most important areas of mathematical physics and various technical problems are linked into application of Bessel functions. Bessel function theory is often used when solving, for example, problems of hydrodynamics, acoustics, radio physics, atomic and nuclear physics, information theory. These functions are also an effective tool for problem solving in areas of wave mechanics and elasticity theory. Bessel functions are an inexhaustible subject, there are always more useful properties than one knows.

On the first page of the famous book (second edition) of G.N. Watson [227] we can read the following:

> The theory of Bessel functions is intimately connected with the theory of a certain type of differential equation of the first order, known as Ricatti's equation. In fact a Bessel function is usually defined as a particular solution of a linear differential equation of the second order (known as Bessel's equation) which is derived from Ricatti's equation by an elementary transformation.

Let us consider the second-order differential equation (see for example the book of G.N. Watson [227, p. 38])

$$z^2 w''(z) + z w'(z) + (z^2 - p^2) w(z) = 0, \tag{1.4}$$

which is called Bessel's equation, where p is an unrestricted (real or complex) number. The function J_p, which is called the Bessel function of the first kind of order p, is defined as a particular solution of (1.4). This function has the form [227, p. 40]

$$J_p(z) = \sum_{n \geq 0} \frac{(-1)^n}{n! \cdot \Gamma(p+n+1)} \left(\frac{z}{2}\right)^{2n+p} \quad \text{for all } z \in \mathbb{C}. \tag{1.5}$$

The modified Bessel differential equation [227, p. 77]

$$z^2 w''(z) + z w'(z) - (z^2 + p^2) w(z) = 0, \tag{1.6}$$

which differs from Bessel's equation (1.4) only in the coefficient of w, is of frequent occurrence in problems of mathematical physics (see for example the function v_p defined by (3.71), which is of special interest in finite elasticity). A particular solution of (1.6) is called the modified Bessel function (or Bessel function with imaginary argument) of the first kind of order p, and is defined by the formula [227, p. 77]

$$I_p(z) = \sum_{n \geq 0} \frac{1}{n! \cdot \Gamma(p+n+1)} \left(\frac{z}{2}\right)^{2n+p} \quad \text{for all } z \in \mathbb{C}. \tag{1.7}$$

The second-order differential equation (see the book of M. Abramowitz and I.A. Stegun [1, p. 437])

$$z^2 w''(z) + 2 z w'(z) + [z^2 - p(p+1)] w(z) = 0, \tag{1.8}$$

which differs from equations (1.4) and (1.6) in the coefficient of w' and w, is called the spherical Bessel equation. One of its particular solutions is called the spherical Bessel function of the first kind of order p and is defined by the formula [1, p. 437]

$$j_p(z) = \sqrt{\frac{\pi}{2z}} J_{p+\frac{1}{2}}(z) = \sqrt{\frac{\pi}{2}} \sum_{n \geq 0} \frac{(-1)^n}{n! \cdot \Gamma\left(p+n+\frac{3}{2}\right)} \left(\frac{z}{2}\right)^{2n+p} \quad \text{for all } z \in \mathbb{C}. \quad (1.9)$$

Finally, let us consider the second-order differential equation [1, p. 443]

$$z^2 w''(z) + 2zw'(z) - [z^2 + p(p+1)]w(z) = 0, \quad (1.10)$$

which differs from equation (1.8) only in the coefficient of w, and is called the modified spherical Bessel equation. One of its particular solutions is called the modified spherical Bessel function of the first kind of order p and is defined by the formula (see the book of M. Abramowitz and I.A. Stegun [1, p. 437])

$$i_p(z) = \sqrt{\frac{\pi}{2z}} I_{p+\frac{1}{2}}(z) = \sqrt{\frac{\pi}{2}} \sum_{n \geq 0} \frac{1}{n! \cdot \Gamma\left(p+n+\frac{3}{2}\right)} \left(\frac{z}{2}\right)^{2n+p} \quad \text{for all } z \in \mathbb{C}. \quad (1.11)$$

Our goal is to study the Bessel functions, the modified Bessel functions, the spherical Bessel functions and the modified spherical Bessel functions together. For this we consider the linear differential equation

$$z^2 w''(z) + bzw'(z) + \left(cz^2 + d\right) w(z) = 0, \quad (1.12)$$

where $b, c \in \mathbb{C}$ and $d = d_1 p^2 + d_2 p + d_3$, with $d_1, d_2, d_3, p \in \mathbb{C}$. Due to our notations by using the Frobenius method we can seek the solution of equation (1.12) in the following form:

$$w(z) = z^p \left(\sum_{n \geq 0} a_n z^n\right),$$

where $a_n \in \mathbb{C}$ for all $n \geq 0$. It is easy to show that we have the following recursion between the coefficients a_n and a_{n-2} of the above infinite power series for all $n \geq 2$:

$$a_n \cdot n\left(n+2p+b-1\right) + a_n \left[(d_1+1)p^2 + (b+d_2-1)p + d_3\right] = -c \cdot a_{n-2}.$$

By putting $d_1 = -1$, $d_2 = 1-b$ and $d_3 = 0$, the recursion becomes

$$a_n \cdot n[n+2p+b-1] = -c \cdot a_{n-2} \quad \text{for all } n \geq 2, \quad (1.13)$$

and the differential equation (1.12) is

$$z^2 w''(z) + bzw'(z) + \left[cz^2 - p^2 + (1-b)p\right] w(z) = 0. \quad (1.14)$$

This equation generalizes the equations (1.4), (1.6), (1.8) and (1.10), and permits the study of Bessel, modified Bessel, spherical Bessel and modified spherical Bessel functions together. Using the recursive relation (1.13), we obtain a particular solution of (1.14) for all $z \in \mathbb{C}$

$$w(z) = a_0(p) \sum_{n \geq 0} \frac{(-c/4)^n z^{2n+p}}{n! \cdot \prod_{m=1}^{n} \left(m+p+\frac{b-1}{2}\right)}$$

$$= a_0(p) \sum_{n \geq 0} \frac{(-c/4)^n \Gamma\left(p+\frac{b+1}{2}\right)}{n! \cdot \Gamma\left(n+p+\frac{b+1}{2}\right)} z^{2n+p},$$

where $a_0(p) \neq 0$. If $a_0(p) = \left[2^p \Gamma\left(p+\frac{b+1}{2}\right)\right]^{-1}$, we denote the above particular solution with w_p, so we obtain

$$w_p(z) = \sum_{n \geq 0} \frac{(-c)^n}{n! \cdot \Gamma\left(p+n+\frac{b+1}{2}\right)} \left(\frac{z}{2}\right)^{2n+p} \quad \text{for all } z \in \mathbb{C}. \qquad (1.15)$$

We note that by the ratio test, the radius of convergence of the series $w_p(z)$ is infinity, and thus $w_p(z)$ is convergent for all $b, c, p, z \in \mathbb{C}$.

Definition 1.1. (Á. Baricz [52]) Any solution of the linear differential equation (1.14) is said to be a generalized Bessel function of order p. The particular solution w_p defined by (1.15) is called the generalized Bessel function of the first kind of order p.

It is clear that for $c = 1$ and $b = 1$ the function w_p reduces to J_p, when $c = -1$ and $b = 1$ the function w_p becomes I_p. Similarly, when $c = 1$ and $b = 2$ the function w_p reduces to $2j_p/\sqrt{\pi}$, while if $c = -1$ and $b = 2$, then w_p becomes $2i_p/\sqrt{\pi}$. Further, from (1.15) we have $w_p(0) = 0$. From (1.15) it follows that for all $z \in \mathbb{C}$

$$w_p(z) = \left[2^p \Gamma\left(p+\frac{b+1}{2}\right)\right]^{-1} z^p \sum_{n \geq 0} \frac{(-c/4)^n \Gamma\left(p+\frac{b+1}{2}\right)}{n! \Gamma\left(n+p+\frac{b+1}{2}\right)} z^{2n}. \qquad (1.16)$$

Set

$$u_p(z) = \sum_{n \geq 0} b_n z^n \quad \text{for all } z \in \mathbb{C}, \qquad (1.17)$$

where

$$b_n = \frac{(-c/4)^n \Gamma\left(p+\frac{b+1}{2}\right)}{n! \cdot \Gamma\left(n+p+\frac{b+1}{2}\right)} \quad \text{for all } n \geq 0. \qquad (1.18)$$

Then (1.16) yields

$$w_p(z) = \left[2^p \Gamma \left(p + \frac{b+1}{2} \right) \right]^{-1} z^p u_p(z^2) \quad \text{for all } z \in \mathbb{C}. \tag{1.19}$$

Taking into consideration that $(a)_n = \Gamma(a+n)/\Gamma(a)$, we obtain for the function u_p the following form

$$u_p(z) = \sum_{n \geq 0} \frac{(-c/4)^n}{(\kappa)_n} \frac{z^n}{n!}, \tag{1.20}$$

where b, p, c are complex numbers such that $\kappa = p + (b+1)/2 \neq 0, -1, -2, \ldots$. The function u_p is called the generalized and "normalized" Bessel function[1] of the first kind of order p (cf. Á. Baricz [52]). In fact this function is an elementary transform of the hypergeometric function $_0F_1(\kappa, \cdot)$, defined by (1.3), i.e. it is easy to see that $u_p(-4z/c) = {}_0F_1(\kappa, z)$. We note that by the ratio test, the radius of convergence of the series $u_p(z)$ is infinity. Moreover, the function u_p is analytic in $\mathbb{C}$ and satisfies the differential equation

$$4z^2 u''(z) + 4\kappa z u'(z) + czu(z) = 0. \tag{1.21}$$

In what follows we present some basic properties of generalized and normalized Bessel functions of the first kind u_p like recursive relations, differentiation formula and integral representations. These results are natural generalizations of some known formulas for Bessel and modified Bessel functions of the first kind and will be used frequently in the following chapters.

The next lemma will be used to prove a recursive relation for the function u_p, which will be applied for the study of the univalence of this function (see the proof of Theorem 2.2).

Lemma 1.1. (Á. Baricz [52]) *For the generalized Bessel function w_p of the first kind of order p the following recursive relations are true for all $b, p, c \in \mathbb{C}$ such that $\kappa \neq 0, -1, -2, \ldots$ and for all $z \in \mathbb{C}$:*

(a) $zw_{p-1}(z) + czw_{p+1}(z) = (2p+b-1)w_p(z)$.
(b) $zw_p'(z) + (p+b-1)w_p(z) = zw_{p-1}(z)$.
(c) $zw_p'(z) + czw_{p+1}(z) = pw_p(z)$.
(d) $[z^{-p}w_p(z)]' = -cz^{-p}w_{p+1}(z)$, *where both sides of the above relation are replaced by their limiting values when $z = 0$.*

---

[1] It is known that a function $f(z) = z + \sum_{n \geq 2} a_n z^n$, which is analytic, is called a normalized function. That is we have $f(0) = 0$ and $f'(0) = 1$. Now for the function u_p it is clear that $u_p(0) = b_0 = 1$ and $u_p'(0) = 0$. In fact the function $(u_p(z) - b_0)/b_1$ or zu_p is normalized and not the function u_p, that is why the word normalized appears between quotation marks. Therefore throughout this paper when we refer to the generalized and normalized Bessel functions, we always think about the function $_0F_1(\kappa, -cz/4) = u_p(z)$.

Proof. (a) Computing the expression $w_{p-1}(z) + w_{p+1}(z)$ we have

$$\sum_{n\geq 0} \frac{(-1)^n c^n}{n!\,\Gamma(\kappa+n-1)} \cdot \left(\frac{z}{2}\right)^{2n+p-1} + \sum_{n\geq 0} \frac{(-1)^n c^n}{n!\,\Gamma(\kappa+n+1)} \cdot \left(\frac{z}{2}\right)^{2n+p+1}$$

$$= \frac{1}{\Gamma(\kappa-1)}\left(\frac{z}{2}\right)^{p-1} + \sum_{n\geq 1}\left[\frac{(-1)^n c^n}{n!\,\Gamma(\kappa+n-1)} + \frac{(-1)^{n-1} c^{n-1}}{(n-1)!\,\Gamma(\kappa+n)}\right]\cdot\left(\frac{z}{2}\right)^{2n+p-1}$$

$$= \frac{2(\kappa-1)}{z}\left[\frac{(z/2)^p}{\Gamma(\kappa)} + \sum_{n\geq 1}\frac{(-1)^n c^n (z/2)^{2n+p}}{n!\,\Gamma(\kappa+n)}\right.$$

$$\left. + \frac{1}{\kappa-1}\sum_{n\geq 1}\frac{(-1)^n c^{n-1}(c-1)}{n!\,\Gamma(\kappa+n)}\cdot\left(\frac{z}{2}\right)^{2n+p}\right].$$

Thus we obtain

$$w_{p-1}(z) + w_{p+1}(z)$$

$$= \frac{2(\kappa-1)}{z}\left[w_p(z) + \frac{1}{\kappa-1}\cdot\sum_{n\geq 0}\frac{(-1)^{n+1}c^n(c-1)}{n!\,\Gamma(\kappa+n+1)}\cdot\left(\frac{z}{2}\right)^{2n+p+2}\right]$$

$$= \frac{2(\kappa-1)}{z}\left[w_p(z) + \frac{z}{2(\kappa-1)}(1-c)w_{p+1}(z)\right],$$

which implies $z[w_{p-1}(z) + cw_{p+1}(z)] = (2p+b-1)w_p(z)$.

(b) Analogously, if we compute the expression $w_{p-1}(z) - w_{p+1}(z)$, we obtain

$$\frac{2p+b-1}{2\Gamma(\kappa)}\left(\frac{z}{2}\right)^{p-1} + \sum_{n\geq 1}\frac{(-1)^n c^n\left(\kappa+n-1+\frac{n}{c}\right)}{n!\,\Gamma(\kappa+n)}\cdot\left(\frac{z}{2}\right)^{2n+p-1}$$

$$= \frac{p}{2\Gamma(\kappa)}\left(\frac{z}{2}\right)^{p-1} + \frac{p+b-1}{2\Gamma(\kappa)}\left(\frac{z}{2}\right)^{p-1} + \sum_{n\geq 1}\frac{(-1)^n c^n\left(n+\frac{p}{2}\right)}{n!\,\Gamma(\kappa+n)}\cdot\left(\frac{z}{2}\right)^{2n+p-1}$$

$$+ \sum_{n\geq 1}\frac{(-1)^n c^n(p+b-1)}{2\cdot n!\,\Gamma(\kappa+n)}\left(\frac{z}{2}\right)^{2n+p-1} + \sum_{n\geq 1}\frac{(-1)^n c^{n-1}}{(n-1)!\,\Gamma(\kappa+n)}\cdot\left(\frac{z}{2}\right)^{2n+p-1}$$

$$= w_p'(z) + \frac{p+b-1}{z}w_p(z) - w_{p+1}(z),$$

and this is equivalent to the second recursive relation.

(c) Combining the recursive relations (a) and (b), we obtain

$$zw_p'(z) + (p+b-1)w_p(z) + czw_{p+1}(z) = (2p+b-1)w_p(z),$$

which leads to the desired equality.

(d) Clearly, when $z = 0$, from (1.16) we have

$$\left[z^{-p} w_p(z) \right]' \Big|_{z=0} = \left[\sum_{n \geq 0} a_0(p) b_n(p) z^{2n} \right]' \Big|_{z=0}$$

$$= \left[2 \sum_{n \geq 1} a_0(p) b_n(p) n z^{2n-1} \right] \Big|_{z=0} = 0$$

and

$$\left[z^{-p} w_{p+1}(z) \right] \Big|_{z=0} = \left[\sum_{n \geq 0} a_0(p) b_n(p+1) z^{2n+1} \right] \Big|_{z=0} = 0,$$

where in view of (1.18)

$$b_n(p) = \frac{(-c/4)^n \Gamma\left(p + \frac{b+1}{2} \right)}{n! \cdot \Gamma\left(n + p + \frac{b+1}{2} \right)} \quad \text{for all} \quad n \geq 0.$$

Now, suppose that $z \neq 0$. Then, using the third recursive relation, we obtain

$$\left[z^{-p} w_p(z) \right]' = z^{-p-1} \left[z w_p'(z) - p w_p(z) \right] = -c z^{-p} w_{p+1}(z).$$

$\square$

Remark 1.1. Since for $c = b = 1$ ($c = -1$, $b = 1$ respectively) the function w_p reduces to J_p (I_p respectively), we get from Lemma 1.1 the following recurrence formulae (for all $p \neq -1, -2, -3, \ldots$ and $z \in \mathbb{C}$) which were discovered for integer values of p by F.W. Bessel and were generalized for real values of p by E.C.J. von Lommel (see the pages 45 and 79 from the book of G.N. Watson [227]):

(a) $z J_{p-1}(z) + z J_{p+1}(z) = 2 p J_p(z)$.
(b) $z J_p'(z) + p J_p(z) = z J_{p-1}(z)$.
(c) $z J_p'(z) + z J_{p+1}(z) = p J_p(z)$.
(d) $\left[z^{-p} J_p(z) \right]' = -z^{-p} J_{p+1}(z)$.
(e) $z I_{p-1}(z) - z I_{p+1}(z) = 2 p I_p(z)$.
(f) $z I_p'(z) + p I_p(z) = z I_{p-1}(z)$.
(g) $z I_p'(z) - z I_{p+1}(z) = p I_p(z)$.
(h) $\left[z^{-p} I_p(z) \right]' = z^{-p} I_{p+1}(z)$.

Remark 1.2. It is also important to note here that in particular Bessel functions of the first kind reduces to some elementary functions related to trigonometric functions, like cosine and sine, while modified Bessel functions of the first kind in particular are related to the hyperbolic sine and cosine. More precisely, we have the following formulae:

$$J_{-1/2}(z) = \sqrt{\frac{2}{\pi z}} \cos z, \quad J_{1/2}(z) = \sqrt{\frac{2}{\pi z}} \sin z,$$

$$J_{3/2}(z) = \sqrt{\frac{2}{\pi z}} \left(\frac{\sin z}{z} - \cos z \right),$$

$$I_{-1/2}(z) = \sqrt{\frac{2}{\pi z}} \cosh z, \; I_{1/2}(z) = \sqrt{\frac{2}{\pi z}} \sinh z$$

and

$$I_{3/2}(z) = -\sqrt{\frac{2}{\pi z}} \left(\frac{\sinh z}{z} - \cosh z \right).$$

Finally, we note that the graphs of Bessel functions of the first kind of real variable J_p look roughly like oscillating sine or cosine functions that decay proportionally to $1/\sqrt{x}$, although their roots are not generally periodic, except asymptotically for large x. The graph of the functions $J_{-1/2}, J_0$ and $J_{1/2}$ are indicated in Fig. 1.1. Unlike the ordinary Bessel functions, which are oscillating as functions of a real argument, the modified Bessel functions of the first kind of real variable I_p are exponentially growing functions. The graph of the functions $I_{-1/2}, I_0$ and $I_{1/2}$ are indicated in Fig. 1.2.

Lemma 1.2. (Á. Baricz [52]) *If $b, p, c \in \mathbb{C}$ and $\kappa \neq 0, -1, -2, \ldots$, then the function u_p satisfies the recursive relation $4\kappa u_p'(z) = -c u_{p+1}(z)$ for all $z \in \mathbb{C}$.*

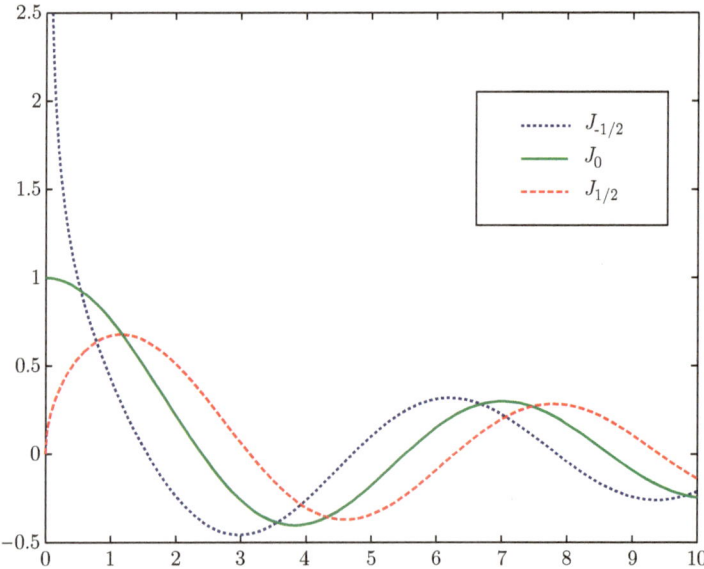

Fig. 1.1 The graph of the functions $J_{-1/2}, J_0$ and $J_{1/2}$

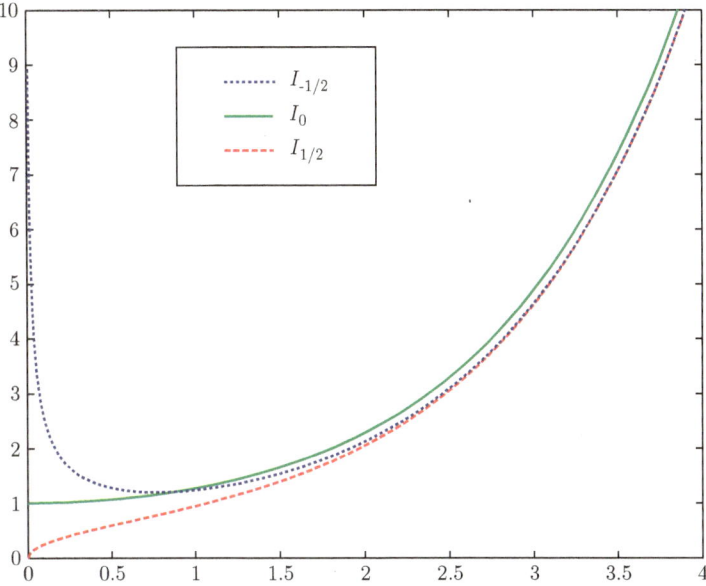

Fig. 1.2 The graph of the functions $I_{-1/2}, I_0$ and $I_{1/2}$

Proof. According to part (d) of Lemma 1.1 we have $[z^{-p}w_p(z)]' = -cz^{-p}w_{p+1}(z)$. Due to relation (1.19), i.e. $w_p(z) = [a_0(p)]z^p u_p(z^2)$ we obtain

$$
\begin{aligned}
2zu'_p(z^2) &= [u_p(z^2)]' \\
&= [a_0(p)]^{-1}[z^{-p}w_p(z)]' \\
&= -[a_0(p)]^{-1}cz^{-p}w_{p+1}(z).
\end{aligned}
$$

But $w_{p+1}(z) = [a_0(p+1)]z^{p+1}u_{p+1}(z^2)$, and therefore

$$2zu'_p(z^2) = -cz[a_0(p)]^{-1}[a_0(p+1)]u_{p+1}(z^2). \tag{1.22}$$

Now if we compute the expression $[a_0(p+1)]/[a_0(p)]$, by using the recursive relation $\Gamma(z+1) = z\Gamma(z)$, we obtain $(2\kappa)a_0(p+1) = a_0(p)$. Thus, by relation (1.22) the proof is complete. □

The function

$$z \mapsto \mathscr{J}_p(z) = 2^p\Gamma(p+1)z^{-p}J_p(z) \tag{1.23}$$

was studied by R. Askey [28], E. Neuman [161] and V. Selinger [208], while the function

$$z \mapsto \mathscr{I}_p(z) = 2^p\Gamma(p+1)z^{-p}I_p(z) \tag{1.24}$$

was studied for example by C. Giordano et al. [100], M.E.H. Ismail [119] and E. Neuman [159]. The function $\lambda_p : \mathbb{C} \to \mathbb{C}$, defined by $\lambda_p(z) = u_p(z^2)$, for $c = b = 1$ becomes $\mathscr{J}_p$ and for $c = -1, b = 1$ it reduces to $\mathscr{I}_p$. Note that when $b = 2$ and $c = 1$ then w_p becomes $2j_p/\sqrt{\pi}$ and in this case λ_p reduces to the function

$$z \mapsto \mathscr{J}_{p+\frac{1}{2}}(z) = 2^{p+\frac{1}{2}} \Gamma\left(p+\frac{3}{2}\right) z^{-(p+\frac{1}{2})} J_{p+\frac{1}{2}}(z).$$

Similarly, when $b = 2$ and $c = -1$ then w_p reduces to $2i_p/\sqrt{\pi}$, and λ_p in this case becomes

$$z \mapsto \mathscr{I}_{p+\frac{1}{2}}(z) = 2^{p+\frac{1}{2}} \Gamma\left(p+\frac{3}{2}\right) z^{-(p+\frac{1}{2})} I_{p+\frac{1}{2}}(z).$$

For later use we note that for the Bessel function of the first kind J_p the following infinite product formula (see G.N. Watson [227, p. 498]) is valid for all $p \neq -1, -2, \ldots$ and arbitrary $z \in \mathbb{C}$:

$$\mathscr{J}_p(z) = 2^p \Gamma(p+1) z^{-p} J_p(z) = \prod_{n \geq 1} \left(1 - \frac{z^2}{j_{p,n}^2}\right),$$

where $j_{p,n}$ denotes the nth zero of $J_p(z)$, the zeros being numbered in increasing order of their real parts. By the celebrated Lommel theorem "$J_p(z)$ has an infinity of real zeros, for any given real value of p" (see G.N. Watson [227] for further details). Therefore, here and in what follows, we will assume that for p real $0 = j_{p,0} < j_{p,1} < j_{p,2} < \ldots < j_{p,n} < \ldots$.

Clearly the function λ_p has the infinite series representation

$$\lambda_p(z) = \sum_{n \geq 0} \frac{(-c/4)^n}{n! \cdot (\kappa)_n} z^{2n}, \tag{1.25}$$

where $\kappa \neq 0, -1, -2, \ldots$ and the following lemma gives an integral representation.

Lemma 1.3. (Á. Baricz and E. Neuman [58]) *Suppose that $c \in \mathbb{R}$ and $p, b \in \mathbb{C}$ are such that $\operatorname{Re} \kappa > 1/2$. Then the function $\lambda_p : \mathbb{C} \to \mathbb{C}$, defined by $\lambda_p(z) = u_p(z^2)$, admits the following integral representation:*

$$\lambda_p(z) = \int_0^1 \cos(tz\sqrt{c}) d\mu_{p,b}(t), \tag{1.26}$$

where $d\mu_{p,b}(t) = \mu_{p,b}(t) \, dt$ with

$$\mu_{p,b}(t) = \frac{2\Gamma(\kappa)(1-t^2)^{\kappa-3/2}}{\sqrt{\pi} \cdot \Gamma(\kappa-1/2)} = \frac{2(1-t^2)^{\kappa-3/2}}{B(\kappa-1/2, 1/2)}.$$

In particular, for $z = x \in \mathbb{R}$ we have

$$
\lambda_p(x) = \begin{cases} \displaystyle\int_0^1 \cos(tx\sqrt{c})d\mu_{p,b}(t), & \text{if } c \geq 0 \\ \displaystyle\int_0^1 \cosh(tx\sqrt{-c})d\mu_{p,b}(t), & \text{if } c \leq 0. \end{cases}
\tag{1.27}
$$

Here $\mu_{p,b}$ is a probability measure on the interval $[0,1]$, i.e.

$$
\int_0^1 d\mu_{p,b}(t) = \int_0^1 \mu_{p,b}(t)\,dt = 1,
$$

and for all $\operatorname{Re}\alpha, \operatorname{Re}\beta > 0$

$$
B(\alpha,\beta) = \frac{\Gamma(\alpha)\Gamma(\beta)}{\Gamma(\alpha+\beta)} = \int_0^1 t^{\alpha-1}(1-t)^{\beta-1}\,dt
$$

is the Euler beta function.

Proof. Note that if $c = 0$ clearly $\lambda_p(z) \equiv 1$. So we assume that $c > 0$. First we prove that the generalized Bessel function w_p admits the following integral representation

$$
w_p(z) = 2\left(\frac{z}{2}\right)^p \frac{1}{\sqrt{\pi}\cdot\Gamma(p+b/2)} \int_0^{\pi/2} (\sin^{2p+b-1}\theta)\cos(\sqrt{cz}\cos\theta)d\theta. \tag{1.28}
$$

Substituting the function $\cos(\sqrt{cz}\cos\theta)$ with the corresponding MacLaurin series and integrating term by term, we obtain

$$
\int_0^{\pi/2} (\sin\theta)^{2p+b-1} \cos(\sqrt{cz}\cos\theta)d\theta
$$

$$
= \sum_{n\geq 0} \frac{(-1)^n c^n}{(2n)!} z^{2n} \int_0^{\pi/2} (\sin\theta)^{2p+b-1}(\cos\theta)^{2n}d\theta
$$

$$
= \frac{1}{2}\sum_{n\geq 0} \frac{(-1)^n c^n}{(2n)!} z^{2n} B\left(p+\frac{b}{2}, n+\frac{1}{2}\right).
$$

On the other hand, we know that

$$
B\left(p+\frac{b}{2}, n+\frac{1}{2}\right) = \frac{\Gamma(p+b/2)\Gamma(n+1/2)}{\Gamma(p+(b+1)/2+n)}, \quad \Gamma\left(\frac{1}{2}\right) = \sqrt{\pi},
$$

$$
\Gamma\left(n+\frac{1}{2}\right) = \frac{1\cdot 3\cdot 5\cdot\ldots\cdot(2n-1)}{2^n}\sqrt{\pi} = \frac{(2n)!}{2^{2n}n!}\sqrt{\pi}
$$

for all $n \in \{1,2,3,\ldots\}$. So we obtain

$$\int_0^{\pi/2} (\sin\theta)^{2p+b-1}\cos(\sqrt{cz}\cos\theta)d\theta$$

$$= \frac{\sqrt{\pi}}{2}\Gamma\left(p+\frac{b}{2}\right)\sum_{n\geq 0}\frac{(-1)^n c^n}{n!\Gamma(p+(b+1)/2+n)}\left(\frac{z}{2}\right)^{2n}$$

$$= \frac{\sqrt{\pi}}{2}\Gamma\left(p+\frac{b}{2}\right)w_p(z)\left(\frac{2}{z}\right)^p.$$

Thus we have the representation (1.28). By using the substitution $\cos\theta = t$, we get

$$w_p(z) = 2\left(\frac{z}{2}\right)^p\frac{1}{\sqrt{\pi}\cdot\Gamma(\kappa-1/2)}\int_0^1 (1-t^2)^{\kappa-3/2}\cos(\sqrt{czt})\,dt.$$

According to (1.19), it follows that

$$u_p(z^2) = \frac{2\Gamma(\kappa)}{\sqrt{\pi}\Gamma(\kappa-1/2)}\int_0^1 (1-t^2)^{\kappa-3/2}\cos(\sqrt{czt})\,dt. \tag{1.29}$$

The situation is similar in the case when $c < 0$, because

$$\cosh(\sqrt{-cz}\cos\theta) = \sum_{n\geq 0}\frac{(-1)^n c^n}{(2n)!}z^{2n}(\cos\theta)^{2n}.$$

Thus we have

$$w_p(z) = 2\left(\frac{z}{2}\right)^p\frac{1}{\sqrt{\pi}\cdot\Gamma(p+b/2)}\int_0^{\pi/2}(\sin^{2p+b-1}\theta)\cosh(\sqrt{-cz}\cos\theta)d\theta$$

$$= 2\left(\frac{z}{2}\right)^p\frac{1}{\sqrt{\pi}\cdot\Gamma(\kappa-1/2)}\int_0^1 (1-t^2)^{\kappa-3/2}\cosh(\sqrt{-czt})\,dt.$$

Consequently, in this case we obtain

$$u_p(z^2) = \frac{2\Gamma(\kappa)}{\sqrt{\pi}\Gamma(\kappa-1/2)}\int_0^1 (1-t^2)^{\kappa-3/2}\cosh(\sqrt{-czt})\,dt. \tag{1.30}$$

Since $\cos z = \cosh(iz)$, for complex argument the formulas (1.29) and (1.30) are the same. But for real argument $z = x \in \mathbb{R}$ (since $\sqrt{c}$ is in integral representations) we have two formulas, namely (1.27). Finally we just need to prove that $\int_0^1 d\mu_{p,b}(t) = 1$. Clearly the function in question is nonnegative on the indicated interval. Moreover after some computations we get that

$$\int_0^1 d\mu_{p,b}(t) = 2A^{-1} \int_0^1 (1-t^2)^{\kappa-3/2} dt$$

$$= A^{-1} \int_0^1 r^{-1/2}(1-r)^{\kappa-3/2} dr$$

$$= A^{-1} \cdot A = 1,$$

where $A = B(\kappa - 1/2, 1/2)$. Here we have used the substitution $t = r^{1/2}$. Thus the proof is complete.   □

When $b = c = 1$, formula (1.26) simplifies to equation (9.1.20) from the handbook of mathematical functions edited by M. Abramowitz and I.A. Stegun [1], which is in fact Lommel's expression of $J_p(z)$ by an integral of Poisson's type (see also the equation (1) on page 47 in the book of G.N. Watson [227]).

The next lemma is the key tool in the proof of Theorem 2.22.

Lemma 1.4. (S. András and Á. Baricz [24]) *Let b,c,p,q,z be arbitrary complex numbers and let $2\kappa_p = 2p+b+1, 2\kappa_q = 2q+b+1$. If $\operatorname{Re}\kappa_p > \operatorname{Re}\kappa_q > 0$, then the following integral representation formula holds:*

$$u_p(z) = \int_0^1 u_q(tz)d\mu_{p,q,b}(t), \tag{1.31}$$

where $d\mu_{p,q,b}(t) = \mu_{p,q,b}(t) dt$ and

$$\mu_{p,q,b}(t) = \frac{t^{\kappa_q-1}(1-t)^{\kappa_p-\kappa_q-1}}{B(\kappa_q, \kappa_p - \kappa_q)}$$

is a probability measure on $[0,1]$.

Proof. If we use (1.20) to represent $u_q(tz)$ and multiply both sides of (1.20) by $t^{q+\frac{b-1}{2}}(1-t)^{p-q-1}$, then after integration on $[0,1]$, we obtain

$$\int_0^1 u_q(tz)t^{q+\frac{b-1}{2}}(1-t)^{p-q-1} dt = \sum_{n\geq 0} \frac{(-c/4)^n \Gamma(\kappa_q)}{\Gamma(\kappa_q+n)} \frac{z^n}{n!} \int_0^1 t^{q+n+\frac{b-1}{2}}(1-t)^{p-q-1} dt. \tag{1.32}$$

Since

$$\int_0^1 t^{q+n+\frac{b-1}{2}}(1-t)^{p-q-1} dt = B(\kappa_q+n, \kappa_p-\kappa_q)$$

$$= \frac{\Gamma(\kappa_q+n)\Gamma(\kappa_p-\kappa_q)}{\Gamma(\kappa_p+n)},$$

the equality (1.32) implies

$$\int_0^1 u_q(tz)t^{q+\frac{b-1}{2}}(1-t)^{p-q-1}\,dt$$

$$= \frac{\Gamma(\kappa_p-\kappa_q)\Gamma(\kappa_q)}{\Gamma(\kappa_p)} \sum_{n\geq 0} \frac{(-c/4)^n}{(\kappa_p)_n} \frac{z^n}{n!}$$

$$= \frac{\Gamma(\kappa_p-\kappa_q)\Gamma(\kappa_q)}{\Gamma(\kappa_p)} u_p(z).$$

Using this we deduce that

$$u_p(z) = \frac{\Gamma(\kappa_p)}{\Gamma(\kappa_p-\kappa_q)\Gamma(\kappa_q)} \int_0^1 u_q(tz)t^{q+\frac{b-1}{2}}(1-t)^{p-q-1}\,dt,$$

i.e. (1.31) holds.
   Since

$$B(\kappa_q,\kappa_p-\kappa_q) = \int_0^1 t^{\kappa_q-1}(1-t)^{\kappa_p-\kappa_q-1}\,dt,$$

it follows that $\int_0^1 d\mu_{p,q,b}(t) = 1$, i.e. $\mu_{p,q,b}$ is a probability measure on $[0,1]$. Thus the proof is complete. □

Remark 1.3. When $\mathrm{Re}\,\kappa_p > \mathrm{Re}\,\kappa_q > 0$, by using the integral representation formula (1.31), we obtain easily the following equality:

$$\lambda_p(z) = \int_0^1 \lambda_q(tz)dv_{p,q,b}(t), \tag{1.33}$$

where $dv_{p,q,b}(t) = v_{p,q,b}(t)\,dt$ and

$$v_{p,q,b}(t) = \frac{2t^{2q+b}(1-t^2)^{p-q-1}}{B\left(q+\frac{b+1}{2},p-q\right)}$$

is also a probability measure on $[0,1]$. Since $v_{p,-b/2,b} = \mu_{p,b}$ and $\lambda_{-b/2}(tz) = \cos(t\sqrt{cz})$, it is clear that (1.33) is a generalization of (1.26). In particular, when $c = 1$ and $b = 1$, the function λ_p reduces to the function $\mathscr{J}_p$. Thus, when $\mathrm{Re}\,p > \mathrm{Re}\,q > -1$, we obtain that

$$\mathscr{J}_p(z) = \int_0^1 \mathscr{J}_q(tz)dv_{p,q}(t), \tag{1.34}$$

where $dv_{p,q}(t) = v_{p,q}(t)\,dt$ with

$$v_{p,q}(t) = v_{p,q,1}(t) = \frac{2t^{2q+1}(1-t^2)^{p-q-1}}{B(q+1,p-q)}$$

being also a probability measure on $[0,1]$. Finally observe that in fact (1.34) is a generalization of (2.32). For this we take $q = -1/2$ in (1.34) obtaining that $v_{p,-1/2} = \mu_p$ and we use (2.28).

Finally, for later use let us mention that the function λ_p in particular reduces to some elementary functions like cosine, hyperbolic cosine, sine and hyperbolic sine. Further properties, especially geometric properties, monotonicity and convexity properties of the function λ_p, various kind of inequalities for this function will be studied in details in the following chapters.

1.3 Classical Inequalities

In what follows we list some known classical inequalities which will be used in the sequel. The first classical inequality which is used in Chap. 2 is the Cauchy–Bunyakovsky–Schwarz inequality (see D.S. Mitrinović [156, p. 32]).

Lemma 1.5. (Cauchy–Bunyakovsky–Schwarz inequality) *If $a_1, \ldots, a_n$ and $b_1, \ldots, b_n$ are arbitrary real sequences, then we have*

$$\left| \sum_{k=1}^{n} a_k b_k \right| \leq \sqrt{\sum_{k=1}^{n} |a_k|^2} \sqrt{\sum_{k=1}^{n} |b_k|^2} \tag{1.35}$$

with equality if and only if the sequences $a_1, \ldots, a_n$ and $b_1, \ldots, b_n$ are proportional.

Its integral version, which is used in Sect. 3.5.2, reads as follows (see the book of D.S. Mitrinović [156, p. 43]).

Lemma 1.6. (Cauchy–Bunyakovsky–Schwarz inequality) *If f and g are real and integrable functions on $[a,b]$, then*

$$\int_{a}^{b} |f(t)g(t)| \, dt \leq \sqrt{\int_{a}^{b} |f(t)|^2 \, dt} \sqrt{\int_{a}^{b} |g(t)|^2 \, dt} \tag{1.36}$$

with equality holding if and only if f and g are linearly dependent functions.

The classical Jensen inequality is useful in the study of functional inequalities for hypergeometric functions and generalized Bessel functions. This inequality reads as follows (see D.S. Mitrinović [156, p. 12]).

Lemma 1.7. (Jensen inequality) *If the function $f : [a,b] \to \mathbb{R}$ is convex, then for all $a_1, \ldots, a_n \in [a,b]$ and all nonnegative numbers $\alpha_1, \ldots, \alpha_n$ such that $\alpha_1 + \ldots + \alpha_n = 1$ we have*

$$f\left(\sum_{k=1}^{n} \alpha_k a_k \right) \leq \sum_{k=1}^{n} \alpha_k f(a_k). \tag{1.37}$$

Notice that if f is strictly convex, then inequality in (1.37) is strict, except when $a_1 = \ldots = a_n$.

Hölder-Rogers inequality is also useful in our study, mainly in the investigation of the convexity of the zero-balanced Gaussian hypergeometric function with respect to power (or Hölder) means (see Sect. 3.4). For this inequality see D.S. Mitrinović [156, Theorem 1, p. 50].

Lemma 1.8. (Hölder-Rogers inequality) *If $a_1,\ldots,a_n \geq 0, b_1,\ldots,b_n \geq 0$ and $1/\alpha + 1/\beta = 1$ with $\alpha > 1$, then*

$$\sum_{k=1}^{n} a_k b_k \leq \left(\sum_{k=1}^{n} a_k^{\alpha}\right)^{1/\alpha} \left(\sum_{k=1}^{n} b_k^{\beta}\right)^{1/\beta} \tag{1.38}$$

with equality holding if and only if $\mu a_k^{\alpha} = v b_k^{\beta}$ for all $k \in \{1,2,\ldots,n\}$, where μ and v are real nonnegative constants with $\mu^2 + v^2 > 0$.

At the end of Sect. 3.6 we use the classical Hermite-Hadamard inequality in order to extend van der Corput's inequality to modified Bessel functions of the first kind. The Hermite-Hadamard inequality (see the paper of D.S. Mitrinović and I. Lacković [157]) reads as follows.

Lemma 1.9. (Hermite-Hadamard inequality) *If $f : [a,b] \to \mathbb{R}$ is a convex function, then*

$$f\left(\frac{a+b}{2}\right) \leq \frac{1}{b-a} \int_a^b f(t)\,dt \leq \frac{f(a)+f(b)}{2}. \tag{1.39}$$

Finally, let us recall the classical Chebyshev integral inequality. For this inequality see the book of D.S. Mitrinović [156, p. 40].

Lemma 1.10. (Chebyshev integral inequality) *If $f,g : [a,b] \to \mathbb{R}$ are integrable functions, both increasing or both decreasing and $p : [a,b] \to \mathbb{R}$ is a positive integrable function, then*

$$\int_a^b p(t)f(t)\,dt \int_a^b p(t)g(t)\,dt \leq \int_a^b p(t)\,dt \int_a^b p(t)f(t)g(t)\,dt. \tag{1.40}$$

Note that if one of the functions f or g is decreasing and the other is increasing, then (1.40) is reversed.

Chapter 2
Geometric Properties of Generalized Bessel Functions

Abstract The goal of the present chapter is to study some geometric properties (like univalence, starlikeness, convexity, close-to-convexity) of generalized Bessel functions of the first kind. In order to achieve our goal we use several methods: differential subordinations technique, Alexander transform, results of L. Fejér, W. Kaplan, S. Owa and H.M. Srivastava, S. Ozaki, S. Ponnusamy and M. Vuorinen, H. Silverman, and Jack's lemma. Moreover, we present some immediate applications of univalence and convexity involving generalized Bessel functions associated with the Hardy space and a monotonicity property of generalized and normalized Bessel functions of the first kind.

2.1 Univalence of Generalized Bessel Functions

Our aim in this section is to give sufficient conditions for the function u_p, defined in (1.20), to be univalent. The results of this section were picked up from the papers of Á. Baricz [38, 42, 43, 52].

Before we state our main results of this section we recall some basic facts. A single-valued function $f : \mathbb{E} \subseteq \mathbb{C} \to \mathbb{C}$ is said to be univalent (or schlicht) in a domain $\mathbb{E}$ if it never takes the same value twice, that is, if $f(z_1) = f(z_2)$ for $z_1, z_2 \in \mathbb{E}$ implies that $z_1 = z_2$. The function $f_1(z) = \bar{z}$ is nowhere analytic, but it is univalent in the complex plane $\mathbb{C}$. A necessary condition for an analytic function $f : \mathbb{E} \subseteq \mathbb{C} \to \mathbb{C}$ to be univalent in $\mathbb{E}$ is that $f'(z) \neq 0$ in $\mathbb{E}$, but this condition is not sufficient; for example e^z is not univalent in $\mathbb{C}$ though its derivative never vanishes in $\mathbb{C}$.

A domain $\mathbb{E} \subseteq \mathbb{C}$ is said to be starlike domain with respect to $z_0 \in \mathbb{E}$ if the line segment joining z_0 to every other point $z \in \mathbb{E}$ lies entirely in $\mathbb{E}$. If a function f maps $\mathbb{E}$ onto a domain that is starlike with respect to z_0, then f is said to be starlike with respect to z_0. In particular, if z_0 is the origin, then we say that f is a starlike function.

Further a domain $\mathbb{E} \subseteq \mathbb{C}$ is said to be convex if the line segment joining any two points of $\mathbb{E}$ lies entirely in $\mathbb{E}$. If a function f maps $\mathbb{E}$ onto a convex domain, then we say that f is a convex function in $\mathbb{E}$.

Due to the famous Riemann mapping theorem, any simply connected domain in the complex plane $\mathbb{C}$ which is not the whole plane, can be mapped by an analytic

Á. Baricz, *Generalized Bessel Functions of the First Kind*,
Lecture Notes in Mathematics 1994, DOI 10.1007/978-3-642-12230-9_2,
© Springer-Verlag Berlin Heidelberg 2010

function conformally onto the unit disk $\mathbb{D} = \{z \in \mathbb{C} : |z| < 1\}$. More precisely, the Riemann mapping theorem states (see the book of P.L. Duren [90, p. 11]): if $\mathbb{U}$ is a simply connected domain which is a proper subset of the complex plane and ζ is a given point in $\mathbb{U}$, then there is a unique function f which maps $\mathbb{U}$ conformally onto the unit disk $\mathbb{D}$ and has the properties $f(\zeta) = 0$ and $f'(\zeta) > 0$. Thus, the investigation of the analytic functions which are univalent in a simply connected domain with more than one boundary point can be confined to the investigation of analytic functions which are univalent in $\mathbb{D}$. Hence without loss of generality we assume that $\mathbb{E}$ is the unit disk $\mathbb{D}$. This is to simplify and to give short and elegant formulae. Note that for functions defined in a domain bordered by an ellipse (see the paper of N.N. Pascu et al. [169]), by a closed Lamé curve or a so-called spindle curve (see the paper of Á. Baricz [40]), the conditions (2.1) and (2.3) become more complicated, thus this also justifies why we need to work on the unit disk.

It is known that an analytic function $f : \mathbb{D} \to \mathbb{C}$, which satisfies $f(0) = 0$ and $f'(0) \neq 0$, is starlike (see the book of P.L. Duren [90]) if and only if

$$\mathrm{Re}\left[\frac{zf'(z)}{f(z)}\right] > 0 \quad \text{for all} \quad z \in \mathbb{D}. \tag{2.1}$$

Moreover, the function f is said (see the papers of B.C. Carlson and D.B. Shaffer [80], I.S. Jack [124] and M.S. Robertson [200]) to be starlike of order α, where $\alpha \in [0, 1)$, if

$$\mathrm{Re}\left[\frac{zf'(z)}{f(z)}\right] > \alpha \quad \text{for all} \quad z \in \mathbb{D}. \tag{2.2}$$

We denote by $\mathscr{S}^*(\alpha)$ the class of all functions which are starlike of order α. Throughout this book, it should be understood that functions such as zf'/f, which have removable singularities at $z = 0$, have had these singularities removed in statements like (2.1), (2.2). It is known that $g : \mathbb{D} \to \mathbb{C}$, which satisfies $g'(0) \neq 0$, is convex (see the book of P.L. Duren [90]) if and only if

$$\mathrm{Re}\left[1 + \frac{zg''(z)}{g'(z)}\right] > 0 \quad \text{for all} \quad z \in \mathbb{D}. \tag{2.3}$$

If in addition

$$\mathrm{Re}\left[1 + \frac{zg''(z)}{g'(z)}\right] > \alpha \quad \text{for all} \quad z \in \mathbb{D},$$

where $\alpha \in [0, 1)$, then g is called convex of order α. We denote the class of convex functions of order α by $\mathscr{C}(\alpha)$. We remark that for all $\alpha \in [0, 1)$ we have

$$\mathscr{S}^*(\alpha) \subseteq \mathscr{S}^*(0) = \mathscr{S}^*, \quad \mathscr{C}(\alpha) \subseteq \mathscr{C}(0) = \mathscr{C}$$

and

$$\mathscr{C}(\alpha) \subset \mathscr{S}^*(\alpha).$$

The classes $\mathscr{S}^*(\alpha)$ and $\mathscr{C}(\alpha)$ were first introduced by M.S. Robertson [200], and were studied subsequently by I.S. Jack [124], T.H. MacGregor [142], B. Pinchuk [172], A. Schild [206], and others.

Let $\mathscr{A}$ denote the class of normalized functions of the form

$$f(z) = z + \sum_{n \geq 2} a_n z^n,$$

where $a_n \in \mathbb{C}$ for all $n \geq 2$, which are analytic in the unit disk $\mathbb{D}$. Further let $\mathscr{S}$ denote the class of all functions in $\mathscr{A}$, which are univalent in the unit disk $\mathbb{D}$. It is known that for normalized functions belonging to the class $\mathscr{A}$ we have $\mathscr{C}(\alpha) \subset \mathscr{S}^*(\alpha) \subset \mathscr{S}$.

An analytic function $f : \mathbb{D} \to \mathbb{C}$ is said to be close-to-convex with respect to a convex function $\varphi : \mathbb{D} \to \mathbb{C}$ if

$$\mathrm{Re}\left[\frac{f'(z)}{\varphi'(z)}\right] > 0 \quad \text{for all} \;\; z \in \mathbb{D}.$$

If there exists a convex function $\varphi : \mathbb{D} \to \mathbb{C}$ such that f is close-to-convex with respect to φ, then we say that f is close-to-convex. We note that f is not required a priori to be univalent (cf. Lemma 2.3 below), and the associated function φ need not be a function belonging to the class $\mathscr{A}$. Moreover, every starlike (and hence convex) function is close-to-convex. However, if f is starlike then it is not necessary that it will be close-to-convex with respect to a particular convex function. For instance, it is known that if $f \in \mathscr{A}$ is convex, then f need not be close-to-convex with respect to the identity function.

Given a number $\alpha \in [0, 1)$, we say that $f : \mathbb{D} \to \mathbb{C}$ is close-to-convex of order α with respect to a convex function $\varphi : \mathbb{D} \to \mathbb{C}$ if

$$\mathrm{Re}\left[\frac{f'(z)}{\varphi'(z)}\right] > \alpha \quad \text{for all} \;\; z \in \mathbb{D}.$$

If there exists a convex function $\varphi : \mathbb{D} \to \mathbb{C}$ such that f is close-to-convex of order α with respect to φ, then we say that f is close-to-convex of order α. For a general reference for the special classes of univalent functions we refer to the books of P.L. Duren [90], A.W. Goodman [101, 102] and C. Pommerenke [173, 174].

In order to prove our results the following preliminary results will be helpful. The first result is due to J.W. Alexander [5] and is called Alexander's duality theorem, and the second one is due to I.S. Jack [124].

Lemma 2.1. *(J.W. Alexander [5]) Let $f : \mathbb{D} \to \mathbb{C}$ be an analytic function. Then f is convex if and only if zf' is starlike.*

Lemma 2.2. *(I.S. Jack [124]) Let $f : \mathbb{D} \to \mathbb{C}$ be an analytic function. Further, consider the function $g : \mathbb{D} \to \mathbb{C}$, defined by $g(z) = z[f'(z)]^{1/(1-\alpha)}$, where $\alpha \in [0, 1)$. Then $f \in \mathscr{C}(\alpha)$ if and only if $g \in \mathscr{S}^*(\alpha)$. Moreover, $f \in \mathscr{C}(\alpha)$ if and only if $zf' \in \mathscr{S}^*(\alpha)$.*

Now, we state the following known condition of univalence (see also Theorem 2.17 in the book of P.L. Duren [90]).

Lemma 2.3. (W. Kaplan [129] and S. Ozaki [167]) *If $f : \mathbb{D} \to \mathbb{C}$ is a close-to-convex function, then it is univalent in $\mathbb{D}$.*

We note that there are other known univalence criteria in the literature, which may be useful to solve similar problems such as treated in this book. For example, the known Nehari's criterion for univalence states (see the paper of Z. Nehari [158]): if $f : \mathbb{D} \to \mathbb{C}$ is analytic and locally univalent in $\mathbb{D}$ and its Schwarzian derivative S_f, defined by

$$S_f(z) = \left[\frac{f''(z)}{f'(z)} \right]' - \frac{1}{2} \left[\frac{f''(z)}{f'(z)} \right]^2,$$

satisfies

$$\left(1 - |z|^2\right)^2 |S_f(z)| \leq 2 \quad \text{for all} \quad z \in \mathbb{D},$$

then f is univalent in $\mathbb{D}$. The constant 2 in Nehari's criterion is best possible and cannot be replaced by any larger number, as it was shown by E. Hille [115] as an addendum to Nehari's paper [158]. In addition to the above condition, in his original paper Z. Nehari [158] showed that $|S_f(z)| \leq \pi^2/2$ for all $z \in \mathbb{D}$ is also a sufficient condition for the function f to be univalent in $\mathbb{D}$. Here the constant $\pi^2/2$ is sharp. However, Nehari's conditions may be awkward to verify because it requires the computation of the Schwarzian derivative. Thus, it is often simpler to work directly with the logarithmic derivative f''/f' of f', called sometimes as the pre-Schwarzian derivative. Such univalence criterion, which involves the pre-Schwarzian derivative is due to J. Becker [66], who proved that the condition

$$\left| \frac{z f''(z)}{f'(z)} \right| \leq \frac{1}{1 - |z|^2},$$

where $z \in \mathbb{D}$, is a sufficient condition for the univalence of f in $\mathbb{D}$. The constant 1 is sharp, as was proved later by J. Becker and C. Pommerenke [67]. Various other conditions of Nehari and Becker type are discussed in the survey article of F.G. Avhadiev and L.A. Aksentev [31]. For more details we refer also to the books of P.L. Duren [90], A.W. Goodman [101, 102] and C. Pommerenke [173, 174].

2.1.1 Sufficient Conditions Involving Jack's Lemma

S. Owa and H. M. Srivastava in [168] proved several interesting results concerning univalent, starlike of order α and convex of order α generalized hypergeometric functions. An important result which was used by S. Owa and H. M. Srivastava is the Jack lemma, i.e.

Lemma 2.4. (I.S. Jack [124]) *Let* f *be a nonconstant analytic function in the unit disk* $\mathbb{D}$ *with* $f(0) = 0$. *If* $|f(z)|$ *attains its maximum value on the circle* $|z| = r \in [0,1)$ *at a point* z_0, *then there exists a real number* $m \geq 1$ *such that* $z_0 f'(z_0) = m f(z_0)$.

By using their technique mutatis mutandis we can obtain easily the following result.

Theorem 2.1. (Á. Baricz [54]) *Let* $u_p(z) = \sum_{n \geq 0} b_n z^n$ *be the generalized and normalized Bessel function, where* b_n *is defined by* (1.18). *Suppose that* $b_1 > 0$ *and that* u_p *satisfies the condition*

$$|u'_p(z) - b_1|^{1-\beta} \cdot \left| \frac{z u''_p(z)}{u'_p(z)} \right|^{\beta} < (b_1)^{1-\beta} \left(\frac{1}{2} \right)^{\beta} \tag{2.4}$$

for all $z \in \mathbb{D}$ *and for some fixed* $\beta \geq 0$. *Then* u_p *is univalent in* $\mathbb{D}$.

Proof. For the function $h : \mathbb{D} \to \mathbb{C}$, defined by $h(z) = (u_p(z) - 1)/b_1$, the condition (2.4) implies

$$|h'(z) - 1|^{1-\beta} \cdot \left| \frac{z h''(z)}{h'(z)} \right|^{\beta} < \left(\frac{1}{2} \right)^{\beta}. \tag{2.5}$$

Further, it is clear that $h \in \mathscr{A}$. Now we define the function $f : \mathbb{D} \to \mathbb{C}$ by $f(z) = h'(z) - 1$. Then it follows that f is analytic in the unit disk $\mathbb{D}$ with $f(0) = 0$. From (2.5) we get

$$|f(z)|^{1-\beta} \left| \frac{z f'(z)}{1 + f(z)} \right|^{\beta} < \left(\frac{1}{2} \right)^{\beta} \tag{2.6}$$

or equivalently

$$|f(z)| \cdot \left| \frac{z f'(z)}{f(z)} \cdot \frac{1}{1 + f(z)} \right|^{\beta} < \left(\frac{1}{2} \right)^{\beta}$$

for all $z \in \mathbb{D}$, where the comment about removable singularities applies just as in (2.1) and (2.2). Now assume that there exists a point $z_0 \in \mathbb{D}$ such that

$$\max\{|f(z)| : |z| \leq |z_0|\} = |f(z_0)| = 1.$$

Then we can put $z_0 f'(z_0)/f(z_0) = m \geq 1$ by means of Lemma 2.4. Therefore, we obtain

$$|f(z_0)| \cdot \left| \frac{z_0 f'(z_0)}{f(z_0)} \cdot \frac{1}{1 + f(z_0)} \right|^{\beta} \geq \left(\frac{m}{2} \right)^{\beta} \geq \left(\frac{1}{2} \right)^{\beta},$$

which contradicts the condition (2.6), and so also (2.4).

This shows that $|f(z)| = |h'(z) - 1| < 1$, which implies $\operatorname{Re} h'(z) > 0$ for all $z \in \mathbb{D}$. But this means that $\operatorname{Re}[u'_p(z)/b_1] > 0$ for all $z \in \mathbb{D}$. Note that $\varphi(z) = b_1 z$ is convex in $\mathbb{D}$. For this φ, the function u_p satisfies $\mathrm{Re}[u_p'(z)/\varphi'(z)] > 0$ for all $z \in \mathbb{D}$. Consequently, by Lemma 2.3 we deduce that u_p is univalent[1] in $\mathbb{D}$.   □

By setting $\beta = 0$ and $\beta = 1$ in Theorem 2.1, we obtain the following corollaries:

Corollary 2.1. (Á. Baricz [54]) *If $u_p(z) = \sum_{n \geq 0} b_n z^n$, where $b_1 > 0$, satisfies the condition $|u_p'(z) - b_1| < b_1$ for all $z \in \mathbb{D}$, then it is univalent in $\mathbb{D}$.*

Corollary 2.2. (Á. Baricz [54]) *If $u_p(z) = \sum_{n \geq 0} b_n z^n$, where $b_1 > 0$, satisfies the following condition $|z u_p''(z)/u_p'(z)| < 1/2$ for all $z \in \mathbb{D}$, then it is univalent in $\mathbb{D}$.*

Remark 2.1. In fact the results of Theorem 2.1 and Corollaries 2.1, 2.2 hold in the disk $|z| < 4/|c|$, which for $0 < |c| < 4$ is larger than the unit disk. For this we remind the reader that the function $_0F_1(\kappa, \cdot)$ is a special case of the function $_qF_r(a_1, \ldots, a_q, b_1, \ldots, b_r, \cdot)$ (see Sect. 1.1). By applying Theorem 1, Corollaries 1, 2 due to S. Owa and H.M. Srivastava [168] to the function $F(z) = _0F_1(\kappa, z)$, we know that if

$$|F'(z) - b_1|^{1-\beta} \cdot \left| \frac{zF''(z)}{F'(z)} \right|^{\beta} < (b_1)^{1-\beta} \left(\frac{1}{2} \right)^{\beta} \text{ for all } z \in \mathbb{D} \qquad (2.7)$$

and some fixed $\beta \geq 0$, then F is univalent in $\mathbb{D}$. Now using the fact that $F(z) = u_p(-4z/c)$ and changing z with $-cz/4$ the inequality (2.7) becomes (2.4), which means that u_p is univalent in the disk $|-cz/4| < 1$.

2.1.2 Sufficient Conditions Involving the Admissible Function Method

The next lemma will be used to prove several theorems.

Lemma 2.5. (S.S. Miller and P.T. Mocanu [153, 155]) *Let $\mathbb{E} \subseteq \mathbb{C}$ be a set in the complex plane $\mathbb{C}$ and $\psi : \mathbb{C}^3 \times \mathbb{D} \mapsto \mathbb{C}$ a function, that satisfies the admissibility condition $\psi(\rho i, \sigma, \mu + vi; z) \notin \mathbb{E}$, where $z \in \mathbb{D}$, $\rho, \sigma, \mu, v \in \mathbb{R}$ with $\mu + \sigma \leq 0$ and $\sigma \leq -(1 + \rho^2)/2$. If h is analytic in the unit disk $\mathbb{D}$, with $h(0) = 1$ and $\psi(h(z), zh'(z), z^2h''(z); z) \in \mathbb{E}$ for all $z \in \mathbb{D}$, then $\mathrm{Re}\, h(z) > 0$ for all $z \in \mathbb{D}$. In particular, if we only have $\psi : \mathbb{C}^2 \times \mathbb{D} \mapsto \mathbb{C}$, the admissibility condition reduces to $\psi(\rho i, \sigma; z) \notin \mathbb{E}$ for all $z \in \mathbb{D}$ and $\rho, \sigma \in \mathbb{R}$ with $\sigma \leq -(1 + \rho^2)/2$.*

Using the above lemma we can prove the following theorem.[2]

---

[1] We note that since $b_1 > 0$ here can be applied another univalence criteria. Namely, due to K. Noshiro [166] and S.E. Warschawski [226] it is known that if $f : \mathbb{D} \to \mathbb{C}$ is analytic and $\mathrm{Re}\, f'(z) > 0$ for all $z \in \mathbb{D}$, then f is univalent in $\mathbb{D}$. Now, since $\mathrm{Re}\, u_p'(z) > 0$ for all $z \in \mathbb{D}$, we conclude that u_p is univalent in $\mathbb{D}$.

[2] Note that there is a mistake in the mentioned paper [38] of the author, namely the expression "$u_p(z) = z^{-p/2} v_p(z^{1/2})$" should be changed with "$u_p(z) = 2^p \cdot \Gamma(\kappa) z^{-p/2} v_p(z^{1/2})$."

Theorem 2.2. (Á. Baricz [38]) *If $b, p, c \in \mathbb{R}$ are such that $\kappa \geq |c|/4 + 1$, then $\operatorname{Re} u_p(z) > 0$ for all $z \in \mathbb{D}$. Further if $\kappa \geq |c|/4$ and $c \neq 0$, then u_p is univalent in $\mathbb{D}$.*

Proof. When $c = 0$, then we have $u_p(z) \equiv 1$, thus $\operatorname{Re} u_p(z) > 0$ for all $z \in \mathbb{D}$. Next suppose that $c \neq 0$. Put $h = u_p$. Since h satisfies (1.21), we have

$$4z^2 h''(z) + 4\kappa z h'(z) + czh(z) = 0. \tag{2.8}$$

If we consider $\psi(r, s, t; z) = 4t + 4\kappa s + czr$ and $\mathbb{E} = \{0\}$, then the equation (2.8) implies $\psi(h(z), zh'(z), z^2 h''(z); z) \in \mathbb{E}$ for all $z \in \mathbb{D}$. Next we will use Lemma 2.5 to prove that $\operatorname{Re} h(z) > 0$ for all $z \in \mathbb{D}$. If we put $z = x + iy$, where $x, y \in \mathbb{R}$, then

$$\operatorname{Re} \psi(\rho i, \sigma, \mu + vi; x + iy) = 4(\mu + \sigma) + 4(\kappa - 1)\sigma - c\rho y \quad \text{for all } \rho, \sigma, \mu, v \in \mathbb{R}.$$

Let $\rho, \sigma, \mu, v \in \mathbb{R}$ satisfy $\mu + \sigma \leq 0$ and $\sigma \leq -(1 + \rho^2)/2$. Since $\kappa > 1$, we have

$$\operatorname{Re} \psi(\rho i, \sigma, \mu + vi; x + iy) \leq -2(\kappa - 1)\rho^2 - cy\rho - 2(\kappa - 1).$$

Set $Q_1(\rho) = -2(\kappa - 1)\rho^2 - cy\rho - 2(\kappa - 1)$. This value will be strictly negative for all real ρ, because the discriminant Δ of $Q_1(\rho)$ satisfies

$$\Delta = c^2 y^2 - 16(\kappa - 1)^2 < c^2 - 16(\kappa - 1)^2 \leq 0$$

whenever $y \in (-1, 1)$. Consequently ψ satisfies the admissibility condition of Lemma 2.5. Hence by Lemma 2.5 we conclude $\operatorname{Re} h(z) = \operatorname{Re} u_p(z) > 0$ for all $z \in \mathbb{D}$.

When $\kappa \geq |c|/4$ and $c \neq 0$, then the above result implies $\operatorname{Re} u_{p+1}(z) > 0$ for all $z \in \mathbb{D}$. Using Lemma 1.2 we conclude that

$$\operatorname{Re}\left[-\frac{4\kappa}{c} u_p'(z) \right] = \operatorname{Re} u_{p+1}(z) > 0 \quad \text{for all } z \in \mathbb{D}.$$

This means that u_p is close-to-convex with respect to the convex function $\varphi : \mathbb{D} \to \mathbb{C}$, defined by $\varphi(z) = -(cz)/(4\kappa)$. By Lemma 2.3 it follows that u_p is univalent in $\mathbb{D}$. $\square$

Note that similar results as in Theorem 2.2 for Gaussian and confluent hypergeometric functions were obtained by S.S. Miller and P.T. Mocanu [154].

In fact Theorem 2.2 may be extended for complex parameters, as we can see in the following result.

Theorem 2.3. (Á. Baricz [42]) *If $b, p, c \in \mathbb{C}$ are such that $\operatorname{Re} \kappa \geq |c|/4 + 1$, then $\operatorname{Re} u_p(z) > 0$ for all $z \in \mathbb{D}$. Further if $\operatorname{Re} \kappa \geq |c|/4$ and $c \neq 0$, then u_p is univalent in $\mathbb{D}$.*

Proof. The proof of this theorem is similar to the proof of the previous theorem. If $c = 0$, then $u_p(z) \equiv 1$, and consequently $\operatorname{Re} u_p(z) > 0$ for all $z \in \mathbb{D}$. Suppose that $\operatorname{Re} \kappa \geq |c|/4 + 1$ and $c \neq 0$. Again we denote $h = u_p$, $\psi(r, s, t; z) = 4t + 4\kappa s + czr$

and $\mathbb{E} = \{0\}$. Then we have $\psi\left(h(z), zh'(z), z^2h''(z); z\right) \in \mathbb{E}$ for all $z \in \mathbb{D}$. Now we will use Lemma 2.5 to prove that $\operatorname{Re} h(z) > 0$. If we put $z = x + iy$ and $c = c_1 + ic_2$, where $x, y, c_1, c_2 \in \mathbb{R}$, then

$$\operatorname{Re} \psi(\rho i, \sigma, \mu + vi; x + iy) = 4(\mu + \sigma) + 4(\operatorname{Re}\kappa - 1)\sigma - (c_1 y + c_2 x)\rho$$

for all $\rho, \sigma, \mu, v \in \mathbb{R}$. Let $\rho, \sigma, \mu, v \in \mathbb{R}$ satisfy $\mu + \sigma \leq 0$ and $\sigma \leq -(1 + \rho^2)/2$. Since $\operatorname{Re}\kappa > 1$, we have

$$\operatorname{Re} \psi(\rho i, \sigma, \mu + vi; x + iy) \leq -2(\operatorname{Re}\kappa - 1)\rho^2 - (c_1 y + c_2 x)\rho - 2(\operatorname{Re}\kappa - 1).$$

Set $Q_1(\rho) = -2(\operatorname{Re}\kappa - 1)\rho^2 - (c_1 y + c_2 x)\rho - 2(\operatorname{Re}\kappa - 1)$. This value will be strictly negative for all real ρ, because the discriminant Δ of $Q_1(\rho)$ satisfies

$$\begin{aligned}
\Delta &= (c_1 y + c_2 x)^2 - 16(\operatorname{Re}\kappa - 1)^2 \\
&\leq |c|^2 |z|^2 - 16(\operatorname{Re}\kappa - 1)^2 \\
&< |c|^2 - 16(\operatorname{Re}\kappa - 1)^2 \leq 0
\end{aligned}$$

whenever $z \in \mathbb{D}$. Consequently ψ satisfies the admissibility condition of Lemma 2.5. Hence by Lemma 2.5 we conclude $\operatorname{Re} h(z) = \operatorname{Re} u_p(z) > 0$ for all $z \in \mathbb{D}$. The second part of the theorem is proved just like in the case of Theorem 2.2.   □

Note that results similar to those given in Theorem 2.3 were obtained for confluent hypergeometric functions by S. Kanas and J. Stankiewicz [128].

Another extension of Theorem 2.2 for real parameters is the following theorem.

Theorem 2.4. (Á. Baricz [52]) *Let* $\alpha \in [0, 1/2)$, *and let* b, p, c *be arbitrary real numbers. If* $\kappa \geq (1 - \alpha)(1 - 2\alpha)^{-1/2}|c|/4 + 1$, *then* $\operatorname{Re} u_p(z) > \alpha$ *for all* $z \in \mathbb{D}$. *Moreover, if* $c \neq 0$ *and* $\kappa \geq (1 - \alpha)(1 - 2\alpha)^{-1/2}|c|/4$, *then* u_p *is close-to-convex of order* α *in* $\mathbb{D}$.

Proof. First assume that $c = 0$. Then $u_p(z) \equiv 1$, and consequently $\operatorname{Re} u_p(z) > \alpha$ for all $z \in \mathbb{D}$. Now suppose that $\kappa \geq (1 - \alpha)(1 - 2\alpha)^{-1/2}|c|/4 + 1$ and $c \neq 0$. Define the function $h : \mathbb{D} \to \mathbb{C}$ by

$$h(z) = \frac{u_p(z) - \alpha}{1 - \alpha}.$$

Since u_p satisfies (1.21), h will satisfy the following differential equation:

$$4z^2 h''(z) + 4\kappa z h'(z) + cz\left[h(z) + \frac{\alpha}{1 - \alpha}\right] = 0. \tag{2.9}$$

Using $\psi(r, s, t; z) = 4t + 4\kappa s + cz[r + \alpha/(1 - \alpha)]$ and $\mathbb{E} = \{0\}$, we see that equation (2.9) implies $\psi\left(h(z), zh'(z), z^2 h''(z); z\right) \in \mathbb{E}$ for all $z \in \mathbb{D}$. Next we use Lemma 2.5 to prove that $\operatorname{Re} h(z) > 0$ for all $z \in \mathbb{D}$. For $z = x + iy$, where $x, y \in \mathbb{R}$, we have

$$\operatorname{Re} \psi(\rho i, \sigma, \mu + vi; x + iy) = 4(\mu + \sigma) + 4(\kappa - 1)\sigma - c\rho y + \alpha c x/(1 - \alpha)$$

for all $\rho, \sigma, \mu, \nu \in \mathbb{R}$. Let $\rho, \sigma, \mu, \nu \in \mathbb{R}$ satisfy $\mu + \sigma \leq 0$ and $\sigma \leq -(1+\rho^2)/2$. Since $\kappa - 1 \geq (1-\alpha)(1-2\alpha)^{-1/2}|c|/4 > 0$, we obtain

$$\text{Re}\, \psi\,(\rho\mathrm{i}, \sigma, \mu + \nu\mathrm{i}; x + \mathrm{i}y) \leq -2(\kappa-1)\rho^2 - cy\rho - 2(\kappa-1) + \alpha cx/(1-\alpha).$$

Set $Q_1(\rho) = -2(\kappa-1)\rho^2 - cy\rho - 2(\kappa-1) + \alpha cx/(1-\alpha)$. This value is strictly negative for all real ρ, because the discriminant Δ_1 of $Q_1(\rho)$ satisfies

$$\Delta_1 = c^2 y^2 - 16(\kappa-1)^2 + 8\alpha cx(\kappa-1)/(1-\alpha)$$
$$< c^2(1-x^2) - 16(\kappa-1)^2 + 8\alpha cx(\kappa-1)/(1-\alpha) = Q_2(x) \leq 0,$$

whenever $x^2 + y^2 < 1$ (because z is in the unit disk) and the discriminant Δ_2 of $Q_2(x)$ is negative. Δ_2 has the form

$$\Delta_2 = 4c^2 \left[-16\frac{1-2\alpha}{(1-\alpha)^2}(\kappa-1)^2 + c^2 \right]$$

and this is negative if and only if we have $\kappa \geq (1-\alpha)(1-2\alpha)^{-1/2}|c|/4 + 1$. Hence by Lemma 2.5 we conclude

$$\text{Re}\, h(z) = \text{Re}\left[\frac{1}{1-\alpha}(u_p(z) - \alpha) \right] > 0 \ \text{ for all } \ z \in \mathbb{D},$$

and this means that $\text{Re}\, u_p(z) > \alpha$ for all $z \in \mathbb{D}$.

Now suppose that $\kappa \geq (1-\alpha)(1-2\alpha)^{-1/2}|c|/4$ and $c \neq 0$. Then the above result implies $\text{Re}\, u_{p+1}(z) > \alpha$ for all $z \in \mathbb{D}$. Using again Lemma 1.2 we conclude that

$$\text{Re}\left[\left(-\frac{4\kappa}{c} \right) u_p'(z) \right] = \text{Re}\, u_{p+1}(z) > \alpha \ \text{ for all } \ z \in \mathbb{D}.$$

This means that u_p is close-to-convex of order α in $\mathbb{D}$ with respect to the function $\varphi(z) = -(cz)/(4\kappa)$.   □

Note that results similar to those given in Theorem 2.4 were obtained by J.H. Choi et al. [84] and by S. Ponnusamy and M. Vuorinen [187, 188] for Gaussian and confluent hypergeometric functions.

2.1.3 Sufficient Conditions Involving the Alexander Transform

The following results of L. Fejér and S. Ozaki will be used to prove another sufficient condition for the univalence of the function u_p, defined by (1.20). As we can see this result in particular is sharper than the result of Theorem 2.2. We note that similar results for Gaussian and confluent hypergeometric functions were obtained by S. Ponnusamy [176].

Lemma 2.6. (L. Fejér [98]) *If the function* $f(z) = \sum_{n\geq 1} A_n z^n$, *where* $A_1 = 1$ *and* $A_n \geq 0$ *for all* $n \geq 2$, *is analytic in* $\mathbb{D}$, *and the sequences* $\{nA_n - (n+1)A_{n+1}\}_{n\geq 1}$, $\{nA_n\}_{n\geq 1}$ *both are decreasing, then* f *is starlike in* $\mathbb{D}$.

Lemma 2.7. (L. Fejér [98]) *If the function* $g(z) = \sum_{n\geq 1} C_n z^{n-1}$, *where* $C_1 = 1$ *and* $C_n \geq 0$ *for all* $n \geq 2$, *is analytic in* $\mathbb{D}$ *and if* $\{C_n\}_{n\geq 1}$ *is a convex decreasing sequence, i.e.,* $C_n - 2C_{n+1} + C_{n+2} \geq 0$ *and* $C_n - C_{n+1} \geq 0$ *for all* $n \geq 1$, *then*

$$\mathrm{Re}\, g(z) > 1/2 \ \textit{for all} \ z \in \mathbb{D}.$$

Lemma 2.8. (S. Ozaki [167]) *If the function* $f(z) = z + \sum_{n\geq 2} A_n z^n$ *is analytic in* $\mathbb{D}$ *and if* $1 \geq 2A_2 \geq \ldots \geq nA_n \geq \ldots \geq 0$ *or* $1 \leq 2A_2 \leq \ldots \leq nA_n \leq \ldots \leq 2$, *then* f *is close-to-convex with respect to* $-\mathrm{Log}(1-z)$.

We note that some generalizations of the coefficient conditions for close-to-convex functions from Lemma 2.8 were obtained recently by S. Ponnusamy [176]. Moreover, it is worth to mention here that the coefficient conditions from Lemma 2.8 does not imply necessarily that the function f is starlike in $\mathbb{D}$. For example, the co-efficients of the function $f(z) = z + z^2/2 + z^3/3$ satisfies $1 \geq 2A_2 \geq \ldots \geq nA_n \geq \ldots \geq 0$, but f is not starlike in $\mathbb{D}$. Indeed, for $z = e^{i\theta}$ we have

$$\mathrm{Re}\left[\frac{f(z)}{zf'(z)}\right] = \frac{3+8\cos\theta}{6(1+2\cos\theta)},$$

which is negative for some values of θ, and thus f is not starlike in $\mathbb{D}$. For more details see the paper of I.R. Nezhmetdinov and S. Ponnusamy [162].

It is also important to note here that I.R. Nezhmetdinov and S. Ponnusamy [162] using the duality technique have obtained other sufficient conditions over the MacLaurin coefficients of an analytic and normalized function f that imply its starlikeness. More precisely, I.R. Nezhmetdinov and S. Ponnusamy [162] in particular proved that if the function $f(z) = z + \sum_{n\geq 2} a_n z^n$ is analytic in $\mathbb{D}$ and in addition $2 \leq 3a_2 \leq \ldots \leq (n+1)a_n \leq \ldots$ and $na_n \leq 2$ for all $n \geq 2$, or $2/3 \geq a_2 \geq 2a_3 \geq \ldots \geq (n-1)a_n \geq \ldots \geq 0$ and $na_n \geq a_2$ for all $n \geq 3$, then f is starlike in $\mathbb{D}$.

For convenience, as in the paper of S. Ponnusamy and M. Vuorinen [187], we denote by $\mathscr{CS}^*$ the family of all functions in $\mathscr{A}$, which are close-to-convex with respect to $-\mathrm{Log}(1-z)$ and also starlike in $\mathbb{D}$.

Now we state one of our main results which gives a sufficient condition for the Alexander transform Λ_f of $f(z) = zu_p(z)$ to be in the family of starlike functions. The Alexander transform of f is defined on $\mathbb{D}$ by

$$\Lambda_f(z) = \int_0^z \frac{(f*l)(t)}{t}\, \mathrm{d}t = \int_0^z \frac{f(t)}{t}\, \mathrm{d}t = z + \sum_{n\geq 2} \frac{\alpha_n}{n} z^n = (f*h)(z), \qquad (2.10)$$

where the functions f, l and h are defined by

$$f(z) = z + \sum_{n \geq 2} a_n z^n, \quad l(z) = \frac{z}{1-z} = \sum_{n \geq 1} z^n$$

and

$$h(z) = -\text{Log}(1-z) = \sum_{n \geq 1} \frac{z^n}{n}.$$

We note that the convolution $f * g$, or Hadamard product (see the book of P.L. Duren [90]), of two power series

$$f(z) = z + \sum_{n \geq 2} a_n z^n$$

and

$$g(z) = z + \sum_{n \geq 2} \beta_n z^n$$

is defined as the power series

$$(f * g)(z) = z + \sum_{n \geq 2} a_n \beta_n z^n.$$

Theorem 2.5. (Á. Baricz [52]) *Let b, p be arbitrary real numbers and let $c < 0$. If $N(c) = [-(c+2) + \sqrt{c^2/2 - 4c + 4}]/2$, $\kappa \geq N(c)/2$ and $f(z) = zu_p(z)$, then $\Lambda_f \in \mathcal{CS}^*$. Moreover, we have $\text{Re}\, u_p(z) > 1/2$ for all $z \in \mathbb{D}$.*

Proof. From (1.20) we have

$$f(z) = z + \sum_{n \geq 2} b_{n-1} z^n = \sum_{n \geq 1} \frac{(-c/4)^{n-1} z^n}{(\kappa)_{n-1}(n-1)!}.$$

So in this case the corresponding Alexander transform takes the form $\Lambda_f(z) = \sum_{n \geq 1} A_n z^n$, where $A_n = b_{n-1}/n$ for all $n \geq 1$, i.e.

$$A_n = \frac{(-c/4)^{n-1}}{(\kappa)_{n-1} n!}.$$

Obviously we have $A_1 = 1$. Because $c < 0$ and $4\kappa \geq 2N(c) > -c > 0$, we also have $A_n > 0$ for all $n \geq 2$. Next we prove that the sequence $\{n A_n\}_{n \geq 1}$ is decreasing. Fix any $n \geq 1$. From the definition of the Pochhammer symbol it follows

$$(n+1)A_{n+1} = -\frac{c}{4(\kappa+n-1)} \cdot A_n. \tag{2.11}$$

Using (2.11) we have

$$nA_n - (n+1)A_{n+1} = \frac{U_1(n) \cdot A_n}{4(\kappa+n-1)}, \tag{2.12}$$

where $U_1(n) = 4n^2 + 4(\kappa - 1)n + c$. Since $n^2 \geq 2n - 1$ and $4\kappa > -c$, we have

$$U_1(n) \geq 4(\kappa + 1)n + c - 4 \geq U_1(1) = 4\kappa + c > 0.$$

Consequently, (2.12) yields $nA_n > (n+1)A_{n+1}$. This shows that the sequence $\{nA_n\}_{n \geq 1}$ is strictly decreasing.

Next, we show that the sequence $\{nA_n - (n+1)A_{n+1}\}_{n \geq 1}$ is also decreasing. For convenience we denote $B_n = nA_n - (n+1)A_{n+1}$ for each $n \geq 1$. Fix any $n \geq 1$. Using (2.12), we find that

$$B_n - B_{n+1} = \frac{U_2(n) \cdot A_n}{2(n+1)(\kappa+n)(\kappa+n-1)},$$

where

$$U_2(n) = 2n^4 + 4\kappa n^3 + D_1 n^2 + D_2 n + D_3;$$
$$D_1 = 2\kappa^2 + 2\kappa + c - 2;$$
$$D_2 = 2\kappa^2 + (c-2)\kappa + c;$$
$$D_3 = (c+8\kappa)c/8.$$

Our aim is to show that $U_2(n) > 0$. First we observe that $n^4 \geq 4n^3 - 6n^2 + 4n - 1$ holds. By using this inequality we obtain $U_2(n) \geq V(n)$, where

$$V(n) = 4(\kappa+2)n^3 + (D_1 - 12)n^2 + (D_2 + 8)n + D_3 - 2.$$

Clearly, the coefficient of n^3 in the above expression is nonnegative, since $\kappa > 0$. Therefore using that $n^3 \geq 3n^2 - 3n + 1$, we obtain $V(n) \geq W(n)$, where

$$W(n) = D_4 n^2 + D_5 n + D_6;$$
$$D_4 = 2\kappa^2 + 14\kappa + c + 10;$$
$$D_5 = 2\kappa^2 + (c-14)\kappa + c - 16;$$
$$D_6 = c^2/8 + (c+4)\kappa + 6.$$

Now, we observe that D_4 is also nonnegative, because

$$\kappa \geq [N(c)]/2 > -c/4 > [-7 + \sqrt{29 - 2c}]/2$$

(the value $[-7 + \sqrt{29 - 2c}]/2$ is the greatest root of the equation $D_4 = 0$). Similarly $n^2 \geq 2n - 1$, therefore $W(n) \geq X(n)$, where $X(n) = D_7 n + D_8$, $D_7 = 2D_4 + D_5$ and $D_8 = D_6 - D_4$. Analogously, by the hypothesis, we can deduce easily that

$$D_7 = 6\kappa^2 + (c+14)\kappa + 3c + 4 > 0.$$

Indeed, the relation

$$\kappa \geq [N(c)]/2 > -c/4 > [-(c+14) + \sqrt{c^2 - 44c + 100}]/12 = \kappa_c$$

(here κ_c is the greatest root of the equation $D_7 = 0$) implies that D_7 is nonnegative, and leads to $X(n) \geq X(1)$. In this case

$$X(1) = D_4 + D_5 + D_6 = 4\kappa^2 + 2(c+2)\kappa + c^2/8 + 2c$$

is also positive, because $\kappa \geq N(c)/2 > -c/4 > 0$. Thus, we have proved a chain of inequalities

$$U_2(n) \geq V(n) \geq W(n) \geq X(n) \geq X(1) > 0,$$

which implies $B_n - B_{n+1} > 0$. Thus the sequence $\{nA_n - (n+1)A_{n+1}\}_{n \geq 1}$ is strictly decreasing. By Lemma 2.6 we deduce that Λ_f is starlike in $\mathbb{D}$.

The sequence $\{nA_n\}_{n \geq 1}$ is strictly decreasing and $2A_2 = b_1 = -c/(4\kappa) < 1$. Thus it follows by Lemma 2.8 that Λ_f is close-to-convex with respect to $-\mathrm{Log}(1-z)$. Consequently we have $\Lambda_f \in \mathscr{CS}^*$.

Now, we apply Lemma 2.7 to prove that $\mathrm{Re}\, u_p(z) > 1/2$ for all $z \in \mathbb{D}$. For this consider $g = u_p$. Therefore we have $C_n = b_{n-1} = nA_n$ for all $n \geq 1$ and thus the sequence $\{C_n\}_{n \geq 1}$ is strictly decreasing. In addition we have $C_n - 2C_{n+1} + C_{n+2} = B_n - B_{n+1} > 0$ for all $n \geq 1$. Now Lemma 2.7 yields the asserted property and this completes the proof.   □

Taking $c = -1$ and $b = 1$ in Theorem 2.5, we obtain:

Corollary 2.3. (Á. Baricz [52]) *If* $p \geq [-5 + \sqrt{17/2}]/4 \simeq -0.5211310131\ldots$, *then we have that* $\int_0^z t^{-p/2} J_p(\mathrm{i}\sqrt{t})\, \mathrm{d}t$ *is in* $\mathscr{CS}^*$. *Moreover,* $\mathrm{Re}\, \mathscr{I}_p(\sqrt{z}) > 1/2$ *for all* $z \in \mathbb{D}$.

Proof. We have $N(-1) = [-1 + \sqrt{17/2}]/2 \simeq 0.9577379740\ldots$. Therefore by Theorem 2.5 we have that Λ_f is in $\mathscr{CS}^*$, where $f(z) = zu_p(z) = z\mathscr{I}_p(z^{1/2})$. But from (1.24) it follows that $\mathscr{I}_p(z^{1/2}) = 2^p \Gamma(p+1) z^{-p/2} I_p(z^{1/2})$. On the other hand, (1.5) and (1.7) imply $I_p(z) = \mathrm{i}^{-p} J_p(\mathrm{i}z)$. Thus

$$\Lambda_f(z) = \int_0^z \mathscr{I}_p(t^{1/2})\, \mathrm{d}t = (-2\mathrm{i})^p \Gamma(p+1) \int_0^z t^{-p/2} J_p(\mathrm{i}\sqrt{t})\, \mathrm{d}t$$

and this completes the proof.   □

Another consequence of Theorem 2.5 is obtained by using the recursive relation $4\kappa u_p'(z) = -cu_{p+1}(z)$ (see Lemma 1.2), namely

Corollary 2.4. (Á. Baricz [52]) *Let* b *and* p *be arbitrary real numbers. If* $c < 0$, $\kappa \geq -c/4 - 1$ *and* $\kappa \neq 0$, *then* u_p *is univalent in* $\mathbb{D}$.

Proof. By the proof of Theorem 2.5 the Alexander transform $\int_0^z u_{p+1}(t)\,dt$ is close-to-convex with respect to the convex function $-\mathrm{Log}(1-z)$ if $\kappa + 1 \geq -c/4$, and therefore, in particular, it is univalent. Using the relation $4\kappa u'_p(z) = -cu_{p+1}(z)$, we have

$$\int_0^z u_{p+1}(t)\,dt = -\frac{4\kappa}{c}\int_0^z u'_p(t)\,dt = -\frac{4\kappa}{c}[u_p(z) - 1].$$

Consequently $-4\kappa[u_p(z) - 1]/c$ is univalent in $\mathbb{D}$. Since the addition of a constant and the multiplication by a nonzero quantity do not disturb the univalence, we immediately deduce that u_p is univalent in $\mathbb{D}$. This completes the proof. □

Remark 2.2. By Theorem 2.2 u_p is univalent in $\mathbb{D}$ when $\kappa \geq |c|/4$ and $c \neq 0$. If we consider $c < 0$, then the above conditions become $\kappa \geq -c/4$. Since $-c/4 > -c/4 - 1$, it follows that the result of Corollary 2.4 is better than the result of Theorem 2.2.

Using the idea of the proof of Corollary 2.4, we deduce the following corollary, which is also an immediate consequence of Theorem 2.5.

Corollary 2.5. (Á. Baricz [52]) *If for $b, p \in \mathbb{R}$, $c < 0$ we have $\kappa + 1 \geq N(c)/2$, where $N(c)$ is defined in Theorem 2.5, then the function $(-4\kappa/c)[u_p(z) - 1]$ is in $\mathscr{CS}^*$.*

2.1.4 Sufficient Conditions Involving Results of L. Fejér

As we can see in this section a direct application of Lemma 2.7 combined with Lemma 2.3 gives in some particular cases a sharper result than Corollary 2.4.

Lemma 2.9. (Á. Baricz [43]) *Let $b, p \in \mathbb{R}$, $c < 0$ and $\alpha \in [0, 1)$ be fixed numbers and let*

$$c_0 = \frac{c}{8(\alpha - 1)};$$

$$c_1 = \frac{-13 + \sqrt{77 - 2c}}{2};$$

$$c_2 = \frac{-(c/2 + 27) + \sqrt{c^2/4 - 23c + 169}}{10}.$$

If $\kappa \geq \max\{c_0, c_1, c_2\}$, then the sequence $\{A_n\}_{n\geq 1}$, defined by

$$A_1 = 1 \quad and \quad A_n = \frac{b_{n-1}}{2(1 - \alpha)} \quad for\ all\ n \geq 2,$$

is a nonnegative convex decreasing sequence, where b_n is defined by (1.18).

Proof. First we prove that the sequence $\{A_n\}_{n\geq 1}$ is decreasing. The inequality $A_2 \leq A_1$ is equivalent to $\kappa \geq c_0 = c/[8(\alpha - 1)]$. Now we prove that $A_n - A_{n+1} \geq 0$

for all $n \geq 2$. Let $n \geq 2$ be fixed. By definition we have $A_n - A_{n+1} = [b_{n-1} - b_n]/[2(1-\alpha)]$. Therefore using the recursive relation $4n(\kappa + n - 1)b_n = -cb_{n-1}$, we obtain

$$A_n - A_{n+1} = \frac{[4n^2 + 4(\kappa - 1)n + c] \cdot b_{n-1}}{8(1-\alpha)n(\kappa + n - 1)}.$$

Let us denote $M_1(n) = 4n^2 + 4(\kappa - 1)n + c$. From $(n-2)^2 \geq 0$, we obtain $n^2 \geq 4n - 4$. Therefore $M_1(n) \geq 4(\kappa + 3)n + c - 16$. By the hypothesis $\kappa > 0$, and therefore $M_1(n) \geq 8\kappa + c + 8$ and this value is nonnegative, because $8\kappa \geq c/(\alpha - 1) \geq -c - 8$. Consequently we have shown that $A_n - A_{n+1} \geq 0$.

The assumptions of the lemma imply that the sequence $\{A_n\}_{n \geq 1}$ is nonnegative. Therefore, we need only to show that this sequence is convex, i.e.

$$A_n - 2A_{n+1} + A_{n+2} \geq 0 \quad \text{for all } n \geq 1. \tag{2.13}$$

Observe that $A_1 - 2A_2 + A_3 \geq 0$ is equivalent to the inequality $16\kappa + c + 16 \geq 0$. But $16\kappa + c + 16 \geq 8(\kappa + 1) > 0$ and therefore the inequality (2.13) holds for $n = 1$. Next we verify inequality (2.13) for $n \geq 2$. Let $n \geq 2$ be fixed. From the definition of A_n, we find that

$$A_n - 2A_{n+1} + A_{n+2} = \frac{b_{n-1}M_2(n)}{4(1-\alpha)n(n+1)(\kappa + n - 1)(\kappa + n)},$$

where

$$M_2(n) = 2n^4 + 4\kappa n^3 + D_1 n^2 + D_2 n + D_3;$$
$$D_1 = 2\kappa^2 + 2\kappa + c - 2;$$
$$D_2 = 2\kappa^2 + (c - 2)\kappa + c;$$
$$D_3 = (8\kappa + c)c/8.$$

With some computation we get

$$A_n - 2A_{n+1} + A_{n+2} = \frac{b_{n-1}M_3(n)}{2(1-\alpha)n(n+1)(\kappa + n - 1)(\kappa + n)},$$

where

$$M_3(n) = (n-2)^4 + 2(\kappa + 4)(n-2)^3 + E_1(n-2)^2 + E_2(n-2) + E_3;$$
$$E_1 = \kappa^2 + 13\kappa + c/2 + 23;$$
$$E_2 = 5\kappa^2 + (c/2 + 27)\kappa + 5c/2 + 28;$$
$$E_3 = 9\kappa^2 + (3c/2 + 57)\kappa + c^2/16 + 9c/2 + 81.$$

If $c \in [-37, 0)$, then the greatest real root $c_1 = [-13 + \sqrt{77 - 2c}]/2$ of the equation $E_1 = 0$ is negative. From the condition $c_1 < c/[8(\alpha - 1)] \leq \kappa$ we deduce that E_1 is nonnegative. Analogously $c_2 = [-(c/2 + 27) + \sqrt{c^2/4 - 23c + 169}]/10$ is the greatest real root of the equation $E_2 = 0$ and c_2 is also negative (when $c \in [-37, 0)$). This and the condition $c_2 < \kappa$, imply that E_2 is nonnegative. In this case ($c \in [-37, 0)$) the greatest root of the equation $E_3 = 0$ is $c_3 = [-(c/2 + 19) + \sqrt{c + 37}]/6$ and this root is nonnegative. Moreover $c_3 \le c/[8(\alpha - 1)]$, therefore from the condition $c_0 \le \kappa$ we deduce that E_3 is also nonnegative.

For $c < -37$ the expression E_3 is nonnegative and because of $\kappa \ge \max\{c_0, c_1, c_2\}$ it follows that E_1 and E_2 are also positive.

Therefore all the numbers E_i ($i \in \{1, 2, 3\}$) are nonnegative. From this observation we deduce that $M_3(n) \ge 0$. Thus $\{A_n\}_{n \ge 1}$ is a convex sequence.   □

Using Lemma 2.9 we obtain the next result.

Theorem 2.6. (Á. Baricz [43]) *Let* $\alpha \in [0, 1)$, $b, p \in \mathbb{R}$, $c < 0$. *If* $\kappa \ge \max\{c_0, c_1, c_2\}$ *(where c_0, c_1, c_2 are defined in Lemma 2.9), then* $\operatorname{Re} u_p(z) > \alpha$ *for all* $z \in \mathbb{D}$.

Proof. For all $n \ge 1$ let A_n be defined as in Lemma 2.9. Then the conclusion $\operatorname{Re} u_p(z) > \alpha$ for all $z \in \mathbb{D}$ is equivalent to

$$\operatorname{Re}\left[1 + \sum_{n \ge 2} A_n z^{n-1}\right] > 1/2 \ \text{ for all } \ z \in \mathbb{D}.$$

By Lemma 2.9 and the hypotheses, we observe that the sequence $\{A_n\}_{n \ge 1}$ is nonnegative, convex and decreasing. Therefore the conclusion follows from Lemma 2.7.
                                                                                            □

In the proof of Lemma 2.9 we have seen that $\max\{c_0, c_1, c_2\} = c_0$ for $c \in [-37, 0)$. Therefore by taking $\alpha - 0$ in Theorem 2.6 we immediately obtain the following result:

Theorem 2.7. (Á. Baricz [43]) *If* $\kappa \ge -c/8 - 1$ *for* $b, p \in \mathbb{R}$, $c \in [-37, 0)$, *then* u_p *is univalent in* $\mathbb{D}$.

Proof. We apply Theorem 2.6 and conclude that $\operatorname{Re} u_{p+1}(z) > 0$ for all $z \in \mathbb{D}$. Using Lemma 1.2, it follows that

$$\operatorname{Re}\left[-\frac{4\kappa}{c} u_p'(z)\right] = \operatorname{Re} u_{p+1}(z) > 0 \ \text{ for all } \ z \in \mathbb{D}.$$

Thus, u_p is close-to-convex with respect to the function $\varphi(z) = -(cz)/(4\kappa)$. By Lemma 2.3 it follows that u_p is univalent in $\mathbb{D}$.   □

Remark 2.3. In Corollary 2.4 we have seen that u_p is univalent in $\mathbb{D}$ for arbitrary $b \in \mathbb{R}$, $c < 0$, $\kappa \ge -c/4 - 1$. Clearly we have $-c/4 - 1 > -c/8 - 1$. Hence for $c \in [-37, 0)$ the result of Theorem 2.7 is better than the result of Corollary 2.4.

Now we end this section with the following consequence of Theorem 2.6.

Corollary 2.6. (Á. Baricz [43]) *If* $\alpha \in [0,1)$, $c \in [-37,0)$, $b \in \mathbb{R}$ *and in addition* $8(1-\alpha)(\kappa+1)+c \geq 0$, *then* $\mathrm{Re}[(-4\kappa/c)u_p'(z)] > \alpha$ *for all* $z \in \mathbb{D}$.

Proof. By Lemma 1.2 we know that $4\kappa u_p'(z) = -cu_{p+1}(z)$ for all $z \in \mathbb{D}$. Therefore we have $u_{p+1}(z) = (-4\kappa/c)u_p'(z)$ for all $z \in \mathbb{D}$. Using Theorem 2.6 for $p+1$ the inequality follows. □

2.2 Starlikeness and Convexity Properties of Generalized Bessel Functions

Our aim in this section is to find sufficient conditions for the function u_p, i.e. for the hypergeometric function $_0F_1(\kappa, -cz/4)$, to be starlike, convex, starlike of order α, convex of order α and close-to-convex with respect to some known functions. Moreover, in the third part of this section we present some immediate applications of convexity and univalence involving Bessel functions associated with the Hardy space of analytic functions.

Note that the results included in this section were proved by the author in the papers [38, 42, 43, 52]. As in the previous section we begin with an application of Jack's lemma.

2.2.1 Sufficient Conditions Involving Jack's Lemma

The following lemma proved by S. Owa and H.M. Srivastava [168] will be useful in this section to find sufficient conditions for the functions u_p, zu_p to be starlike, and for the function u_p to be convex of order α.

Lemma 2.10. (S. Owa and H.M. Srivastava [168]) *If* $f \in \mathscr{A}$ *and*

$$\left| \frac{zf'(z)}{f(z)} - 1 \right|^{1-\beta} \cdot \left| \frac{zf''(z)}{f'(z)} \right|^{\beta} < (1-\alpha)^{1-2\beta}(1 - 3\alpha/2 + \alpha^2)^{\beta}$$

for some fixed $\alpha \in [0, 1/2]$ *and* $\beta \geq 0$, *and for all* $z \in \mathbb{D}$, *then* f *is in the class* $\mathscr{S}^*(\alpha)$.

By applying Lemma 2.10, we prove the following theorems concerning the starlikeness of the generalized Bessel functions of order α.

Theorem 2.8. (Á. Baricz [54]) *If the function* u_p, *defined by* (1.20)*, satisfies the condition*

$$\left| \frac{zu_p'(z)}{u_p(z)} \right| < 1 - \alpha,$$

where $\alpha \in [0, 1/2]$ *and* $z \in \mathbb{D}$, *then* $zu_p \in \mathscr{S}^*(\alpha)$.

Proof. Define the function g by $g(z) = zu_p(z)$ for all $z \in \mathbb{D}$. The given condition becomes

$$\left| \frac{zg'(z)}{g(z)} - 1 \right| < 1 - \alpha,$$

where $z \in \mathbb{D}$. By taking $\beta = 0$ in Lemma 2.10, we thus conclude from the previous inequality that $g \in \mathscr{S}^*(\alpha)$, which proves Theorem 2.8.   □

Remark 2.4. Evidently, since $\mathscr{S}^*(\alpha) \subseteq \mathscr{S}^* \subset \mathscr{S}$, the function zu_p is univalent in $\mathbb{D}$ under the hypothesis of Theorem 2.8.

Theorem 2.9. (Á. Baricz [54]) *If the function u_p, defined by (1.20), satisfies the condition*

$$\left| \frac{zu_p''(z)}{u_p'(z)} \right| < \frac{1 - 3\alpha/2 + \alpha^2}{1 - \alpha},$$

where $\alpha \in [0, 1/2]$ and $z \in \mathbb{D}$, then it is starlike of order α with respect to 1.

Proof. We define the function $h : \mathbb{D} \to \mathbb{C}$ by $h(z) = [u_p(z) - b_0]/b_1$. Then $h \in \mathscr{A}$ and

$$\left| \frac{zh''(z)}{h'(z)} \right| = \left| \frac{zu_p''(z)}{u_p'(z)} \right| < \frac{1 - 3\alpha/2 + \alpha^2}{1 - \alpha},$$

where $\alpha \in [0, 1/2]$ and $z \in \mathbb{D}$. Consequently, it follows by Lemma 2.10 (choose $\beta = 1$) that $h \in \mathscr{S}^*(\alpha)$, i.e. h is starlike of order α with respect to the origin for $\alpha \in [0, 1/2]$. Now Theorem 2.9 follows from the definition of the function h, because $b_0 = 1$.   □

Remark 2.5. Since $\mathscr{S}^*(\alpha) \subseteq \mathscr{S}^* \subset \mathscr{S}$, the function $z \mapsto [u_p(z) - b_0]/b_1$ is univalent in $\mathbb{D}$ under the hypothesis of Theorem 2.9.

Corresponding to Theorem 2.8, we have the following result on the convexity of the generalized Bessel functions.

Theorem 2.10. (Á. Baricz [54]) *If for $\alpha \in [0, 1/2]$ and $c \neq 0$ we have*

$$\left| \frac{zu_{p+1}'(z)}{u_{p+1}(z)} \right| < 1 - \alpha$$

for all $z \in \mathbb{D}$, then $u_p \in \mathscr{C}(\alpha)$.

Proof. Theorem 2.8 implies that $zu_{p+1} \in \mathscr{S}^*(\alpha)$. On the other hand, Lemma 1.2 yields $b_1 zu_{p+1}(z) = zu_p'(z)$, where $b_1 = -c/(4\kappa) \neq 0$. Consequently $zu_p' \in \mathscr{S}^*(\alpha)$ and therefore $u_p \in \mathscr{C}(\alpha)$.   □

Remark 2.6. The results of the Theorems 2.8, 2.9 and 2.10 hold in the disk $|z| < 4/|c|$, which for $0 < |c| < 4$ is larger than the unit disk. This can be proved by using the same argument as in Remark 2.1. By applying the Theorems 2, 3, and 5 by S. Owa and H.M. Srivastava [168] to the function $F(z) = {}_0F_1(\kappa, z)$, using the equality $F(z) = u_p(-4z/c)$ and changing z with $-cz/4$, we obtain that the Theorems 2.8, 2.9 and 2.10 hold in the disk $|z| < 4/|c|$.

2.2.2 Sufficient Conditions Involving the Admissible Function Method

In this section we use the admissible function method (or, with other words the technique of differential subordinations), i.e. Lemma 2.5, to give sufficient conditions for the function zu_p to be starlike and for the function u_p to be convex.

Theorem 2.11. (Á. Baricz [38]) *Let $b, c, p \in \mathbb{R}$, of which $c \neq 0$. The functions w_p and u_p, defined by (1.15) and (1.20), respectively, have the following properties:*

(a) *If $\kappa \geq |c|/4 + 1/2$, then $u_p \in \mathscr{C}$.*
(b) *If $\kappa \geq |c|/4 + 3/2$, then $zu_p \in \mathscr{S}^*$.*
(c) *If $\kappa \geq |c|/2 + 1$, then $zu_p \in \mathscr{S}^*(1/2)$.*
(d) *If $\kappa \geq |c|/2 + 1$, then $z^{1-p}w_p \in \mathscr{S}^*$.*

Proof. (a) Since $\kappa > |c|/4$ and $c \neq 0$, Theorem 2.2 implies $\operatorname{Re} u_{p+1}(z) > 0$ for all $z \in \mathbb{D}$. According to Lemma 1.2 it follows that $u'_p(z) \neq 0$ for all $z \in \mathbb{D}$. Define $q : \mathbb{D} \to \mathbb{C}$ by

$$q(z) = 1 + \frac{zu''_p(z)}{u'_p(z)}.$$

The function q is analytic in $\mathbb{D}$ and $q(0) = 1$. Suppose that $z \neq 0$. Since u_p satisfies the differential equation (1.21), we have

$$4zu''_p(z) + 4\kappa u'_p(z) + cu_p(z) = 0.$$

If we differentiate this equation, we obtain

$$4zu'''_p(z) + 4(\kappa + 1)u''_p(z) + cu'_p(z) = 0.$$

We know that $u'_p(z) \neq 0$, therefore if we divide both sides of this equation with $u'_p(z)$, and multiply with z, we obtain

$$4 \left[\frac{zu'''_p(z)}{u''_p(z)} \right] \left[\frac{zu''_p(z)}{u'_p(z)} \right] + 4(\kappa + 1) \left[\frac{zu''_p(z)}{u'_p(z)} \right] + cz = 0. \qquad (2.14)$$

Now we differentiate logarithmically and multiply with z on both sides of the equation $q(z) - 1 = zu''_p(z)/u'_p(z)$. Thus we obtain

$$\frac{zq'(z)}{q(z) - 1} = 1 + \frac{zu'''_p(z)}{u''_p(z)} - [q(z) - 1],$$

and therefore

$$\frac{zu_p'''(z)}{u_p''(z)} = \frac{zq'(z)+q^2(z)-3q(z)+2}{q(z)-1}.$$

In view of (2.14) this result reveals that q satisfies the following differential equation:

$$4zq'(z)+4q^2(z)+4(\kappa-2)q(z)+cz-4(\kappa-1)=0. \qquad (2.15)$$

Obviously, this equation is also valid when $z=0$.

If we use $\psi(r,s;z)=4s+4r^2+4(\kappa-2)r+cz-4(\kappa-1)$ and $\mathbb{E}=\{0\}$, then (2.15) implies $\psi(q(z),zq'(z);z)\in\mathbb{E}$ for all $z\in\mathbb{D}$. Now we use Lemma 2.5 to prove that $\mathrm{Re}\,q(z)>0$ for all $z\in\mathbb{D}$. For $z=x+iy\in\mathbb{D}$ (with $x,y\in\mathbb{R}$) and $\rho,\sigma\in\mathbb{R}$ satisfying $\sigma\le-(1+\rho^2)/2$, we obtain

$$\begin{aligned}\mathrm{Re}\,\psi(\rho i,\sigma;x+iy)&=4\sigma-4\rho^2+cx-4(\kappa-1)\\&\le-6\rho^2+cx-2(2\kappa-1)\\&<|c|-2(2\kappa-1)\le0.\end{aligned}$$

By Lemma 2.5 we conclude that $\mathrm{Re}\,q(z)>0$ for all $z\in\mathbb{D}$, which shows that u_p is convex in $\mathbb{D}$.

(b) According to (a) the function u_{p-1} is convex. By Alexander's duality theorem, i.e. Lemma 2.1 it follows that $zu_{p-1}'\in\mathscr{S}^*$. But, on the other hand, Lemma 1.2 yields

$$czu_p(z)=-4(\kappa-1)zu_{p-1}'(z).$$

Consequently, it results that $zu_p\in\mathscr{S}^*$.

(c) According to Theorem 2.2 we have $\mathrm{Re}\,u_p(z)>0$ for all $z\in\mathbb{D}$, hence $u_p(z)\neq0$ for all $z\in\mathbb{D}$. Define $q:\mathbb{D}\to\mathbb{C}$ by

$$q(z)=1+2\frac{zu_p'(z)}{u_p(z)}.$$

The function q is analytic in $\mathbb{D}$ and $q(0)=1$. Assume that $z\neq0$. Because u_p satisfies the equation (1.21), it satisfies the following equation too:

$$4\left[\frac{zu_p''(z)}{u_p'(z)}\right]\left[\frac{zu_p'(z)}{u_p(z)}\right]+4\kappa\left[\frac{zu_p'(z)}{u_p(z)}\right]+cz=0. \qquad (2.16)$$

We proceed as in part (a), we differentiate logarithmically and multiply with z the expression $[q(z)-1]/2=[zu_p'(z)]/u_p(z)$. We obtain

$$\frac{zu_p''(z)}{u_p'(z)}=\frac{2zq'(z)+q^2(z)-4q(z)+3}{2(q(z)-1)}. \qquad (2.17)$$

In view of (2.16) this result reveals that q satisfies the differential equation

$$2zq'(z) + q^2(z) + 2(\kappa - 2)q(z) + cz - 2(\kappa - 3/2) = 0, \qquad (2.18)$$

which is also valid when $z = 0$.

If $\psi(r,s;z) = 2s + r^2 + 2(\kappa - 2)r + cz - 2(\kappa - 3/2)$ and $\mathbb{E} = \{0\}$, then (2.18) implies $\psi(q(z), zq'(z); z) \in \mathbb{E}$ for all $z \in \mathbb{D}$. We use Lemma 2.5 to prove that $\operatorname{Re} q(z) > 0$ for all $z \in \mathbb{D}$. For $z = x + iy \in \mathbb{D}$ with $x, y \in \mathbb{R}$, and $\rho, \sigma \in \mathbb{R}$ satisfying $\sigma \leq -(1 + \rho^2)/2$, we obtain

$$\operatorname{Re} \psi(\rho i, \sigma; x + iy) = 2\sigma - \rho^2 + cx - 2(\kappa - 3/2)$$
$$\leq -2\rho^2 + cx - 2(\kappa - 1)$$
$$< |c| - 2(\kappa - 1) \leq 0.$$

By Lemma 2.5 we conclude that $\operatorname{Re} q(z) > 0$ for all $z \in \mathbb{D}$. Now consider the function $g_p : \mathbb{D} \to \mathbb{C}$, defined by $g_p(z) = z u_p(z)$. Since

$$\frac{z g_p'(z)}{g_p(z)} = \frac{1}{2} + \frac{1}{2}q(z),$$

it follows that

$$\operatorname{Re} \left[\frac{z g_p'(z)}{g_p(z)} \right] > \frac{1}{2} \quad \text{for all } z \in \mathbb{D},$$

which shows that g_p is starlike of order $1/2$.

(d) Define the function $h_p : \mathbb{D} \to \mathbb{C}$ by $h_p(z) = z^{1-p}w_p(z)$. Since $h_p(z) = a_0(p)z u_p(z^2)$, where $a_0(p) = [2^p \Gamma(\kappa)]^{-1}$ (see (1.19)), it follows that

$$\frac{z h_p'(z)}{h_p(z)} = 2 \left[\frac{z^2 g_p'(z^2)}{g_p(z^2)} - \frac{1}{2} \right].$$

But from part (c) we know that the function $g_p : \mathbb{D} \to \mathbb{C}$, defined by $g_p(z) = z u_p(z)$, is starlike of order $1/2$. Thus we conclude that

$$\operatorname{Re} \left[\frac{z h_p'(z)}{h_p(z)} \right] > 0 \quad \text{for all } z \in \mathbb{D},$$

and hence h_p is starlike.   □

Taking in Theorems 2.2 and 2.11 the values $b = c = 1$, we obtain the following corollary.[3]

---

[3] Note that in the papers of Á. Baricz [38] and V. Selinger [208] the expression "$f_p(z) = z^{-p/2}J_p(z^{1/2})$" should be replaced with "$f_p(z) = 2^p \Gamma(p+1)z^{-p/2}J_p(z^{1/2})$."

Corollary 2.7. (V. Selinger [208]) *Let $p \in \mathbb{R}$. For the function*

$$z \mapsto \mathscr{J}_p(z^{1/2}) = 2^p \Gamma(p+1) z^{-p/2} J_p(z^{1/2}),$$

where J_p is defined by (1.5), the following properties are true:

(a) *If $p \geq 1/4$, then $\operatorname{Re} \mathscr{J}_p(z^{1/2}) > 0$ for all $z \in \mathbb{D}$.*
(b) *If $p \geq -3/4$, then $z \mapsto \mathscr{J}_p(z^{1/2})$ is univalent in $\mathbb{D}$.*
(c) *If $p \geq -1/4$, then $z \mapsto \mathscr{J}_p(z^{1/2})$ is convex in $\mathbb{D}$.*
(d) *If $p \geq 3/4$, then $z \mapsto z \mathscr{J}_p(z^{1/2})$ is starlike in $\mathbb{D}$.*
(e) *If $p \geq 1/2$, then $z \mapsto z \mathscr{J}_p(z^{1/2})$ is starlike of order $1/2$ in $\mathbb{D}$.*
(f) *If $p \geq 1/2$, then $z \mapsto z^{1-p} J_p(z)$ is starlike in $\mathbb{D}$.*

Remark 2.7. We note that R.K. Brown [78, Theorem 3] proved that if $p = p_1 + i p_2$ ($p_1, p_2 \in \mathbb{R}$) is a complex number satisfying one of the following conditions $p_2 \leq p_1 \in [0,1)$ or $p_1 \geq 1$, $2p_1 - 1 > p_2^2$, then the normalized Bessel function $z \mapsto z^{1-p} J_p(z)$ is regular, univalent, and spirallike in every circle $|z| < r = \rho_\mu^*$, where $\mu^2 = \operatorname{Re}[p^2]$, $\mu > 0$, and ρ_μ^* is the smallest positive zero of the function

$$r \mapsto r J_\mu'(r) + \operatorname{Re}[1 - p] J_\mu(r).$$

In the particular case when p is real the function $z \mapsto z^{1-p} J_p(z)$ is starlike in $|z| < \rho_\mu^*$, but it is not univalent in any larger circle. The method used by R.K. Brown is completely different from the method of differential subordinations, but if p is real the inequality $2p_1 - 1 > p_2^2$ becomes $p_1 > 1/2$, which appears in part (f) of Corollary 2.7. Nevertheless the method of differential subordinations gives only fairly weak estimates, because for real values of p we obtain that ρ_μ^* is precisely ρ_p obtained by E. Kreyszig and J. Todd [138], and the circle $|z| < \rho_p$ is larger than the unit disk (we know the inequalities $2\sqrt{(p+1)/3} < \rho_p < \sqrt{12(p+2)/5}$ for all $p > -1$). In fact ρ_p is the smallest positive zero of the function $z \mapsto z^{1-p} J_p(z)$. The authors proved in [138] that for $p > -1$ the radius of univalence ρ_p of the function $z \mapsto z^{1-p} J_p(z)$ increases steadily with p. For more details the interested reader is referred also to the paper of H.S. Wilf [229], where the author proved that

$$\rho_p = \sqrt{2p} \left[1 + 1/(4p) + \mathscr{O}(p^{-2}) \right], \quad p \to \infty.$$

It is also worth mentioning that R.K. Brown [79] proved that if $p \in (-1/2, 0)$, then the normalized Bessel function $z \mapsto z^{1-p} J_p(z)$ is starlike in $\{z \in \mathbb{C} : |z| < \rho_p^*\}$, where ρ_p^* is the smallest positive zero of the function $\rho J_p'(\rho) + (1-p) J_p(\rho)$. This result is sharp and extends the author's previous result [78] obtained for $p \geq 0$.

Remark 2.8. Observe that if we choose $c = -1$ and $b = 1$ in Theorems 2.2 and 2.11, then for the function

$$z \mapsto \mathscr{I}_p(z^{1/2}) = 2^p \Gamma(p+1) z^{-p/2} I_p(z^{1/2}),$$

where I_p is the modified Bessel function of the first kind, defined by (1.7), the properties are the same as for the function $z \mapsto \mathscr{I}_p(z^{1/2})$, because in this case $|c| = 1$.

If we take $b = 2$ and $c = 1$, then the Theorems 2.2 and 2.11 yield.

Corollary 2.8. (Á. Baricz [38]) *Let $p \in \mathbb{R}$. For the function*

$$z \mapsto \mathscr{I}_{p+\frac{1}{2}}(z^{1/2}) = 2^{p+\frac{1}{2}} \Gamma\left(p + \frac{3}{2}\right) z^{-\frac{1}{2}(p+\frac{1}{2})} J_{p+\frac{1}{2}}(z^{1/2}),$$

where J_p is defined by (1.5), the following assertions are true:

(a) *If $p \geq -1/4$, then Re $\mathscr{I}_{p+1/2}(z^{1/2}) > 0$ for all $z \in \mathbb{D}$.*
(b) *If $p \geq -5/4$, then $z \mapsto \mathscr{I}_{p+1/2}(z^{1/2})$ is univalent in $\mathbb{D}$.*
(c) *If $p \geq -3/4$, then $z \mapsto \mathscr{I}_{p+1/2}(z^{1/2})$ is convex in $\mathbb{D}$.*
(d) *If $p \geq 1/4$, then $z \mapsto z\,\mathscr{I}_{p+1/2}(z^{1/2})$ is starlike in $\mathbb{D}$.*
(e) *If $p \geq 0$, then $z \mapsto z\,\mathscr{I}_{p+1/2}(z^{1/2})$ is starlike of order $1/2$ in $\mathbb{D}$.*
(f) *If $p \geq 0$, then $z \mapsto z^{1-p} j_p(z)$ is starlike in $\mathbb{D}$.*

Remark 2.9. If we consider the Bessel functions of the first kind of order p, which can be expressed with elementary functions as $\cos, \sin, \cosh$ and $\sinh$, we may obtain some interesting examples of univalent, starlike and convex functions, using the definitions of the functions $\mathscr{I}_p$ and $\mathscr{I}_p$.

(a) The functions $\cos \sqrt{z}$, $\cosh \sqrt{z}$ and $(\sin \sqrt{z})/\sqrt{z}$ are univalent in $\mathbb{D}$, because the functions

$$\mathscr{I}_{-1/2}(z^{1/2}) = \sqrt{\pi/2} \cdot z^{1/4} J_{-1/2}(\sqrt{z}) = \cos \sqrt{z},$$

$$\mathscr{I}_{-1/2}(z^{1/2}) = \sqrt{\pi/2} \cdot z^{1/4} I_{-1/2}(\sqrt{z}) = \cosh \sqrt{z},$$

$$\mathscr{I}_{1/2}(z^{1/2}) = \sqrt{\pi/2} \cdot z^{-1/4} J_{1/2}(\sqrt{z}) = \frac{\sin \sqrt{z}}{\sqrt{z}}$$

satisfy the conditions of Theorem 2.2.
(b) The functions $(\sin \sqrt{z} - \sqrt{z}\cos \sqrt{z})/(z\sqrt{z})$, $(\sinh \sqrt{z})/\sqrt{z}$ and $(\sin \sqrt{z})/\sqrt{z}$ are convex, because the functions

$$\mathscr{I}_{3/2}(z^{1/2}) = \frac{3}{z}\left(\frac{\sin \sqrt{z}}{\sqrt{z}} - \cos \sqrt{z}\right),$$

$$\mathscr{I}_{1/2}(z^{1/2}) = \sqrt{\pi/2} \cdot z^{-1/4} I_{1/2}(\sqrt{z}) = \frac{\sinh \sqrt{z}}{\sqrt{z}}$$

and $\mathscr{I}_{1/2}(z^{1/2}) = (\sin \sqrt{z})/\sqrt{z}$ satisfy the conditions of Theorem 2.11 (first part).

(c) The functions $\sqrt{z}\sin\sqrt{z}$ and $\sqrt{z}\sinh\sqrt{z}$ are starlike functions of order $1/2$, because the functions

$$z\mathscr{J}_{1/2}(z^{1/2}) = \sqrt{\pi/2}\cdot z^{3/4}J_{1/2}(\sqrt{z}) = \sqrt{z}\sin\sqrt{z}$$

and analogously

$$z\mathscr{I}_{1/2}(z^{1/2}) = \sqrt{\pi/2}\cdot z^{3/4}I_{1/2}(\sqrt{z}) = \sqrt{z}\sinh\sqrt{z}$$

satisfy the conditions of Theorem 2.11 (third part).

(d) The functions $\sin z$, $\sinh z$ and $(\sin\sqrt{z})/\sqrt{z} - \cos\sqrt{z}$ are starlike functions in $\mathbb{D}$, because the functions $z^{1/2}J_{1/2}(z) = \sqrt{2/\pi}\cdot\sin z$, $z^{1/2}I_{1/2}(z) = \sqrt{2/\pi}\cdot\sinh z$ and

$$z\mathscr{J}_{3/2}(z^{1/2}) = 3\left(\frac{\sin\sqrt{z}}{\sqrt{z}} - \cos\sqrt{z}\right)$$

satisfy the conditions of Theorem 2.11 (second and fourth part).

Note that similar results as those given in Theorem 2.11 were obtained by S.S. Miller and P.T. Mocanu [154] for Gaussian and confluent hypergeometric functions.

Theorem 2.11 can be extended for complex parameters, as the following result reveals.

Theorem 2.12. (Á. Baricz [42]) *Let $b, p, c \in \mathbb{C}$, of which $c \neq 0$. Then w_p and u_p, defined by (1.15) and (1.20), respectively, have the following properties*[4]:

(a) *If* $\operatorname{Re}\kappa \geq |c|/4 + (\operatorname{Im}\kappa)^2/6 + 1/2$, *then* $u_p \in \mathscr{C}$.
(b) *If* $\operatorname{Re}\kappa \geq |c|/4 + (\operatorname{Im}\kappa)^2/6 + 3/2$, *then* $zu_p \in \mathscr{S}^*$.
(c) *If* $\operatorname{Re}\kappa \geq |c|/2 + (\operatorname{Im}\kappa)^2/4 + 1$, *then* $zu_p \in \mathscr{S}^*(1/2)$.
(d) *If* $\operatorname{Re}\kappa \geq |c|/2 + (\operatorname{Im}\kappa)^2/4 + 1$, *then* $z^{1-p}w_p \in \mathscr{S}^*$.

Proof. The proof of this theorem is very similar to the proof of Theorem 2.11. For convenience we just sketch the proof.

(a) Since $\operatorname{Re}(\kappa+1) \geq |c|/4 + (\operatorname{Im}\kappa)^2/6 + 3/2 > |c|/4 + 1$ and $c \neq 0$, Theorem 2.3 implies $\operatorname{Re}u_{p+1}(z) > 0$ for all $z \in \mathbb{D}$. According to Lemma 1.2 it follows that $u_p'(z) \neq 0$ for all $z \in \mathbb{D}$. Define $h : \mathbb{D} \to \mathbb{C}$ by

$$h(z) = 1 + \frac{zu_p''(z)}{u_p'(z)}.$$

The function h is analytic in $\mathbb{D}$ and $h(0) = 1$. Just like in the proof of part (a) of Theorem 2.11 it is shown that h satisfies the differential equation

$$4zh'(z) + 4h^2(z) + 4(\kappa-2)h(z) + cz - 4(\kappa-1) = 0. \tag{2.19}$$

---

[4] Note that there is a small error in the mentioned paper [42], namely the expression "$2\operatorname{Im}\kappa - 1$" should be replaced with "$2\operatorname{Im}\kappa$."

If
$$\psi(r,s;z) = 4s + 4r^2 + 4(\kappa - 2)r + cz - 4(\kappa - 1)$$

and $\mathbb{E} = \{0\}$, then (2.19) implies $\psi(h(z), zh'(z); z) \in \mathbb{E}$ for all $z \in \mathbb{D}$. We use Lemma 2.5 to prove that $\operatorname{Re} h(z) > 0$ for all $z \in \mathbb{D}$. Let $z = x + iy \in \mathbb{D}$ and $c = c_1 + ic_2$ (with $x, y, c_1, c_2 \in \mathbb{R}$). For all $\rho, \sigma \in \mathbb{R}$ satisfying $\sigma \leq -(1 + \rho^2)/2$ we obtain

$$\operatorname{Re}\psi(\rho i, \sigma; x + iy) = 4\sigma - 4\rho^2 - 4\rho\operatorname{Im}\kappa + c_1 x - c_2 y - 4(\operatorname{Re}\kappa - 1)$$
$$\leq -6\rho^2 - 4(\operatorname{Im}\kappa)\rho + c_1 x - c_2 y - 2(2\operatorname{Re}\kappa - 1) = Q_1(\rho).$$

The discriminant Δ_1 of the quadratic form $Q_1(\rho)$ is

$$\Delta_1 = 4[4(\operatorname{Im}\kappa)^2 + 6(c_1 x - c_2 y) - 12(2\operatorname{Re}\kappa - 1)].$$

By the Cauchy–Bunyakovsky–Schwarz inequality (1.35) we have

$$c_1 x - c_2 y \leq |c_1 x - c_2 y| \leq \sqrt{c_1^2 + c_2^2}\sqrt{x^2 + y^2} < |c|.$$

Therefore we have

$$\Delta_1/4 < 4(\operatorname{Im}\kappa)^2 + 6|c| - 12(2\operatorname{Re}\kappa - 1) \leq 0.$$

Thus, the quadratic form $Q_1(\rho)$ is strictly negative, and consequently we have $\operatorname{Re}\psi(\rho i, \sigma; x + iy) < 0$. By Lemma 2.5 we conclude that $\operatorname{Re} h(z) > 0$ for all $z \in \mathbb{D}$, which shows that u_p is convex in $\mathbb{D}$.

(b) Since $\operatorname{Re}(\kappa - 1) \geq |c|/4 + (\operatorname{Im}\kappa)^2/6 + 1/2 = |c|/4 + [\operatorname{Im}(\kappa - 1)]^2/6 + 1/2$, it follows from (a) that u_{p-1} is convex. By applying Alexander's duality theorem, i.e. Lemma 2.1 we conclude that $zu'_{p-1} \in \mathcal{S}^*$. But, on the other hand, Lemma 1.2 yields $czu_p(z) = -4(\kappa - 1)zu'_{p-1}(z)$. Consequently, it results that $zu_p \in \mathcal{S}^*$.

(c) Since the condition of the first part of Theorem 2.3 holds, i.e. we have $\operatorname{Re}\kappa > |c|/4 + 1$, we deduce that $u_p(z) \neq 0$ for all $z \in \mathbb{D}$. Define $h : \mathbb{D} \to \mathbb{C}$ by

$$h(z) = 1 + 2\frac{zu'_p(z)}{u_p(z)}.$$

The function h is analytic in $\mathbb{D}$ and $h(0) = 1$. Just like in the proof of part (c) of Theorem 2.11 it is shown that h satisfies the differential equation

$$2zh'(z) + h^2(z) + 2(\kappa - 2)h(z) + cz - (2\kappa - 3) = 0. \tag{2.20}$$

If
$$\psi(r,s;z) = 2s + r^2 + 2(\kappa - 2)r + cz - (2\kappa - 3)$$

and $\mathbb{E} = \{0\}$, then (2.20) implies $\psi(h(z), zh'(z); z) \in \mathbb{E}$ for all $z \in \mathbb{D}$. We use Lemma 2.5 to prove that $\operatorname{Re} h(z) > 0$ for all $z \in \mathbb{D}$. If $z = x + iy \in \mathbb{D}$ and $c = c_1 + ic_2$ (with $x, y, c_1, c_2 \in \mathbb{R}$), we obtain for all $\sigma, \rho \in \mathbb{R}$ satisfying $\sigma \leq -(1 + \rho^2)/2$ that

$$\operatorname{Re}\psi(\rho i, \sigma; x + iy) = 2\sigma - \rho^2 - 2(\operatorname{Im}\kappa)\rho + c_1 x - c_2 y - (2\operatorname{Re}\kappa - 3)$$
$$\leq -2\rho^2 - 2(\operatorname{Im}\kappa)\rho + c_1 x - c_2 y - (2\operatorname{Re}\kappa - 2) = Q_2(\rho).$$

The discriminant Δ_2 of the quadratic form $Q_2(\rho)$ is

$$\Delta_2 = 4(\operatorname{Im}\kappa)^2 + 8(c_1 x - c_2 y) - 8(2\operatorname{Re}\kappa - 2).$$

By the Cauchy–Bunyakovsky–Schwarz inequality (1.35), we know that $c_1 x - c_2 y < |c|$. Therefore we have $\Delta_2 < 4(\operatorname{Im}\kappa)^2 + 8|c| - 8(2\operatorname{Re}\kappa - 2) \leq 0$. Thus $Q_2(\rho) < 0$, and consequently $\operatorname{Re}\psi(\rho i, \sigma; x + iy) < 0$. By Lemma 2.5 we conclude that $\operatorname{Re} h(z) > 0$ for all $z \in \mathbb{D}$, and this inequality shows that $g_p : \mathbb{D} \to \mathbb{C}$, defined by $g_p(z) = zu_p(z)$, is starlike of order $1/2$.

(d) The proof of this assertion coincides with the proof of (d) in Theorem 2.11. □

Note that similar results as those given in Theorem 2.12 were obtained by S. Kanas and J. Stankiewicz [128] for confluent hypergeometric functions. Moreover, it is worth mentioning here that part (d) of Theorem 2.12 slightly improves the result of R.K. Brown [78, Theorem 3]. More precisely, recall that R.K. Brown proved (see Remark 2.7) that if $p = \operatorname{Re} p + i\operatorname{Im} p$ satisfies one of the following conditions $\operatorname{Im} p \leq \operatorname{Re} p \in [0, 1)$ or $\operatorname{Re} p \geq 1$, $2\operatorname{Re} p - 1 > (\operatorname{Im} p)^2$, then the normalized Bessel function $z \mapsto z^{1-p}J_p(z)$ is univalent in every disk $\mathbb{D}_r = \{z \in \mathbb{C} : |z| < r = \rho_\mu^*\}$, where $\mu^2 = \operatorname{Re}[p^2]$, $\mu > 0$, and ρ_μ^* is the smallest positive zero of the function

$$r \mapsto rJ_\mu'(r) + \operatorname{Re}[1 - p]J_\mu(r).$$

Now, in the case of the unit disk, i.e. for $z \in \mathbb{D} \cap \mathbb{D}_r$ part (d) of Theorem 2.12 states that if $\operatorname{Re} p \geq 1/2 + (\operatorname{Im} p)^2/4$, then the normalized Bessel function $z \mapsto z^{1-p}J_p(z)$ is still starlike and hence univalent in $\mathbb{D}$. And this slightly improves the above result of R.K. Brown because we have that $\operatorname{Re} p \geq 1/2 + (\operatorname{Im} p)^2/2 \geq 1/2 + (\operatorname{Im} p)^2/4$.

Another extension of Theorem 2.11 for real parameters is included in the next theorem.

Theorem 2.13. (Á. Baricz [52]) *If $\alpha \in [0, 1)$ and b, p, c are real numbers such that $c \neq 0$ and $4\alpha^2 + (|c| - 6)\alpha + 2 \geq 0$, then the functions w_p and u_p, defined by (1.15) and (1.20), respectively, have the following properties:*

(a) *If $\kappa \geq \dfrac{|c| + 2(1 - \alpha)(1 - 2\alpha)}{4(1 - \alpha)}$, then $u_p \in \mathscr{C}(\alpha)$. Moreover, we have that*

$$\operatorname{Re}\left[u_p'(z)\right]^{\frac{1}{2(1-\alpha)}} > \frac{1}{2} \text{ for all } z \in \mathbb{D}.$$

(b) If $\kappa \geq \dfrac{|c| + 2(1-\alpha)(3-2\alpha)}{4(1-\alpha)}$, then $zu_p \in \mathscr{S}^*(\alpha)$ and

$$z\left[-\frac{c}{4(\kappa-1)}u_p(z)\right]^{\frac{1}{1-\alpha}} \in \mathscr{S}^*(\alpha).$$

(c) If $\kappa \geq \dfrac{|c| + 2(1-\alpha)(3-2\alpha)}{4(1-\alpha)}$ and $\alpha \neq 0$, then $z^{\frac{2(1-\alpha)-p}{2\alpha}} w_p\left(z^{\frac{1}{2\alpha}}\right) \in \mathscr{S}^*$.

Proof. (a) The equality

$$\frac{|c| + 2(1-\alpha)(1-2\alpha)}{4(1-\alpha)} = \frac{|c|}{4} + \frac{4\alpha^2 + (|c| - 6)\alpha + 2}{4(1-\alpha)}$$

implies $\kappa \geq |c|/4$. By applying Theorem 2.2 we conclude that $\operatorname{Re} u_{p+1}(z) > 0$ for all $z \in \mathbb{D}$. According to Lemma 1.2 it follows that $u'_p(z) \neq 0$ for all $z \in \mathbb{D}$. Define $q : \mathbb{D} \to \mathbb{C}$ by

$$q(z) = 1 + \frac{zu''_p(z)}{(1-\alpha)u'_p(z)}.$$

The function q is analytic in $\mathbb{D}$ and $q(0) = 1$. Since u_p satisfies the differential equation (1.21) it can be shown, as in the proof of Theorem 2.11, that q satisfies the following differential equation:

$$4(1-\alpha)zq'(z) + 4(1-\alpha)^2 q^2(z) + 2(1-\alpha)e_1 q(z) + cz - 2(1-\alpha)e_2 = 0, \tag{2.21}$$

where $e_1 = 2\kappa + 4(\alpha - 1)$ and $e_2 = 2\kappa + 2(\alpha - 1)$.

If

$$\psi(r,s;z) = 4(1-\alpha)s + 4(1-\alpha)^2 r^2 + 2(1-\alpha)e_1 r + cz - 2(1-\alpha)e_2$$

and $\mathbb{E} = \{0\}$, then (2.21) implies $\psi(q(z), zq'(z); z) \in \mathbb{E}$ for all $z \in \mathbb{D}$. We use Lemma 2.5 to prove that $\operatorname{Re} q(z) > 0$ for all $z \in \mathbb{D}$. For $z = x + iy \in \mathbb{D}$ (with $x, y \in \mathbb{R}$) and $\rho, \sigma \in \mathbb{R}$ satisfying $\sigma \leq -(1+\rho^2)/2$, we obtain

$$\operatorname{Re} \psi(\rho i, \sigma; x+iy) = 4(1-\alpha)\sigma - 4(1-\alpha)^2\rho^2 + cx - 2(1-\alpha)e_2$$
$$\leq -2(1-\alpha)(3-2\alpha)\rho^2 + cx - 2(1-\alpha)(1+e_2)$$
$$< |c| + 2(1-\alpha)(1-2\alpha) - 4(1-\alpha)\kappa \leq 0.$$

By Lemma 2.5 we conclude that $\operatorname{Re} q(z) > 0$ for all $z \in \mathbb{D}$. This result implies

$$\operatorname{Re}\left[1 + \frac{zu''_p(z)}{u'_p(z)}\right] = (1-\alpha)\operatorname{Re} q(z) + \alpha > \alpha \quad \text{for all } z \in \mathbb{D},$$

which shows that u_p is convex of order α in $\mathbb{D}$. On the other hand, it is known (see the paper of I.S. Jack [124, p. 473]) that if $f \in \mathscr{C}(\alpha)$, then

$$\mathrm{Re}\left[f'(z)\right]^{\frac{1}{2(1-\alpha)}} > \frac{1}{2} \text{ for all } z \in \mathbb{D}.$$

Now, using the fact that u_p is convex of order α in $\mathbb{D}$, the asserted inequality follows.

(b) Since

$$\kappa - 1 \geq \frac{|c| + 2(1-\alpha)(3-2\alpha)}{4(1-\alpha)} - 1 = \frac{|c| + 2(1-\alpha)(1-2\alpha)}{4(1-\alpha)},$$

it follows from (a) that u_{p-1} is convex of order α. By applying the general version of Alexander's duality theorem, i.e. Lemma 2.2 we conclude that $zu'_{p-1} \in \mathscr{S}^*(\alpha)$ and $z[u'_{p-1}]^{1/(1-\alpha)} \in \mathscr{S}^*(\alpha)$. According to Lemma 1.2 these imply that

$$-\frac{c}{4(\kappa-1)}zu_p(z) \in \mathscr{S}^*(\alpha)$$

and

$$z\left[-\frac{c}{4(\kappa-1)}u_p(z)\right]^{\frac{1}{1-\alpha}} \in \mathscr{S}^*(\alpha).$$

(c) Define the functions $g_p : \mathbb{D} \to \mathbb{C}$ and $h_p : \mathbb{D} \to \mathbb{C}$ by

$$g_p(z) = zu_p(z) \text{ and } h_p(z) = z^{[2(1-\alpha)-p]/(2\alpha)}w_p\left(z^{1/(2\alpha)}\right),$$

respectively. Since $h_p(z) = a_0(p)z^{(1-\alpha)/\alpha}u_p\left(z^{1/\alpha}\right)$, where $a_0(p) = [2^p\Gamma(\kappa)]^{-1}$ (see (1.19)), it follows that

$$\frac{zh'_p(z)}{h_p(z)} = \frac{1}{\alpha}\left[\frac{z^{1/\alpha}g'_p(z^{1/\alpha})}{g_p(z^{1/\alpha})} - \alpha\right].$$

Finally because $g_p \in \mathscr{S}^*(\alpha)$, we deduce that $h_p \in \mathscr{S}^*$.   □

2.2.3 Sufficient Conditions Involving Results of H. Silverman

In this section we place conditions on b, p and c to guarantee that zu_p is in various subclasses of starlike and convex functions, where u_p is defined by (1.20). Further constraints lead to coefficient characterizations of the families. An integral operator related to the generalized Bessel function is also examined. For similar results involving Gaussian hypergeometric functions we refer to the paper of H. Silverman [212], which is the starting point of this section.

Denote by $\mathscr{S}_1^*(\alpha)$, where $\alpha \in [0,1)$, the subclass of $\mathscr{S}^*(\alpha)$ consisting of functions f for which

$$\left| \frac{z f'(z)}{f(z)} - 1 \right| < 1 - \alpha \quad \text{for all } z \in \mathbb{D}.$$

A function f is said to be in $\mathscr{C}_1(\alpha)$ if $z f' \in \mathscr{S}_1^*(\alpha)$. Coefficient bounds and other extremal properties for $\mathscr{S}_1^*(\alpha)$ were found in papers of P.J. Eenigenburg [92] and H. Silverman [211].

We will make use of the following lemmata.

Lemma 2.11. (H. Silverman [210]) *Let $\alpha \in [0,1)$. A sufficient condition for $f(z) = z + \sum_{n \geq 2} a_n z^n$ to be in $\mathscr{S}_1^*(\alpha)$ and $\mathscr{C}_1(\alpha)$, respectively, is that*

$$\sum_{n \geq 2} (n - \alpha)|a_n| \leq 1 - \alpha,$$

$$\sum_{n \geq 2} n(n - \alpha)|a_n| \leq 1 - \alpha$$

respectively.

Lemma 2.12. (E.P. Merkes, M.S. Robertson and W.T. Scott [150], H. Silverman [210]) *Let $\alpha \in [0,1)$. Suppose that $f(z) = z - \sum_{n \geq 2} a_n z^n$, $a_n \geq 0$. Then a necessary and sufficient condition for f to be in $\mathscr{S}_1^*(\alpha)$ and $\mathscr{C}_1(\alpha)$, respectively, is that*

$$\sum_{n \geq 2} (n - \alpha)a_n \leq 1 - \alpha,$$

$$\sum_{n \geq 2} n(n - \alpha)a_n \leq 1 - \alpha$$

respectively. In addition, $f \in \mathscr{S}_1^*(\alpha) \Longleftrightarrow f \in \mathscr{S}^*(\alpha)$, $f \in \mathscr{C}_1(\alpha) \Longleftrightarrow f \in \mathscr{C}(\alpha)$, and $f \in \mathscr{S}^* \Longleftrightarrow f \in \mathscr{S}$.

Our main results of this section are the following theorems.

Theorem 2.14. (Á. Baricz [54]) *If $\alpha \in [0,1)$, $c < 0$ and $\kappa > 0$, then a sufficient condition for $z u_p$ to be in $\mathscr{S}_1^*(\alpha)$ is*

$$u_p(1) + u_p'(1)/(1 - \alpha) \leq 2. \tag{2.22}$$

Moreover, (2.22) is necessary and sufficient for $\psi(z) = z[2 - u_p(z)]$ to be in $\mathscr{S}_1^*(\alpha)$.

Proof. Since $z u_p(z) = z + \sum_{n \geq 2} b_{n-1} z^n$, according to Lemma 2.11 we need only show that

$$\sum_{n \geq 2} (n - \alpha)b_{n-1} \leq 1 - \alpha.$$

We notice that

$$\sum_{n\geq 2}(n-\alpha)b_{n-1}=\sum_{n\geq 2}(n-1)b_{n-1}+\sum_{n\geq 2}(1-\alpha)b_{n-1}$$

$$=\sum_{n\geq 2}\frac{(-c/4)^{n-1}}{(\kappa)_{n-1}(n-2)!}+(1-\alpha)[u_p(1)-1].$$

Taking into consideration that $(\kappa)_{n-1}=\kappa(\kappa+1)_{n-2}$ we may write the above expression as follows

$$\sum_{n\geq 2}(n-\alpha)b_{n-1}=-\frac{c}{4\kappa}\sum_{n\geq 2}\frac{(-c/4)^{n-2}}{(\kappa+1)_{n-2}(n-2)!}+(1-\alpha)[u_p(1)-1]$$

$$=(1-\alpha)[u_p(1)-1]+\left(-\frac{c}{4\kappa}\right)u_{p+1}(1).$$

Using Lemma 1.2 we obtain

$$\sum_{n\geq 2}(n-\alpha)b_{n-1}=(1-\alpha)[u_p(1)-1]+u_p'(1).$$

This sum is bounded above by $1-\alpha$ if and only if (2.22) holds. Since

$$z[2-u_p(z)]=z-\sum_{n\geq 2}b_{n-1}z^n,$$

the necessity of (2.22) for ψ to be in $\mathscr{S}_1^*(\alpha)$ follows from Lemma 2.12.   □

Remark 2.10. Condition (2.22) with $\alpha=0$ is both necessary and sufficient for ψ to be in $\mathscr{S}$. For the convenience throughout this section we denote

$$\varsigma_p(t)=\frac{2(1-t^2)^{\kappa-3/2}}{B(\kappa-1/2,1/2)}\cosh(t\sqrt{-c}). \qquad (2.23)$$

By Lemma 1.3 the condition (2.22) may be written as

$$(1-\alpha)\int_0^1\varsigma_p(t)\,dt-\frac{c}{4\kappa}\int_0^1\varsigma_{p+1}(t)\,dt\leq 2(1-\alpha).$$

In particular when $c=-1$ and $b=1$ (in this case $u_p(z)$ becomes $\mathscr{I}_p(z^{1/2})$, defined by (1.24)) we obtain that condition (2.22) simplifies to

$$2^{p-2}[I_{p+1}(1)+2(1-\alpha)I_p(1)]\cdot\Gamma(p+1)\leq 1-\alpha,$$

which guarantees that

$$z\mathscr{I}_p(z^{1/2}) = 2^p\Gamma(p+1)z^{1-p/2}I_p(z^{1/2}) \in \mathscr{S}_1^*(\alpha).$$

Moreover, the above condition is necessary and sufficient for $z[2 - \mathscr{I}_p(z^{1/2})]$ to be in $\mathscr{S}_1^*(\alpha)$.

Our next theorem is similar to Theorem 2.14, but it deals with the convex case.

Theorem 2.15. (Á. Baricz [54]) *If $\alpha \in [0,1)$, $c < 0$ and $\kappa > 0$, then a sufficient condition for zu_p to be in $\mathscr{C}_1(\alpha)$ is*

$$u_p''(1) + (3 - \alpha)u_p'(1) + (1 + \alpha)u_p(1) \leq 2. \tag{2.24}$$

Moreover, this condition is necessary and sufficient for $\psi(z) = z[2 - u_p(z)]$ to be in $\mathscr{C}_1(\alpha)$.

Proof. In view of Lemma 2.11, we need only show that

$$\sum_{n\geq 2} n(n - \alpha)b_{n-1} \leq 1 - \alpha.$$

We notice that

$$\sum_{n\geq 2} n(n - \alpha)b_{n-1} = \sum_{n\geq 0} (n+2)(n+2-\alpha)b_{n+1} = A - \alpha B,$$

where

$$A = \sum_{n\geq 0} (n+2)^2 b_{n+1} \quad \text{and} \quad B = \sum_{n\geq 0} (n+2)b_{n+1}.$$

Thus, we have to prove that $A - \alpha B \leq 1 - \alpha$. Computing A and B, we obtain:

$$A = \sum_{n\geq 0} (n+1)\frac{(-c/4)^{n+1}}{(\kappa)_{n+1}n!} + 2\sum_{n\geq 0} \frac{(-c/4)^{n+1}}{(\kappa)_{n+1}n!} + \sum_{n\geq 0} \frac{(-c/4)^{n+1}}{(\kappa)_{n+1}(n+1)!}$$

$$= \sum_{n\geq 1} \frac{(-c/4)^{n+1}}{(\kappa)_{n+1}(n-1)!} + 3\sum_{n\geq 0} \frac{(-c/4)^{n+1}}{(\kappa)_{n+1}n!} + \sum_{n\geq 0} \frac{(-c/4)^{n+1}}{(\kappa)_{n+1}(n+1)!}$$

$$= \frac{(-c/4)^2}{\kappa(\kappa+1)} \sum_{n\geq 1} \frac{(-c/4)^{n-1}}{(\kappa+2)_{n-1}(n-1)!} + 3\frac{(-c/4)}{\kappa} \sum_{n\geq 0} \frac{(-c/4)^n}{(\kappa+1)_n n!} + \sum_{n\geq 0} \frac{(-c/4)^{n+1}}{(\kappa)_{n+1}(n+1)!}$$

$$= \frac{(-c/4)^2}{\kappa(\kappa+1)} u_{p+2}(1) - 3\frac{c}{4\kappa} u_{p+1}(1) + u_p(1) - 1$$

$$= u_p''(1) + 3u_p'(1) + u_p(1) - 1,$$

where we used that $(a)_n = a(a+1)_{n-1}$ and Lemma 1.2, i.e. the relations

$$u'_p(z) = -\frac{c}{4\kappa}u_{p+1}(z), \quad u''_p(z) = \frac{(-c/4)^2}{\kappa(\kappa+1)}u_{p+2}(z).$$

Analogously, we obtain that

$$B = \sum_{n\geq 0}\frac{(-c/4)^{n+1}}{(\kappa)_{n+1}n!} + \sum_{n\geq 0}\frac{(-c/4)^{n+1}}{(\kappa)_{n+1}(n+1)!} = u'_p(1) + u_p(1) - 1.$$

Therefore it is clear that the expression

$$A - \alpha B = u''_p(1) + (3-\alpha)u'_p(1) + (1+\alpha)u_p(1) - (1+\alpha)$$

is bounded above by $1-\alpha$ if and only if (2.24) holds. Lemma 2.12 implies that (2.24) is also necessary for ψ to be in $\mathscr{C}_1(\alpha)$.   □

Remark 2.11. Using (2.23) and Lemma 1.3 we see that condition (2.24) may be written as follows:

$$\frac{(-c/4)^2}{\kappa(\kappa+1)}\int_0^1\varsigma_{p+2}(t)\,dt + (3-\alpha)\left(-\frac{c}{4\kappa}\right)\int_0^1\varsigma_{p+1}(t)\,dt + (1+\alpha)\int_0^1\varsigma_p(t)\,dt \leq 2.$$

In particular, when $c = -1$ and $b = 1$, the condition (2.24) becomes

$$2^{p-3}[I_{p+2}(1) + (3-\alpha)I_{p+1}(1) + 4(1+\alpha)I_p(1)]\cdot\Gamma(p+1) \leq 2,$$

which guarantees that

$$z\mathscr{I}_p(z^{1/2}) = 2^p\Gamma(p+1)z^{1-p/2}I_p(z^{1/2}) \in \mathscr{C}_1(\alpha).$$

Moreover, the above condition is necessary and sufficient for $z[2 - \mathscr{I}_p(z^{1/2})]$ to be in $\mathscr{C}_1(\alpha)$.

As in the paper of H. Silverman [212] one can look at other linear operators acting on u_p to obtain similar results. We illustrate this idea in the case of a particular integral operator.

Theorem 2.16. (Á. Baricz [54]) *If $c < 0$, $\kappa > 0$ and $u_p(1) \leq 2$, then $\int_0^z u_p(t)\,dt \in \mathscr{S}^*$.*

Proof. Since

$$\int_0^z u_p(t)\,dt = \sum_{n\geq 0}\frac{b_n}{n+1}z^{n+1} = z + \sum_{n\geq 2}\frac{b_{n-1}}{n}z^n,$$

we note that

$$\sum_{n\geq 2} n \cdot \frac{b_{n-1}}{n} = \sum_{n\geq 2} b_{n-1} = u_p(1) - 1 \leq 1$$

if and only if $u_p(1) \leq 2$. □

Remark 2.12. Comparable bounds to Theorem 2.16 may be obtained for positive order of starlikeness too. Denoting $f(z) = zu_p(z)$ and $g(z) = \int_0^z u_p(t)\,dt$, it is clear that $g \in \mathscr{C}_1(\alpha)$ ($\mathscr{C}(\alpha)$) if and only if $f \in \mathscr{S}_1^*(\alpha)$ ($\mathscr{S}^*(\alpha)$). This follows observing that $g'(z) = u_p(z)$, $g''(z) = u_p'(z)$, and so $1 + zg''(z)/g'(z) = 1 + zu_p'(z)/u_p(z) = zf'(z)/f(z)$. Thus any starlikeness type result concerning zu_p leads to a convexity result concerning g, for example condition (2.22) guarantees that $g \in \mathscr{C}_1(\alpha)$.

2.2.4 Close-to-Convexity with Respect to Certain Functions

Motivated by the papers of S. Ponnusamy and M. Vuorinen [187, 188], we discuss in this section a few conditions concerning the parameters of u_p, which guarantee the close-to-convexity with respect to the functions

$$-\text{Log}(1-z) \quad \text{and} \quad \frac{1}{2}\text{Log}\left(\frac{1+z}{1-z}\right).$$

For this we need the following result of S. Ozaki [167].

Lemma 2.13. (S. Ozaki [167]) *If* $f(z) = z + \sum_{n\geq 2} B_{2n-1}z^{2n-1}$ *is analytic in* $\mathbb{D}$ *and if* $1 \geq 3B_3 \geq \ldots \geq (2n-1)B_{2n-1} \geq \ldots \geq 0$ *or* $1 \leq 3B_3 \leq \ldots \leq (2n-1)B_{2n-1} \leq \ldots \leq 2$, *then* f *is univalent in* $\mathbb{D}$.

We note that, as S. Ponnusamy and M. Vuorinen pointed out in [187], proceeding exactly as in the proof of Lemma 2.8 (also due to S. Ozaki [167]), one can verify directly that if a function $f : \mathbb{D} \to \mathbb{C}$ satisfies the hypothesis of Lemma 2.13, then it is close-to-convex with respect to the convex function

$$z \mapsto \frac{1}{2}\text{Log}\left(\frac{1+z}{1-z}\right).$$

Theorem 2.17. (Á. Baricz [52]) *If* $c < 0$ *and* $b, p \in \mathbb{R}$, *then* u_p, *defined by* (1.20), *has the following properties:*

(a) *If* $\kappa \geq -c/2$, *then* $z \mapsto zu_p(z)$ *is close-to-convex with respect to the function* $-\text{Log}(1-z)$, *and consequently it is univalent in* $\mathbb{D}$.

(b) *If* $\kappa \geq -3c/4$, *then* $z \mapsto zu_p(z^2)$ *is close-to-convex with respect to the function* $\frac{1}{2}\text{Log}\left(\frac{1+z}{1-z}\right)$, *and consequently it is univalent in* $\mathbb{D}$.

Proof. (a) Set

$$f(z) = z u_p(z) = z + \sum_{n \geq 2} b_{n-1} z^n.$$

We have $b_{n-1} > 0$ for all $n \geq 2$ and $2b_1 = -c/(2\kappa) \leq 1$. From the definition of the ascending factorial notation (we use the formula $(\kappa)_n = (\kappa + n - 1)(\kappa)_{n-1}$) we observe that

$$b_n = -\frac{c}{4n(\kappa + n - 1)} b_{n-1} \quad \text{for all } n \geq 2.$$

We use Lemma 2.8 to prove that f is close-to-convex with respect to the function $-\mathrm{Log}(1 - z)$. Therefore, we show that $\{n b_{n-1}\}_{n \geq 2}$ is a decreasing sequence. By a short computation we obtain

$$n b_{n-1} - (n+1)b_n = b_{n-1} \left[n + \frac{c(n+1)}{4n(\kappa + n - 1)} \right] = \frac{b_{n-1} \cdot U_1(n)}{4n(\kappa + n - 1)},$$

where $U_1(n) = 4n^3 + 4(\kappa - 1)n^2 + cn + c$. Using the inequalities $n^3 \geq 3n^2 - 3n + 1$, $n^2 \geq 2n - 1$ and $8\kappa + c + 4 > 0$, we obtain

$$U_1(n) \geq 4(\kappa + 2)n^2 + (c - 12)n + c + 4$$
$$\geq (8\kappa + c + 4)n - 4(\kappa + 2) + c + 4$$
$$\geq 2(2\kappa + c) \geq 0.$$

This implies that $n b_{n-1} - (n+1)b_n \geq 0$ for all $n \geq 2$, thus, $\{n b_{n-1}\}_{n \geq 2}$ is a decreasing sequence. By Lemma 2.8 it follows that f is close-to-convex with respect to the convex function $-\mathrm{Log}(1 - z)$. Consequently, f is univalent in $\mathbb{D}$ (see Lemma 2.3).

(b) Set

$$f(z) = z u_p(z^2) = z + \sum_{n \geq 2} B_{2n-1} z^{2n-1},$$

where $B_{2n-1} = b_{n-1}$ for all $n \geq 2$. Therefore we have $3B_3 = 3b_1 = -(3c)/(4\kappa) \leq 1$ and $B_{2n-1} > 0$ for all $n \geq 2$. We want to show that $\{(2n - 1)B_{2n-1}\}_{n \geq 2}$ is a decreasing sequence. Fix $n \geq 2$. Then we have

$$(2n - 1)B_{2n-1} - (2n + 1)B_{2n+1} = \frac{B_{2n-1} \cdot U_2(n)}{4n(\kappa + n - 1)},$$

where $U_2(n) = 8n^3 + 8(\kappa - 3/2)n^2 - 4(\kappa - c/2 - 1)n + c$. Using the inequalities $n^3 \geq 3n^2 - 3n + 1$ and $n^2 \geq 2n - 1$, we obtain

$$U_2(n) \geq 8(\kappa + 3/2)n^2 - 4(\kappa - c/2 + 5)n + c + 8$$
$$\geq 12(\kappa + c/6 + 1/3)n - 8(\kappa + 3/2) + c + 8$$
$$\geq 4(\kappa + 3c/4) \geq 0.$$

Hence $\{(2n-1)B_{2n-1}\}_{n \geq 2}$ is a decreasing sequence. But $z \mapsto \dfrac{1}{2} \mathrm{Log}\left(\dfrac{1+z}{1-z}\right)$ is convex in $\mathbb{D}$, so by applying Lemma 2.3 the desired conclusion follows. $\square$

In particular if we choose $c = -1$ and $b = 1$, then we obtain the following sufficient condition for the close-to-convexity of the normalized and modified Bessel functions of the first kind of order p.

Corollary 2.9. (Á. Baricz [52]) *If $p \geq -1/2$, then $z \mapsto z\mathscr{I}_p(\sqrt{z})$ is close-to-convex with respect to the function $-\mathrm{Log}(1-z)$, and consequently it is univalent in $\mathbb{D}$. Moreover, if $p \geq -1/4$, then $z \mapsto z\mathscr{I}_p(z)$ is close-to-convex with respect to the function $\dfrac{1}{2}\mathrm{Log}\left(\dfrac{1+z}{1-z}\right)$, and consequently it is univalent in $\mathbb{D}$.*

Remark 2.13. If we consider the Bessel functions which can be expressed with elementary functions as cosh and sinh (see Remark 2.9), we may obtain the following examples of close-to-convex functions.

(a) The functions $z\cosh\sqrt{z}$ and $\sqrt{z}\sinh\sqrt{z}$ are close-to-convex functions with respect to the convex function $-\mathrm{Log}(1-z)$, because $z\mathscr{I}_{-1/2}(z^{1/2}) = \sqrt{\pi/2} \cdot z^{5/4}I_{-1/2}(\sqrt{z}) = z\cosh\sqrt{z}$ and $z\mathscr{I}_{1/2}(z^{1/2}) = \sqrt{\pi/2} \cdot z^{3/4}I_{1/2}(\sqrt{z}) = \sqrt{z}\sinh\sqrt{z}$ satisfy the condition of Theorem 2.17 (first part).

(b) The function $zJ_0(iz)$ is close-to-convex with respect to $z \mapsto \dfrac{1}{2}\mathrm{Log}\left(\dfrac{1+z}{1-z}\right)$, because $z\mathscr{I}_0(z) = zI_0(z) = zJ_0(iz)$ satisfies the condition of Theorem 2.17 (second part). We note that this property of the Bessel function of the first kind of zero order was established also by S. Ponnusamy and M. Vuorinen [187, p. 83].

2.3 Applications Involving Bessel Functions Associated with Hardy Space of Analytic Functions

Our aim in this section is to find conditions on the parameters which guarantee that the generalized and normalized Bessel function belongs to a certain Hardy space. Moreover, we present a monotonicity property of the generalized and normalized Bessel functions. In addition, we find the radius of the smallest disk centered at 1, which contains the image region $u_p(\mathbb{D})$.

2.3.1 Bessel Transforms and Hardy Space of Generalized Bessel Functions

In this section we present some immediate applications of convexity and univalence involving Bessel functions associated with the Hardy space of analytic functions, i.e. we obtain conditions for the function u_p, defined by (1.20), to belong to the Hardy space $\mathscr{H}^\infty$.

Let $\mathscr{H}$ be the set of all analytic functions on $\mathbb{D}$. For any $\mu \in (0, \infty]$, any function $f \in \mathscr{H}$ and any $r \in [0, 1)$ set

$$M_\mu(r, f) = \begin{cases} \left(\dfrac{1}{2\pi} \displaystyle\int_0^{2\pi} \left| f(re^{i\theta}) \right|^\mu d\theta \right)^{1/\mu}, & \text{if } 0 < \mu < \infty \\ \max_{|z| \le r} |f(z)|, & \text{if } \mu = \infty. \end{cases}$$

By definition, the function $f \in \mathscr{H}$ is said to belong to the Hardy space $\mathscr{H}^\mu$, where $0 < \mu \le \infty$, if the set $\left\{ M_\mu(r, f) \,\middle|\, r \in [0, 1) \right\}$ is bounded. We note that for $1 \le \mu \le \infty$, $\mathscr{H}^\mu$ is actually a Banach space with the norm defined by $\|f\|_\mu = \lim_{r \to 1^-} M_\mu(r, f)$. Furthermore, $\mathscr{H}^\infty$ is the class of bounded analytic functions in $\mathscr{H}$. We note that for $0 < \mu \le \nu \le \infty$, it can be shown that $\mathscr{H}^\nu$ is a subset of $\mathscr{H}^\mu$ (see the book of P.L. Duren [89]).

For $\alpha < 1$ we introduce the class

$$\mathscr{P}(\alpha) = \{ f \in \mathscr{H} : f(0) = 1, \exists\, \eta \in \mathbb{R} \text{ such that } \mathrm{Re}\, [e^{i\eta} f(z)] > \alpha \text{ for all } z \in \mathbb{D} \}$$

and define $\mathscr{R}(\alpha) = \{ f \in \mathscr{A} : f' \in \mathscr{P}(\alpha) \}$, i.e.

$$\mathscr{R}(\alpha) = \{ f \in \mathscr{A} : \exists\, \eta \in \mathbb{R} \text{ such that } \mathrm{Re}\, [e^{i\eta} f'(z)] > \alpha \text{ for all } z \in \mathbb{D} \},$$

where we used the fact that $f'(0) = 1$ for every $f \in \mathscr{A}$. When $\eta = 0$ we denote $\mathscr{P}(\alpha)$ and $\mathscr{R}(\alpha)$ simply by $\mathscr{P}_0(\alpha)$ and $\mathscr{R}_0(\alpha)$, respectively; for $\alpha = 0$ we denote $\mathscr{P}_0(\alpha)$ and $\mathscr{R}_0(\alpha)$ simply by $\mathscr{P}$ and $\mathscr{R}$, respectively.

The next lemmata will be used to prove several theorems.

Lemma 2.14. (S. Ponnusamy [180]) *For $\alpha, \beta < 1$ and $\gamma = 1 - 2(1 - \alpha)(1 - \beta)$, we have $\mathscr{R}(\alpha) * \mathscr{R}_0(\beta) \subset \mathscr{R}(\gamma)$, or equivalently $\mathscr{P}(\alpha) * \mathscr{P}_0(\beta) \subset \mathscr{P}(\gamma)$.*

Lemma 2.15. (P.J. Eenigenburg and F.R. Keogh [93]) *Let $\alpha \in [0, 1)$. If the function $f \in \mathscr{C}(\alpha)$ is not of the form*

$$\begin{cases} f(z) = \mu + vz(1 - ze^{i\gamma})^{2\alpha - 1}, & \alpha \ne 1/2 \\ f(z) = \mu + v\,\mathrm{Log}(1 - ze^{i\gamma}), & \alpha = 1/2 \end{cases}$$

for some $\mu, \nu \in \mathbb{C}$ *and* $\gamma \in \mathbb{R}$, *then the following statements hold:*

(a) *There exists* $\delta = \delta(f) > 0$ *such that* $f' \in \mathscr{H}^{\delta + 1/[2(1-\alpha)]}$.
(b) *If* $\alpha \in [0, 1/2)$, *then there exists* $\tau = \tau(f) > 0$ *such that* $f \in \mathscr{H}^{\tau + 1/(1-2\alpha)}$.
(c) *If* $\alpha \geq 1/2$, *then* $f \in \mathscr{H}^{\infty}$.

Our main results read as follows.

Theorem 2.18. (Á. Baricz [43]) *Let* $\alpha \in [0, 1)$ *and* b, p, c *be real numbers such that* $c \neq 0$ *and* $4\alpha^2 + (|c| - 6)\alpha + 2 \geq 0$. *If* $4(1-\alpha)\kappa \geq |c| + 2(1-\alpha)(1-2\alpha)$, *then the following assertions are true:*

(a) *If* $\alpha \in [0, 1/2)$, *then* $u_p \in \mathscr{H}^{1/(1-2\alpha)}$.
(b) *If* $\alpha \geq 1/2$, *then* $u_p \in \mathscr{H}^{\infty}$.

Proof. First we observe that

$$\mu + \frac{\nu z}{(1 - ze^{i\gamma})^{1-2\alpha}} = \mu + \nu z F(1, 1 - 2\alpha, 1, ze^{i\gamma})$$

$$= \mu + \nu \sum_{n \geq 0} \frac{(1-2\alpha)_n}{n!} e^{i\gamma n} z^{n+1}$$

for $\mu, \nu \in \mathbb{C}$, $\alpha \neq 1/2$ and for real γ, where $F(a, b, c; z)$ is the Gaussian hypergeometric series defined in (1.1). On the other hand

$$\mu + \nu \operatorname{Log}(1 - ze^{i\gamma}) = \mu - \nu z F(1, 1, 2, ze^{i\gamma})$$

$$= \mu - \nu \sum_{n \geq 0} \frac{1}{n+1} e^{i\gamma n} z^{n+1}.$$

Therefore, since γ is real, the function u_p cannot be of the form $\mu + \nu z (1 - ze^{i\gamma})^{2\alpha-1}$ (for $\alpha \neq 1/2$) or $\mu + \nu \operatorname{Log}(1 - ze^{i\gamma})$ (for $\alpha = 1/2$). We know that u_p is convex of order α (property (a) in Theorem 2.13). Hence by Lemma 2.15 the proof is completed. $\qquad \square$

Theorem 2.19. (Á. Baricz [43]) *If* $b, p, c \in \mathbb{R}$ *such that* $4\kappa \geq |c| - 2$ *and* $c \neq 0$, *then* $u_p \in \mathscr{H}^{\infty}$.

Proof. Since $\kappa + 1 \geq |c|/4 + 1/2$, it follows by Theorem 2.11 that $u_{p+1} \in \mathscr{C} = \mathscr{C}(0)$. In addition, by the hypergeometric series representation we observe that u_{p+1} is not of the form

$$\mu + \nu z (1 - ze^{i\gamma})^{-1} = \mu + \nu z F(1, 1, 1, ze^{i\gamma})$$

$$= \mu + \nu \sum_{n \geq 0} e^{i\gamma n} z^{n+1},$$

where $\mu, \nu \in \mathbb{C}$ and $\gamma \in \mathbb{R}$. Taking into consideration that $4\kappa u_p'(z) = -cu_{p+1}(z)$ (see Lemma 1.2), we conclude that u_p' is convex in $\mathbb{D}$ and is not of the form $\mu + \nu z(1 - ze^{i\gamma})^{-1}$. Hence by Lemma 2.15 we have $u_p' \in \mathscr{H}^1$. On the other hand, it is known that an analytic function $f : \mathbb{C} \to \mathbb{C}$ is continuous in $\overline{\mathbb{D}} = \{z \in \mathbb{C} : |z| \le 1\}$ and absolutely continuous on $\partial\mathbb{D} = \{z \in \mathbb{C} : |z| = 1\}$ if and only if $f' \in \mathscr{H}^1$ (see the book of P.L. Duren [89, Theorem 3.11]). From this we deduce that u_p is continuous in $\overline{\mathbb{D}}$, so u_p is a bounded analytic function in $\mathbb{D}$. This completes the proof of Theorem 2.19.   □

We note that u_p is in fact an entire function, i.e. it is holomorphic everywhere on the whole complex plane. Thus, since every continuous function on a compact set is bounded, we deduce that $u_p \in \mathscr{H}^\infty$ for all admissible values of the parameters b, c and p.

The main aim in the sequel is to find conditions on α_1 and α_2 and the parameters b, c and p such that $zu_p * f$ maps $\mathscr{R}(\alpha_1)$ into $\mathscr{R}(\alpha_2)$.

Theorem 2.20. (Á. Baricz [43]) *Let $\alpha \in [0, 1/2)$, and let b, p, c be real numbers such that $\kappa \ge (1 - \alpha)(1 - 2\alpha)^{-1/2}|c|/4 + 1$. If $f \in \mathscr{R}(\alpha_1)$, where $\alpha_1 < 1$, then $zu_p * f \in \mathscr{R}(\gamma)$, where $\gamma = 1 - 2(1 - \alpha_1)(1 - \alpha)$.*

Proof. Set $g(z) = zu_p(z) * f(z)$. Then we have $g'(z) = u_p(z) * f'(z)$. By the hypotheses and Theorem 2.4, we have $u_p \in \mathscr{P}_0(\alpha)$. Using Lemma 2.14 and the fact that $f' \in \mathscr{P}(\alpha_1)$, we immediately obtain that the function g' belongs to $\mathscr{P}(\gamma)$, where $\gamma = 1 - 2(1 - \alpha_1)(1 - \alpha)$. But $g' \in \mathscr{P}(\gamma)$ implies $g \in \mathscr{R}(\gamma)$, therefore the proof is complete.   □

As an immediate consequence of Theorem 2.20 we have the following corollary.

Corollary 2.10. (Á. Baricz [43]) *Let α, b, p, c satisfy the hypotheses of Theorem 2.20. If $f \in \mathscr{R}(\alpha_1)$, where $\alpha_1 = (1 - 2\alpha)/(2 - 2\alpha)$, then $zu_p * f \in \mathscr{R}(0)$.*

Taking $\alpha = 0$ in the above Corollary we obtain the next result.

Corollary 2.11. (Á. Baricz [43]) *Let $b, p, c \in \mathbb{R}$ such that $\kappa \ge |c|/4 + 1$. If $f \in \mathscr{R}(1/2)$, then $zu_p * f \in \mathscr{R}(0)$.*

Now we present the analogue of Theorem 2.20, using Lemma 2.9 and Theorem 2.6.

Theorem 2.21. (Á. Baricz [43]) *Let $\alpha_1 < 1, \alpha \in [0, 1), b \in \mathbb{R}, c \in [-37, 0)$ such that $8(1 - \alpha)\kappa + c \ge 0$. If $f \in \mathscr{R}(\alpha_1)$, then $zu_p * f \in \mathscr{R}(\gamma)$, with $\gamma = 1 - 2(1 - \alpha_1)(1 - \alpha)$.*

Proof. Using Theorem 2.6, we have that $u_p \in \mathscr{P}_0(\alpha)$. Therefore the proof is the same as the proof of Theorem 2.20.   □

As an immediate consequence of Theorem 2.21, we have for $\gamma = 0$ and $\alpha = 0$, respectively.

Corollary 2.12. (Á. Baricz [43]) *Let* $\alpha \in [0,1)$, $b \in \mathbb{R}$ *and* $c \in [-37,0)$ *be such that* $8(1 - \alpha)\kappa + c \geq 0$. *If* $f \in \mathscr{R}(\alpha_1)$, *where* $\alpha_1 = (1 - 2\alpha)/(2 - 2\alpha)$, *then* $zu_p * f \in \mathscr{R}(0)$. *In particular, let* $b \in \mathbb{R}$, $c \in [-37,0)$ *be such that* $\kappa \geq -c/8$. *If* $f \in \mathscr{R}(1/2)$, *then* $zu_p * f \in \mathscr{R}(0)$.

Note that it is easy to verify that the results of Corollary 2.12 for $c \in [-37,0)$ are better than the results of Corollaries 2.10 and 2.11. The situation is the same for Theorems 2.21 and 2.20. So in certain cases the "method of sequences" is better than the method of differential subordinations.

For Gaussian and confluent hypergeometric functions (and as well for generalized hypergeometric functions) may be found similar results with those given in this section in the works published by J.H. Choi et al. [84] and S. Ponnusamy [175,176].

We end this section by applying the above presented theorems and corollaries for some particular cases. By applying the Theorems 2.18, 2.19, 2.20, 2.21 and Corollaries 2.10, 2.11, 2.12 to the function

$$z \mapsto \mathscr{I}_p(z^{1/2}) = 2^p \Gamma(p+1) z^{-p/2} I_p(z^{1/2})$$

we obtain the following result.

Corollary 2.13. (Á. Baricz [43]) *Let* $p \in \mathbb{R}$. *For the function* $z \mapsto \mathscr{I}_p(z^{1/2})$ *the following assertions are true:*

(a) *Let* $\alpha \in [0,1)$ *satisfy* $4(1 - \alpha)p \geq 4\alpha^2 - 2\alpha - 1$. *If* $\alpha \geq 1/2$, *then* $\mathscr{I}_p(z^{1/2}) \in \mathscr{H}^\infty$; *if* $\alpha \in [0,1/2)$ *we have* $\mathscr{I}_p(z^{1/2}) \in \mathscr{H}^{1/(1-2\alpha)}$.
(b) *If* $p \geq -5/4$, *then* $\mathscr{I}_p(z^{1/2}) \in \mathscr{H}^\infty$.
(c) *Let* $\alpha_1 < 1$, $4p\sqrt{1 - 2\alpha} \geq 1 - \alpha$, *where* $\alpha \in [0,1/2)$. *If* $f \in \mathscr{R}(\alpha_1)$, *then* $z\mathscr{I}_p(z^{1/2}) * f(z) \in \mathscr{R}(\gamma)$, *with* $\gamma = 1 - 2(1 - \alpha_1)(1 - \alpha)$. *Moreover, if* $\alpha_1 = (1 - 2\alpha)/(2 - 2\alpha)$ *and* $f \in \mathscr{R}(\alpha_1)$, *then* $z\mathscr{I}_p(z^{1/2}) * f(z) \in \mathscr{R}(0)$. *In particular, if* $p \geq 1/4$ *and* $f \in \mathscr{R}(1/2)$, *then* $z\mathscr{I}_p(z^{1/2}) * f(z) \in \mathscr{R}(0)$.
(d) *Let* $\alpha_1 < 1$, $8(1 - \alpha)(p + 1) \geq 1$, *where* $\alpha \in [0,1)$. *If* $f \in \mathscr{R}(\alpha_1)$, *then* $z\mathscr{I}_p(z^{1/2}) * f(z) \in \mathscr{R}(\gamma)$, *with* $\gamma = 1 - 2(1 - \alpha_1)(1 - \alpha)$. *If* $\alpha_1 = (1 - 2\alpha)/(2 - 2\alpha)$ *and* $f \in \mathscr{R}(\alpha_1)$, *then* $z\mathscr{I}_p(z^{1/2}) * f(z) \in \mathscr{R}(0)$. *In particular, if* $p \geq -7/8$ *and* $f \in \mathscr{R}(1/2)$, *then* $z\mathscr{I}_p(z^{1/2}) * f(z) \in \mathscr{R}(0)$.

Remark 2.14. For a normalized function $f \in \mathscr{A}$ define the integral transform

$$V_\mu(f)(z) = \int_0^1 \mu(t) \frac{f(tz)}{t} \, dt, \tag{2.25}$$

where μ is a real-valued, nonnegative weight function normalized so that

$$\int_0^1 \mu(t) \, dt = 1.$$

Now, for $\alpha < 1$ and $0 \leq \beta \leq 1$ we define the class of functions

$$\mathscr{P}_\beta(\alpha) = \left\{ f \in \mathscr{A} : \exists \, \eta \in \mathbb{R} \text{ such that } \operatorname{Re} \Delta_\beta(z) > \alpha \text{ for all } z \in \mathbb{D} \right\},$$

where

$$\Delta_\beta(z) = e^{i\eta} \left((1-\beta) \frac{f(z)}{z} + \beta f'(z) \right).$$

In the last two decades there has been a vivid interest concerning the integral transform (2.25). By using the duality technique among others many authors have found conditions for $f \in \mathscr{P}_\beta(\alpha)$ such that $V_\mu(f) \in \mathscr{S}^*$ or $V_\mu(f) \in \mathscr{P}_\beta(\delta)$. The reason for the considerable interest in finding such conditions is that for appropriate choices of μ the integral transform $V_\mu(f)$ reduces to some known operators like Alexander, Bernardi, Libera and Komatu transforms, respectively. For more details and for related results we refer to the papers of R.M. Ali and V. Singh [6], R. Fournier and S. Ruscheweyh [99], S. Ponnusamy and F. Rønning [181–184], Y.C. Kim and F. Rønning [134], and to references therein. In the study of the above conditions the Gaussian hypergeometric function plays an important role. More precisely, the convolution operator $V_{a,b,c}(f)(z) = zF(a,b,c,z) * f(z)$ for admissible values of a,b and c is exactly an integral operator of the form (2.25), which reduces for $a = 1$ to the Carlson-Shaffer operator

$$V_{1,b,c}(f)(z) = \frac{\Gamma(c)}{\Gamma(b)\Gamma(c-b)} \int_0^1 t^{b-1}(1-t)^{c-b-1} \frac{f(tz)}{t} \, dt.$$

The hypergeometric transform $V_{a,b,c}(f)$ has been studied intensively in the last two decades. For example, S. Ponnusamy [178] has found conditions on α_1, α_2 and on the parameters a,b and c such that the operator $V_{a,b,c}(f)$ maps $\mathscr{R}(\alpha_1)$ into $\mathscr{R}(\alpha_2)$, Y.C. Kim and F. Rønning [134] have found the sharp value of α such that $V_{a,b,c}(f) \in \mathscr{S}^*$, where $f \in \mathscr{P}_\beta(\alpha)$. For other important geometric properties of the integral transform (2.25) and of the hypergeometric transform $V_{a,b,c}(f)$ we refer to the papers of R. Balasubramanian et al. [33–35, 37], R.W. Barnard et al. [65], J.H. Choi et al. [83] and to the references therein.

2.3.2 A Monotonicity Property of Generalized Bessel Functions

Let $z \mapsto F(a,b,c,z)$ be the Gaussian hypergeometric function, defined by (1.1). Based on numerical experiments S. Ponnusamy and M. Vuorinen [188, p. 351] enounced the following interesting open problems (which to our knowledge are still open):

1. There exist positive numbers α_1, α_2 such that for $a \in (0, \alpha_1)$ and $b \in (0, \alpha_2)$ the normalized function $z \mapsto zF(a,b,a+b,z)$ ($z \mapsto zF(a,b,a+b,z^2)$, respectively) maps the $\mathbb{D}$ into a strip domain. For example, the functions

$$-\mathrm{Log}(1-z) = zF(1,1,2,z)$$

and

$$\frac{1}{2}\text{Log}\left(\frac{1+z}{1-z}\right) = zF\left(1,\frac{1}{2},\frac{3}{2},z^2\right)$$

map $\mathbb{D}$ into a strip. Therefore, the problem here is to find the exact range of the constants α_1, α_2 and conditions on a and b such that this property holds.

2. We recall that the Koebe function

$$\frac{z}{(1-z)^2} = zF(1,2,1,z)$$

maps $\mathbb{D}$ into the complement of the ray $\{z = x+iy \in \mathbb{C} : x = 0, y \le -1/4\}$. This function raises the following question: Suppose that $a,b,c > 0$ with $c < a+b$. Do there exist $\alpha_3, \alpha_4 > 0$ such that for $a \in (0,\alpha_3)$ and $b \in (0,\alpha_4)$ the function $z \mapsto zF(a,b,c,z)$ ($z \mapsto zF(a,b,c,z^2)$), respectively) has the property that the image domain is completely contained in a sector type domain where the "angle" depends on $a+b-c$?

Let u_p be the generalized and normalized Bessel function of the first kind of order p depending on the parameters b,c,p and let $\lambda_p(z) = u_p(z^2)$, $z \in \mathbb{D}$. Motivated by the above open problems, we prove in this section that the image region $\lambda_p(\mathbb{D})$ is completely contained in a disk type domain (or egg-shaped domain). We also find the radius of the smallest disk centered at 1 which contains the image region $\lambda_p(\mathbb{D})$. Further, we aim to prove that if $p > q$, then $\lambda_p(\mathbb{D}) \subset \lambda_q(\mathbb{D})$, under certain conditions on the parameters b,c,p,q. The original results of this section were obtained by S. András and Á. Baricz [24].

Due to Lommel's expression of $J_p(z)$ as an integral of Poisson type (see the book of G.N. Watson [227, p. 48]) we know that if $\text{Re } p > -1/2$ and $z \in \mathbb{C}$ then

$$J_p(z) = \frac{(z/2)^p}{\Gamma\left(p+\frac{1}{2}\right)\Gamma\left(\frac{1}{2}\right)}\int_0^\pi e^{iz\cos\theta}\sin^{2p}\theta d\theta. \tag{2.26}$$

If p is real and greater than $-1/2$, then it follows from (2.26) (see G.N. Watson [227, p. 49]) that

$$|J_p(z)| \le \frac{|(z/2)^p|}{\Gamma\left(p+\frac{1}{2}\right)\Gamma\left(\frac{1}{2}\right)}\int_0^\pi e^{|\text{Im} z|}\sin^{2p}\theta d\theta = \frac{|(z/2)^p|}{\Gamma(p+1)}e^{|\text{Im} z|}. \tag{2.27}$$

By using the expression $\sqrt{2/(\pi z)}\cos z$ for $J_{-1/2}(z)$ it may be shown that (2.27) is also valid when $p = -1/2$.

Let us consider now the function $\mathcal{J}_p$ defined by (1.23). Recall that in particular we have

$$\mathcal{J}_{-1/2}(z) = \sqrt{\pi/2}\cdot z^{1/2}J_{-1/2}(z) = \cos z, \tag{2.28}$$

$$\mathcal{J}_{1/2}(z) = \sqrt{\pi/2}\cdot z^{-1/2}J_{1/2}(z) = \frac{\sin z}{z}, \tag{2.29}$$

$$\mathscr{J}_{3/2}(z) = 3\sqrt{\pi/2} \cdot z^{-3/2} J_{3/2}(z) = 3\left(\frac{\sin z}{z^3} - \frac{\cos z}{z^2}\right). \qquad (2.30)$$

From (2.27) it is clear that for $p \geq -1/2$ and $z \in \mathbb{C}$ we have the inequality $|\mathscr{J}_p(z)| \leq e^{|\operatorname{Im} z|}$. Thus if z lies in $\mathbb{D}$ and $p \geq -1/2$, then $|\mathscr{J}_p(z)| \leq e$. Comparing this inequality with the following inequalities (see the books of M. Abramowitz and I.A. Stegun [1, p. 75], D.S. Mitrinović [156, p. 323])

$$|\mathscr{J}_{-1/2}(z)| = |\cos z| < 2, \ |\mathscr{J}_{1/2}(z)| = \left|\frac{\sin z}{z}\right| < \frac{6}{5}, \text{ whenever } z \in \mathbb{D}, \qquad (2.31)$$

it is natural to seek the radius of the smallest disk centered at the origin which contains the image region $\mathscr{J}_p(\mathbb{D})$. In Corollary 2.14 we give the radius of this disk. Moreover, using a disk with center at 1, we obtain a much sharp result.

Remark 2.15. When $\operatorname{Re} p > -1/2$, then the function $\mathscr{J}_p$ under discussion admits the integral representation (see the handbook of M. Abramowitz and I.A. Stegun [1, p. 360] or the relation (1.26) for $c = 1$ and $b = 1$)

$$\mathscr{J}_p(z) = \int_0^1 \cos(tz) d\mu_p(t), \qquad (2.32)$$

where $d\mu_p(t) = \mu_p(t)dt$ with

$$\mu_p(t) = \frac{2(1-t^2)^{p-1/2}}{B\left(p+\frac{1}{2},\frac{1}{2}\right)},$$

μ_p being the probability measure on $[0,1]$. From the first inequality in (2.31) we easily get

$$|\mathscr{J}_p(z)| = \left|\int_0^1 \cos(tz) d\mu_p(t)\right| \leq \int_0^1 |\cos(tz)| d\mu_p(t) < \int_0^1 2 d\mu_p(t) = 2.$$

In other words, the above particular inequality $|\cos z| < 2$ implies $|\mathscr{J}_p(z)| < 2$ for all $z \in \mathbb{D}$, but this upper bound is far from being the best possible.

Remark 2.16. Numerical computations and graphics in Derive6, Maple6 guarantee the following properties of the functions $\mathscr{J}_{-1/2}$, $\mathscr{J}_{1/2}$ and $\mathscr{J}_{3/2}$ defined above by the relations (2.28), (2.29) and (2.30), respectively:

1. $\mathscr{J}_{-1/2}(\mathbb{D})$ is contained in the disk centered at 1, with radius $0.543080635\ldots$, i.e. we have $|\mathscr{J}_{-1/2}(z) - 1| < 0.543080635\ldots$ for all $z \in \mathbb{D}$ (see Fig. 2.1).
2. $\mathscr{J}_{1/2}(\mathbb{D})$ is contained in the disk centered at 1, with radius $0.175201194\ldots$, i.e. we have $|\mathscr{J}_{1/2}(z) - 1| < 0.175201194\ldots$ for all $z \in \mathbb{D}$ (see Fig. 2.2).
3. $\mathscr{J}_{3/2}(\mathbb{D})$ is contained in the disk centered at 1, with radius 0.103638323, i.e. we have $|\mathscr{J}_{3/2}(z) - 1| < 0.103638323\ldots$ for all $z \in \mathbb{D}$.

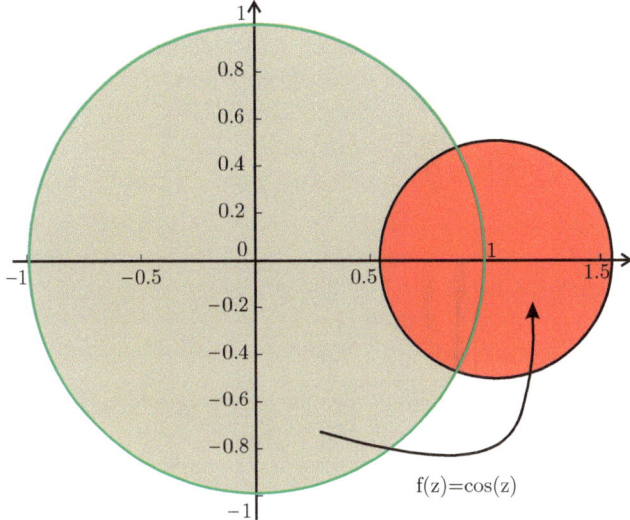

Fig. 2.1 The graph of the function $f : \mathbb{D} \to \mathbb{C}$, defined by $f(z) = \mathcal{J}_{-1/2}(z)$

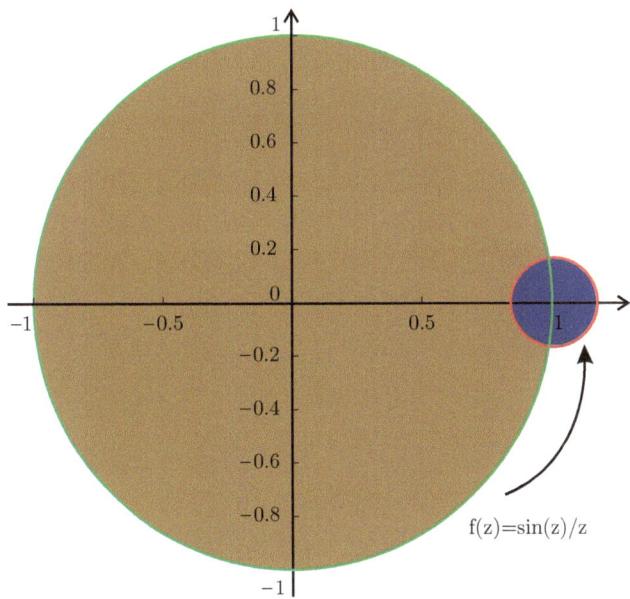

Fig. 2.2 The graph of the function $f : \mathbb{D} \to \mathbb{C}$, defined by $f(z) = \mathcal{J}_{1/2}(z)$

Taking into account the first property we may conclude that in fact $\mathscr{J}_{-1/2}(\mathbb{D})$ is contained in the disk centered at origin with radius $\cosh 1 = 1.543080635\dots$, and clearly this is the smallest disk (centered at origin) which contains $\mathscr{J}_{-1/2}(\mathbb{D})$, because

$$|\cos z| = |\cos(x+iy)| = \sqrt{\cos^2 x + \sinh^2 y} \le \cosh y < \cosh 1.$$

Thus using the same argument as in Remark 2.15 we can conclude that for $\operatorname{Re} p \ge -1/2$, the inequality $|\mathscr{J}_p(z)| < 1.543080635\dots$ holds for all $z \in \mathbb{D}$, and this upper bound is the best possible for $p = -1/2$.

In view of Remark 2.16 and Figs. 2.3 and 2.4 we may ask the following questions:

(a) Is it true that if $p > -1$ increases, then the image region $\mathscr{J}_p(\mathbb{D})$ decreases, more precisely if $p > q > -1$, then $\mathscr{J}_p(\mathbb{D}) \subset \mathscr{J}_q(\mathbb{D})$?
(b) Find the radius r_p (depending on p) of the smallest disk centered at 1 which contains the image region $\mathscr{J}_p(\mathbb{D})$ when $p > -1$.

Our aim in this section is to answer the above questions for generalized and normalized Bessel functions, Theorem 2.22 is the main result of this section. In particular we prove that $r_p = \mathscr{I}_p(1) - 1$, where $\mathscr{I}_p$ is defined by (1.24). Moreover we prove that if $p > q \ge -1/4$, then $\mathscr{J}_p(\mathbb{D}) \subset \mathscr{J}_q(\mathbb{D})$ holds (see Corollary 2.14 below). Part (a) of the above questions for $p > q$ and $q \in (-1, -1/4)$ remains open.

The following result contains conditions on the parameters b, c, p, q such that $\lambda_p(\mathbb{D}) \subset \lambda_q(\mathbb{D})$ holds.

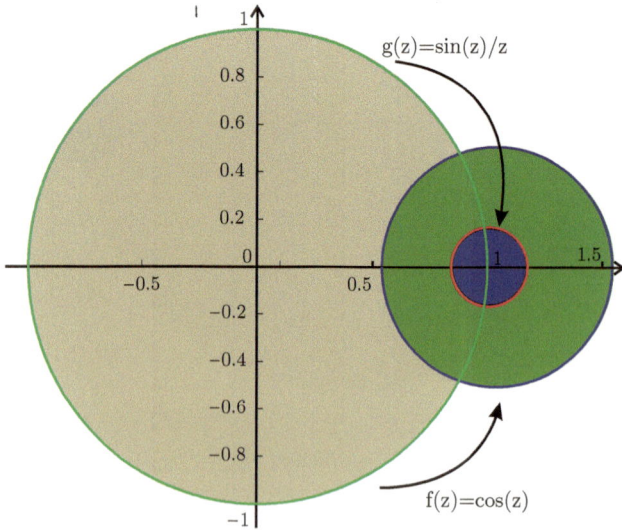

Fig. 2.3 The graph of the functions $f, g : \mathbb{D} \to \mathbb{C}$, defined by $f(z) = \mathscr{J}_{-1/2}(z)$ and $g(z) = \mathscr{J}_{1/2}(z)$

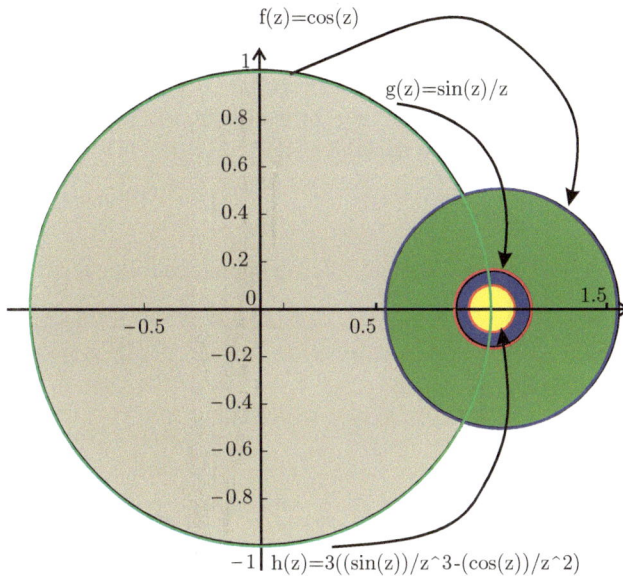

Fig. 2.4 The graph of the functions $f, g, h : \mathbb{D} \to \mathbb{C}$, defined by $f(z) = \mathcal{J}_{-1/2}(z)$, $g(z) = \mathcal{J}_{1/2}(z)$ and $h(z) = \mathcal{J}_{3/2}(z)$

Theorem 2.22. (S. András and Á. Baricz [24]) *Let $p, b, c \in \mathbb{C}$ be such that $2\kappa_p = 2p + b + 1 \neq 0, -2, -4, \dots$. The radius of the smallest disk centered at 1 which contains the domain $\lambda_p(\mathbb{D})$ is*

$$\sum_{n \geq 1} \frac{|c/4|^n}{n! \cdot |(\kappa_p)_n|}.$$

Moreover, if $q \in \mathbb{C}$ such that $\operatorname{Re} \kappa_p > \operatorname{Re} \kappa_q$ and

$$12 \operatorname{Re} \kappa_q \geq 3|c| + 2(\operatorname{Im} \kappa_q)^2 + 6, \tag{2.33}$$

where $2\kappa_q = 2q + b + 1$, then $\lambda_p(\mathbb{D}) \subset \lambda_q(\mathbb{D})$.

Proof. First observe that

$$\left| \sum_{m=1}^{n} \frac{(-c/4)^m}{(\kappa_p)_m} \frac{z^{2m}}{m!} \right| \leq \sum_{m=1}^{n} \left| \frac{(-c/4)^m}{(\kappa_p)_m} \frac{z^{2m}}{m!} \right| < \sum_{m=1}^{n} \left| \frac{(-c/4)^m}{(\kappa_p)_m} \frac{1}{m!} \right| \tag{2.34}$$

holds for all $z \in \mathbb{D}$. For $z \to \pm i$ we obtain equality, so this is the best upper bound for the whole disk. From the D'Alembert ratio test the radius of convergence for the power series $\lambda_p(z) - \lambda_p(0)$ is infinity, so if in (2.34) n tends to infinity, we get the required result.

By the definition of the function λ_p, it is enough to show that under the above assumptions $u_p(\mathbb{D}) \subset u_q(\mathbb{D})$ holds. Due to condition (2.33) from part (a) of Theorem 2.12 we know that u_q is a convex function in $\mathbb{D}$, i.e. $u_q(\mathbb{D})$ is convex. From Lemma 1.4, equation (1.31) we conclude, by using a known integral mean-value formula, that $u_p(\mathbb{D})$ is contained in the convex hull of $\{u_q(t\mathbb{D}) : t \in [0,1]\} = \{u_q(tz) : t \in [0,1], z \in \mathbb{D}\}$. Since $u_q(\mathbb{D})$ is convex we obtain that $u_p(\mathbb{D}) \subset u_q(\mathbb{D})$. Thus the proof is complete. $\square$

Choosing $c = 1$, $b = 1$ in Theorem 2.22 we obtain the following result:

Corollary 2.14. (S. András and Á. Baricz [24]) *Let $p > -1$. The radius r_p of the smallest disk centered at 1 which contains the domain $\mathscr{I}_p(\mathbb{D})$ is $\mathscr{I}_p(1) - 1$. Moreover, if $p, q \in \mathbb{C}$ such that* $\operatorname{Re} p > \operatorname{Re} q \geq -1/4 + (\operatorname{Im} q)^2/6$, *then $\mathscr{I}_p(\mathbb{D}) \subset \mathscr{I}_q(\mathbb{D})$ holds. In particular, if p, q are real numbers such that $p > q \geq -1/4$, then $\mathscr{I}_p(\mathbb{D}) \subset \mathscr{I}_q(\mathbb{D})$.*

Remark 2.17. Let $\mathscr{I}_p$ defined by (1.24). Analogously with the relations (2.28), (2.29) and (2.30) we have

$$\mathscr{I}_{-1/2}(z) = \sqrt{\pi/2} \cdot z^{1/2} I_{-1/2}(z) = \cosh z,$$

$$\mathscr{I}_{1/2}(z) = \sqrt{\pi/2} \cdot z^{-1/2} I_{1/2}(z) = \frac{\sinh z}{z},$$

$$\mathscr{I}_{3/2}(z) = 3\sqrt{\pi/2} \cdot z^{-3/2} I_{3/2}(z) = -3\left(\frac{\sinh z}{z^3} - \frac{\cosh z}{z^2}\right).$$

Thus, we have

$$r_{-1/2} = \mathscr{I}_{-1/2}(1) - 1 = \cosh 1 - 1 \simeq 0.543080635\ldots,$$

$$r_{1/2} = \mathscr{I}_{1/2}(1) - 1 = \sinh 1 - 1 \simeq 0.175201194\ldots$$

and

$$r_{3/2} = \mathscr{I}_{3/2}(1) - 1 = 3(\cosh 1 - \sinh 1) - 1 \simeq 0.103638323\ldots.$$

Remark 2.18. Note that under the condition (2.33) the image region $u_q(\mathbb{D})$ (and consequently $\lambda_q(\mathbb{D})$) is clearly starlike with respect to 1, i.e. u_q is a starlike function with respect to 1. This is justified by convexity of $u_q(\mathbb{D})$ and by the relation $u_q(0) = 1$. On the other hand, $\lambda_p(\mathbb{D})$ is symmetric with respect to the real axis, since according to (1.26) we have

$$\lambda_p(\bar{z}) = \int_0^1 \cos(t\bar{z}\sqrt{c})d\mu_{p,b}(t) = \overline{\int_0^1 \cos(tz\sqrt{c})d\mu_{p,b}(t)} = \overline{\lambda_p(z)} \text{ for all } z \in \mathbb{D}.$$

Moreover, using results of S. Owa and H.M. Srivastava [168, Lemma 3, p. 1062] we can find easily other sufficient conditions for the function u_q to be starlike with respect to 1. For example, if u_q satisfies the condition $|zu_q''(z)/u_q'(z)| < 1$, where $\kappa_q \neq 0, -1, -2, \ldots$ and $z \in \mathbb{D}$, then it is starlike with respect to 1 (see Theorem 2.9).

Remark 2.19. The Schwarz lemma (see the book of P.L. Duren [90, p. 3]) states: if $f : \mathbb{D} \to \mathbb{D}$ is analytic with $f(0) = 0$, then $|f'(0)| \leq 1$ and $|f(z)| \leq |z|$ for all $z \in \mathbb{D}$. On the other hand, an analytic function $f : \mathbb{D} \to \mathbb{C}$ is said to be subordinate (see the book of P.L. Duren [90]) to an analytic function $g : \mathbb{D} \to \mathbb{C}$ (written $f \prec g$) if $f = g \circ \varphi$ for some analytic function $\varphi : \mathbb{D} \to \mathbb{D}$ with $\varphi(0) = 0$. The superordinate function g need not be univalent. From Schwarz's lemma it is clear that if $f \prec g$, then $f(0) = g(0)$ and $f(\mathbb{D}) \subseteq g(\mathbb{D})$. Moreover if the function g is univalent, then the inverse implication also holds. Thus, taking into consideration that $u_p(0) = u_q(0) = 1$ and that the function u_q is univalent when $\operatorname{Re} \kappa_q \geq |c|/4$ (by Theorem 2.3), Theorem 2.22 can be written in terms of subordination, i.e. if the hypotheses of Theorem 2.22 hold, then $u_p \prec u_q$. Finally, note that similar results for confluent and Gaussian hypergeometric functions were obtained by S.S. Miller and P.T. Mocanu [154].

Chapter 3
Inequalities Involving Bessel and Hypergeometric Functions

Abstract In this chapter we deduce several functional inequalities for generalized Bessel functions of the first kind, Gaussian and Kummer hypergeometric functions as well as for general power series with positive coefficients. We present extensions to Bessel functions of some known trigonometric inequalities like Jordan, Cusa, van der Corput, Redheffer, Mahajan, Mitrinović. Moreover, we establish some Grünbaum, Askey and Landen type inequalities for generalized Bessel functions. The methods used to derive these inequalities are based on classical analysis. Among others, we use a criterion for the monotonicity of the quotient of two MacLaurin series and the monotone form of l'Hospital's rule.

This chapter is devoted to the study of the function u_p, defined by (1.20), for real variable, under the assumption that $p, b, c \in \mathbb{R}$. In the first section we treat some interesting functional inequalities involving ratios of the function u_p. In the second section we continue this investigations but only for the function u_p (not for ratios) obtaining analogous functional inequalities. Further relevant connections for hypergeometric functions are given. In the third section we prove a Landen-type inequality for the function λ_p, defined by $\lambda_p(x) = u_p(x^2)$ under certain conditions on the parameters b, c, p. These three sections were motivated by two open problems due to G.D. Anderson et al. [17]. The results may be found in the papers of the author [39, 41, 45]. The fourth section is devoted to the study of the convexity of zero-balanced Gaussian hypergeometric functions and general power series with respect to Hölder means. In the fifth section we prove that the Askey-type inequality [28] holds for the function λ_p under certain conditions on the parameters b, c, p. Moreover, we find a different upper bound for the sum $\lambda_p(x) + \lambda_p(y)$ and we establish computable lower and upper bounds for the function λ_p. The results of the fourth section may be found in [58], which is a joint work of the author with E. Neuman. In the sixth section intrinsic properties, including logarithmic convexity (concavity), of the modified Bessel functions of the first kind and some other related functions are obtained. Some of the results of this section may be found in [59], which is another joint work with E. Neuman. In the seventh section we extend for the function λ_p some classical inequalities, like Mahajan's inequality, Mitrinović's

Á. Baricz, *Generalized Bessel Functions of the First Kind*,
Lecture Notes in Mathematics 1994, DOI 10.1007/978-3-642-12230-9_3,
© Springer-Verlag Berlin Heidelberg 2010

inequality, improvements of Jordan's inequality, Redheffer's inequality, using an adequate integral representation of λ_p and the monotone form of l'Hospital's rule. Moreover, we prove that the integral

$$\varsigma_p(x) = \int_0^x \lambda_p(t)\,dt$$

is sub-additive (super-additive) under certain conditions on the parameters b, c, p.

Finally, in the last section, by using mathematical induction and infinite product representations of the functions $\mathscr{J}_p$ and $\mathscr{I}_p$, we present an extension of Redheffer's inequality for the function $\mathscr{J}_p$ and a Redheffer-type inequality for the function $\mathscr{I}_p$. Moreover, by using some known results on the zeros of Bessel functions of the first kind we establish sharp exponential Redheffer-type inequalities for Bessel and modified Bessel functions of the first kind. The original results of the last two sections of this chapter may be found in the papers of Á. Baricz [47,49], Á. Baricz and S. Wu [62,63].

Before we state our main results of this chapter we recall some basic facts.

A function $f : [a,b] \subseteq \mathbb{R} \to \mathbb{R}$ is said to be convex if for all $r, s \in [a,b]$ and $\lambda \in [0,1]$ we have (see the book of A.W. Roberts and D.E. Varberg [199])

$$f(\lambda r + (1-\lambda)s) \leq \lambda f(r) + (1-\lambda)f(s).$$

In other words, the function f is convex if and only if its epigraph (the set of points lying on or above the graph) is a convex set. If the above inequality is reversed, then f is called concave, in other words f is said to be concave if $-f$ is convex. Moreover, if for $r \neq s$ and $\lambda \in (0,1)$ the above inequality is strict then f is said to be strictly convex. A similar characterization of strictly concave functions is also valid. It is known that if f is differentiable, then f is (strictly) convex (concave) if and only if f' is (strictly) increasing (decreasing). Moreover, if f is twice differentiable, then f is (strictly) convex (concave) if and only if f'' is (strictly) positive (negative). We note that the sum of (strictly) convex (concave) functions is also (strictly) convex (concave) and a convergent sequence of (strictly) convex (concave) functions has a (strictly) convex (concave) limit function.

A function $g : [a,b] \subseteq \mathbb{R} \to (0,\infty)$ is said to be logarithmically convex, or simply log-convex, if its natural logarithm $\log f$ is convex, that is, for all $r, s \in [a,b]$ and $\lambda \in [0,1]$ we have (see the book of A.W. Roberts and D.E. Varberg [199])

$$g(\lambda r + (1-\lambda)s) \leq [g(r)]^\lambda [g(s)]^{1-\lambda}.$$

If the above inequality is reversed, then g is called logarithmically concave, or simply log-concave. In other words g is log-concave if $1/g$ is log-convex. Moreover, if for $r \neq s$ and $\lambda \in (0,1)$ the above inequality is strict then g is said to be strictly log-convex. A similar characterization of strictly log-concave functions is also valid. It is known that if g is differentiable, then g is (strictly) log-convex (log-concave) if and only if its logarithmic derivative g'/g is (strictly) increasing (decreasing). Apparently it would seem that log-concave (log-convex) functions would be unremarkable because they are simply related to concave (convex) functions. But they have some surprising properties. It is known that the product of log-concave (log-convex) functions is also log-concave (log-convex). Moreover, the sum of log-convex functions is also log-convex and a convergent sequence of log-convex (log-concave) functions has a log-convex (log-concave) limit function, provided the limit is positive. However, the sum of log-concave functions is not necessarily log-concave. Due to their interesting properties, the log-convex (log-concave) functions appear frequently in many problems of classical analysis and probability theory. Finally, let us mention that a log-convex function is a convex function, but the converse is not always true. Similarly a concave function is also log-concave, but the converse is not necessarily true.

Now, let us recall that a function $M : (0,\infty) \times (0,\infty) \to (0,\infty)$ is called a mean (function) if

(a) $M(r,s) = M(s,r)$ for all $r,s > 0$;
(b) $M(r,r) = r$ for all $r > 0$;
(c) $r < M(r,s) < s$ whenever $0 < r < s$;
(d) $M(ar,as) = aM(r,s)$ for all $a,r,s > 0$.

By definition, if M and N are arbitrary mean functions, a continuous function $f : [a,b] \subseteq (0,\infty) \to (0,\infty)$ is $MN-$convex if for all $r,s \in [a,b]$ we have

$$f(M(r,s)) \leq N(f(r), f(s)).$$

If the above inequality is reversed then we say that f is $MN-$concave. This concept of generalized convexity is quite old and has been studied extensively in the literature from various points of view. For more details we refer to the papers of J. Aczél [3], G.D. Anderson et al. [21], G. Aumann [30], Á. Baricz [57], J. Matkowski [145], C.P. Niculescu [163] and to the references therein. We note that in Sect. 3.4.2 we consider a particular case of $MN-$convexity, namely when M and N are power (or Hölder) mean functions. Finally, let us mention that the above definition reduces to usual convexity (concavity) when M and N are the arithmetic mean, and similarly reduces to log-convexity (log-concavity) when M is the arithmetic mean and N is the geometric mean.

3.1 Functional Inequalities Involving Quotients of Some Special Functions

Recall that for $a,b,c \in \mathbb{C}$, of which $c \neq 0,-1,-2,\ldots$, the Gaussian hypergeometric series is defined by

$$F(a,b,c,r) = \sum_{n \geq 0} \frac{(a)_n (b)_n}{(c)_n} \frac{r^n}{n!} \quad \text{for all } r \in (-1,1). \tag{3.1}$$

We next recall the behavior of the function $r \in (0,1) \mapsto F(a,b,c,r) \in (0,\infty)$ near 1 when $a,b,c > 0$.

1. For $c > a+b$ (see the book of E.D. Rainville [197, p. 49])

$$\lim_{r \to 1} F(a,b,c,r) = \frac{\Gamma(c)\Gamma(c-a-b)}{\Gamma(c-a)\Gamma(c-b)}.$$

2. For $c = a+b$ we have the formula (see the paper of R.J. Evans [97]):

$$\lim_{r \to 1} \frac{F(a,b,a+b,r)}{\log(1-r)} = -\frac{\Gamma(a+b)}{\Gamma(a)\Gamma(b)}.$$

3. For $c < a+b$, the corresponding formula (see the book of E.T. Whittaker and G.N. Watson [228, p. 299]) is

$$\lim_{r \to 1}(1-r)^{a+b-c} F(a,b,c,r) = \frac{\Gamma(c)\Gamma(a+b-c)}{\Gamma(a)\Gamma(b)}.$$

We note that the second case actually can be rewritten as the asymptotic relation

$$F(a,b,a+b,r) \sim -\frac{1}{B(a,b)}\log(1-r)$$

as $r \to 1$, which is due to C.F. Gauss, and its refined form

$$B(a,b)F(a,b,a+b,r) + \log(1-r) = -\Psi(a) - \Psi(b) - 2\gamma + \mathscr{O}((1-r)\log(1-r))$$

is due to S. Ramanujan (see the papers of R.J. Evans [97, p. 553] and B.C. Berndt [69, p. 71]). Here Ψ is the known psi or digamma function defined as the logarithmic derivative of Euler's gamma function, i.e. $\Psi(r) = \Gamma'(r)/\Gamma(r)$, and

$$\gamma = \lim_{n \to \infty}\left(1 + \frac{1}{2} + \frac{1}{3} + \ldots + \frac{1}{n} - \log n\right) = 0.5772156649\ldots$$

is the Euler-Mascheroni constant. It is important to note here that cases **2** and **3** above have been extended and improved by G.D. Anderson et al. [11], S. Ponnusamy and M. Vuorinen [186], see also the book of G.D. Anderson et al. [19].

Now, we recall the Ramanujan differentiation formula (see the book of B.C. Berndt [69, p. 87]) for the quotient of two Gaussian hypergeometric functions: if $a,b,c,d > 0$ with $a+b+1 = c+d$, then for all $r \in (0,1)$

$$\frac{d}{dr}\left[\frac{F(a,b,d,1-r)}{F(a,b,c,r)}\right] = -\frac{\Gamma(c)\Gamma(d)}{\Gamma(a)\Gamma(b)} \cdot \frac{1}{r^c(1-r)^d [F(a,b,c,r)]^2}, \qquad (3.2)$$

which in particular for $a, r \in (0, 1)$ reduces to Ramanujan's formula (see the book of B.C. Berndt [69, p. 88]) for zero-balanced Gaussian hypergeometric functions

$$\frac{d}{dr} \left[\frac{F(a, 1-a, 1, 1-r)}{F(a, 1-a, 1, r)} \right] = -\frac{\sin(\pi a)}{\pi} \cdot \frac{1}{r(1-r) \left[F(a, 1-a, 1, r) \right]^2}. \tag{3.3}$$

Now, let us recall the Elliott formula (see the paper of E.B. Elliott [96]): if $a, b, c \geq 0$ and $r \in (-1, 1)$, then

$$st + uv - ut = \frac{B(1+a+b, 1+b+c)}{B(3/2+a+b+c, 1/2+b)}, \tag{3.4}$$

where

$$s = F(1/2+a, -1/2-c, 1+a+b, r^2),$$
$$t = F(1/2-a, 1/2+c, 1+b+c, 1-r^2),$$
$$u = F(1/2+a, 1/2-c, 1+a+b, r^2),$$
$$v = F(-1/2-a, 1/2+c, 1+b+c, 1-r^2).$$

A particular case of Elliott's formula is the important identity of Legendre

$$\mathscr{E}(r)\mathscr{K}(\sqrt{1-r^2}) + \mathscr{E}(\sqrt{1-r^2})\mathscr{K}(r) - \mathscr{K}(r)\mathscr{K}(\sqrt{1-r^2}) = \frac{\pi}{2}, \tag{3.5}$$

where $r \in (0, 1)$ and

$$\mathscr{K}(r) = \int_0^{\pi/2} \frac{d\theta}{\sqrt{1 - r^2 \sin^2 \theta}} = \frac{\pi}{2} F\left(\frac{1}{2}, \frac{1}{2}, 1, r^2\right),$$

$$\mathscr{E}(r) = \int_0^{\pi/2} \sqrt{1 - r^2 \sin^2 \theta} \, d\theta = \frac{\pi}{2} F\left(\frac{1}{2}, -\frac{1}{2}, 1, r^2\right)$$

are the complete elliptic integrals of the first and second kinds.

We note that recently R. Balasubramanian et al. [32] showed that in fact Ramanujan's differentiation formula (3.2) for quotients of hypergeometric functions is equivalent to Elliott's identity (3.4). For more details see [32, Corollary 1.8]. Applying this result, R. Balasubramanian et al. [32] obtained a number of identities associated with the complete elliptic integrals of the first and second kinds. For related results and other generalizations of Legendre's identity (3.5) we refer to the papers of Anderson et al. [12], R. Balasubramanian et al. [37], E.A. Karatsuba and M. Vuorinen [130]. For general references on other beautiful properties of the hypergeometric functions we refer also to the books of B.C. Berndt [68–72]. For more details on complete elliptic integrals, especially on new bounds and inequalities we refer to the papers of H. Alzer and S.L. Qiu [9], G.D. Anderson et al. [15, 16],

S. András and Á. Baricz [25], Á. Baricz [50], B.N. Guo and F. Qi [109], H. Kazi and E. Neuman [131, 132], E. Neuman [160] and to the references therein.

G.D. Anderson et al. [17] introduced the homeomorphism $m : (0,1) \to (0,\infty)$ by

$$m(r) = \frac{F(a,b,a+b,1-r^2)}{F(a,b,a+b,r^2)}$$

and posed the following problem [17, p. 80]:

For which $a,b \in (0,1)$ does the functional inequality

$$m(r) + m(s) \le 2m(\sqrt{rs}) \tag{3.6}$$

hold for all $r,s \in (0,1)$?

Some particular cases of this problem were solved by G.D. Anderson et al. [15] and by S.L. Qiu and M. Vuorinen [195]. Recently, S.L. Qiu and M. Vuorinen [195] proved (fifth inequality of first part of Theorem 1.18, p. 112) that

$$\mu_a(r) + \mu_a(s) \le 2\mu_a(\sqrt{rs}) \tag{3.7}$$

for $a \in (0,1/2]$ and $r,s \in (0,1)$, where

$$\mu_a(r) = \frac{\pi}{2\sin \pi a} \frac{F(a,1-a,1,1-r^2)}{F(a,1-a,1,r^2)}.$$

It is also important to note here that the properties of the solutions s of the modular equations $m(s) = pm(r)$ with $a+b=1$ and p, a small positive integer, have been studied by B.C. Berndt et al. [73] to establish a number of identities, which were stated without proof by S. Ramanujan.

Using the Ramanujan differentiation formula (3.3) for zero-balanced hypergeometric functions, R. Balasubramanian et al. [36] proved recently that the inequality (3.6) holds when $a + b = 1$. Moreover, they found a lower bound for the sum $m(r) + m(s)$ for $a \in (0,2)$ and $b \in (0,2-a]$, and proved that

$$m(r) + m(s) \ge 2m\left(\sqrt{1 - \sqrt{(1-r^2)(1-s^2)}}\right) \tag{3.8}$$

for all $r,s \in (0,1)$ if $a \in (0,2)$ and $b \in (0,2-a]$.

In this section we prove that (3.8) holds for all $a,b > 0$ and we prove an analogous inequality for the generalized and normalized Bessel functions. Theorems 3.1 and 3.3 are the main results of this section.

The following lemma due to M. Biernacki and J. Krzyż [74] is one of the crucial facts in the proof of our results in this section. Note that this lemma is a special case of a more general lemma established by S. Ponnusamy and M. Vuorinen [186]. A proof of this lemma may be found in Sect. 3.6 (see the proof of Lemma 3.6).

Lemma 3.1. (M. Biernacki and J. Krzyż [74], S. Ponnusamy and M. Vuorinen [186]) *Suppose that the power series $f(x) = \sum_{n \geq 0} \alpha_n x^n$ and $g(x) = \sum_{n \geq 0} \beta_n x^n$ both converge for $x \in (-1,1)$, where $\alpha_n \in \mathbb{R}$ and $\beta_n > 0$ for all $n \geq 0$. Then f/g is (strictly) increasing (decreasing) on $(0,1)$ if $\{\alpha_n/\beta_n\}_{n \geq 0}$ is (strictly) increasing (decreasing).*

3.1.1 Preliminary Results

Lemma 3.2. (Á. Baricz [41]) *Let $u_p(r) = \sum_{n \geq 0} b_n r^n$, where the coefficients b_n are defined in (1.18). If $c < 0$ and $\kappa > 0$, then $b_n > 0$ for all $n \geq 0$ and the function $r \in (0,1) \mapsto u'_p(r)/u_p(r) \in \mathbb{R}$ is strictly decreasing.*

Proof. Because $c < 0$ and $\kappa > 0$, we have $b_n > 0$ for all $n \geq 0$. This implies $u_p(r) \neq 0$ for all $r \in (0,1)$. Therefore $u'_p(r)/u_p(r)$ makes sense. For the convenience we denote $u'_p(r) = \sum_{n \geq 0} \alpha_n r^n$. By Lemma 3.1 it suffices to show that the sequence $\{\alpha_n/b_n\}_{n \geq 0}$ is strictly decreasing. Since

$$\frac{\alpha_n}{b_n} = \frac{(n+1)b_{n+1}}{b_n} = -\frac{c}{4(\kappa+n)} \quad \text{for all } n \geq 0,$$

we have

$$\frac{\alpha_n}{b_n} - \frac{\alpha_{n+1}}{b_{n+1}} = -\frac{c}{4(\kappa+n)(\kappa+n+1)} > 0 \quad \text{for all } n \geq 0.$$

Hence, $\{\alpha_n/b_n\}_{n \geq 0}$ is strictly decreasing. By Lemma 3.1 it follows that the function $r \in (0,1) \mapsto u'_p(r)/u_p(r) \in \mathbb{R}$ is strictly decreasing. $\square$

Lemma 3.3. (Á. Baricz [41]) *If $c < 0$ and $\kappa > 0$, then the function $Q : (0,\infty) \to (0,\infty)$, defined by $Q(t) = u_p(e^{-t})/u_p(1-e^{-t})$, is strictly decreasing and strictly log-convex.*

Proof. Differentiating the function Q, we obtain

$$Q'(t) = (-e^{-t}) \cdot \frac{u_p(1-e^{-t})u'_p(e^{-t}) + u'_p(1-e^{-t})u_p(e^{-t})}{[u_p(1-e^{-t})]^2} \quad \text{for all } t \in (0,\infty). \quad (3.9)$$

This equality implies $Q'(t) < 0$ for all $t \in (0,\infty)$ because $u_p(r) > 0$ and $u'_p(r) > 0$ for all $r \in (0,1)$. As a consequence, Q is strictly decreasing.

Now we prove that Q is log-convex. For this let us denote $f_1(t) = u_p(e^{-t})$ and $f_2(t) = 1/u_p(1-e^{-t})$ for each $t \in (0,\infty)$. We prove that these functions are log-convex, consequently their product $Q = f_1 f_2$ will be also log-convex. By definition f_1 and f_2 are log-convex if the functions $g_1(t) = \log f_1(t)$ and $g_2(t) = \log f_2(t)$ are convex. We shall use the known fact that a sufficient condition for a real-valued function to be convex is that its second derivative is nonnegative. Differentiating the functions g_1 and g_2, we obtain

$$g_1'(t) = (-e^{-t}) \cdot \frac{u_p'(e^{-t})}{u_p(e^{-t})}, \quad g_2'(t) = (-e^{-t}) \cdot \frac{u_p'(1-e^{-t})}{u_p(1-e^{-t})},$$

$$g_1''(t) = (e^{-t}) \left[\frac{u_p'(e^{-t})}{u_p(e^{-t})} + e^{-t} \frac{u_p''(e^{-t})u_p(e^{-t}) - [u_p'(e^{-t})]^2}{[u_p(e^{-t})]^2} \right],$$

$$g_2''(t) = (e^{-t}) \left[\frac{u_p'(1-e^{-t})}{u_p(1-e^{-t})} - e^{-t} P(t) \right],$$

where

$$P(t) = \frac{u_p''(1-e^{-t})u_p(1-e^{-t}) - [u_p'(1-e^{-t})]^2}{[u_p(1-e^{-t})]^2}.$$

We show that $g_1''(t) \geq 0$ and $g_2''(t) \geq 0$ for all $t \in (0,\infty)$. For the function g_1 we need to prove the inequality

$$u_p'(e^{-t})u_p(e^{-t}) + e^{-t}u_p''(e^{-t})u_p(e^{-t}) \geq e^{-t}[u_p'(e^{-t})]^2 \quad \text{for all } t \in (0,\infty).$$

For convenience, we make the substitution $r = e^{-t} \in (0,1)$. After simplification, we see that the above inequality is equivalent to

$$\frac{1}{r} + \frac{u_p''(r)}{u_p'(r)} - \frac{u_p'(r)}{u_p(r)} \geq 0 \quad \text{for all } r \in (0,1). \tag{3.10}$$

With the notation $h_Q(r) = u_p'(r)/u_p(r)$, (3.10) can be written as

$$1/r + h_Q'(r)/h_Q(r) \geq 0 \quad \text{for all } r \in (0,1).$$

But this is equivalent to

$$[\log(rh_Q(r))]' \geq 0 \quad \text{for all } r \in (0,1). \tag{3.11}$$

On the other hand, we have

$$rh_Q(r) = r\frac{u_p'(r)}{u_p(r)} = \frac{\sum\limits_{n \geq 0} nb_n r^n}{\sum\limits_{n \geq 0} b_n r^n} \quad \text{for all } r \in (0,1).$$

Since $b_n > 0$ for all $n \geq 0$ and the sequence $\{nb_n/b_n\}_{n \geq 0}$ is strictly increasing, it follows by Lemma 3.1 that the function $r \in (0,1) \mapsto rh_Q(r) \in (0,\infty)$ is strictly increasing. Because the function log is also strictly increasing, we obtain that the composite function $r \in (0,1) \mapsto \log(rh_Q(r)) \in \mathbb{R}$ is strictly increasing too. Thus inequality (3.11) holds.

For the function g_2 we need to prove the inequality

$$u'_p(1 - e^{-t})u_p(1 - e^{-t}) - e^{-t}u''_p(1 - e^{-t})u_p(1 - e^{-t}) \geq e^{-t}[u'_p(1 - e^{-t})]^2$$

for all $t \in (0,\infty)$. With the notation $r = 1 - e^{-t} \in (0,1)$, the above inequality is equivalent to

$$\frac{u''_p(r)}{u'_p(r)} - \frac{u'_p(r)}{u_p(r)} - \frac{1}{1-r} \leq 0 \text{ for all } r \in (0,1).$$

Using h_Q, this inequality can be written in the form

$$h'_Q(r)/h_Q(r) - 1/(1-r) \leq 0 \text{ for all } r \in (0,1).$$

For proving this it suffices to show that

$$[\log(1-r)h_Q(r)]' \leq 0 \text{ for all } r \in (0,1).$$

But this is true, because Lemma 3.2 guarantees that the function $h_Q : (0,1) \to (0,\infty)$ is strictly decreasing, the function $r \in (0,1) \mapsto 1 - r \in (0,1)$ is strictly decreasing, and consequently $r \in (0,1) \mapsto \log[(1-r)h_Q(r)] \in \mathbb{R}$ is also a strictly decreasing function on $(0,1)$. This completes the proof of the log-convexity of Q.

Now we prove that Q is actually strictly log-convex. Observe that for this it suffices to show that g_1 or g_2 is strictly convex. Consider the function g_1 and suppose that g_1 is not strictly convex. It is known that if a function $f : I \subseteq \mathbb{R} \to \mathbb{R}$ is convex, then it is strictly convex if and only if there is no subinterval of I on which f is affine (see page 7 on the book of A.W. Roberts and D.E. Varberg [199]). Now, using the above result it follows that there exists an interval $(t_1,t_2) \subseteq (0,\infty)$ on which g_1 is affine, i.e. g_1 has the form $g_1(t) = \alpha t + \beta$ on (t_1,t_2). This implies that $g'_1(t) = \alpha$ for all $t \in (t_1,t_2)$. Consequently we have $rh_Q(r) = ru'_p(r)/u_p(r) = -\alpha$ for all $r \in (r_2,r_1) \subseteq (0,1)$, where $r_1 = e^{-t_1}$ and $r_2 = e^{-t_2}$. But, this contradicts the fact that the function $r \in (0,1) \mapsto rh_Q(r) \in (0,\infty)$ is strictly increasing. Hence g_1 is strictly convex and consequently Q is strictly log-convex. This completes the proof. $\square$

Lemma 3.4. (Á. Baricz [41]) *Let $T(r) = F(a,b,a+b,r)$ be the zero-balanced Gaussian hypergeometric function defined by (3.1). If $a,b > 0$, then the function $K : (0,\infty) \to (0,\infty)$, defined by $K(t) = T(e^{-t})/T(1-e^{-t})$, is strictly decreasing and strictly log-convex.*

Proof. By the proof of Lemma 3.3 clearly it is enough to prove the inequalities

$$[\log(rh_K(r))]' \geq 0$$

and

$$[\log(1-r)h_K(r)]' \leq 0$$

for all $r \in (0,1)$, where $h_K(r) = T'(r)/T(r)$.

For the first inequality we apply the same method as in the proof of Lemma 3.3. So, by (3.1) and Lemma 3.1, we obtain that the function

$$rh_K(r) = r\frac{T'(r)}{T(r)} = \frac{\displaystyle\sum_{n\geq0} nd_n r^n}{\displaystyle\sum_{n\geq0} d_n r^n}, \quad \text{where } d_n = \frac{(a)_n(b)_n}{(a+b)_n}\cdot\frac{1}{n!} \quad \text{for all } n \geq 0,$$

is strictly increasing, because $d_n > 0$ for all $n \geq 0$ and the sequence $\{nd_n/d_n\}_{n\geq0}$ is strictly increasing. Thus the function $r \in (0,1) \mapsto \log(rh_K(r)) \in \mathbb{R}$ is strictly increasing and this implies that the first derivative of this function is nonnegative.

For the second inequality we prove that $\log(1-r)h_K(r)$ is strictly decreasing for $r \in (0,1)$. For this it is enough to show that the function $r \in (0,1) \mapsto (1-r)h_K(r) \in \mathbb{R}$ is strictly decreasing. By (3.1) we obtain that

$$(1-r)h_K(r) = (1-r)\frac{T'(r)}{T(r)} = \frac{\displaystyle\sum_{n\geq0}[(n+1)d_{n+1} - nd_n]r^n}{\displaystyle\sum_{n\geq0} d_n r^n}.$$

We denote $A_n = [(n+1)d_{n+1} - nd_n]/d_n$ for all $n \geq 0$. By the ascending factorial notation we obtain that

$$(n+1)d_{n+1} = \frac{(a)_{n+1}(b)_{n+1}}{(a+b)_{n+1}\cdot n!} = \frac{(a+n)(b+n)}{a+b+n}d_n \quad \text{for all } n \geq 0,$$

so $A_n = ab/(a+b+n)$ for all $n \geq 0$. The sequence $\{A_n\}_{n\geq0}$ is clearly strictly decreasing because $ab > 0$. By Lemma 3.1 we obtain that the function $r \in (0,1) \mapsto (1-r)h_K(r) \in \mathbb{R}$ is strictly decreasing. Thus the proof is complete. □

3.1.2 Inequalities Involving Ratios of Generalized Bessel Functions

In this section our aim is to deduce some inequalities for quotients of generalized Bessel functions.

Theorem 3.1. (Á. Baricz [41]) *If $c < 0$, $\kappa > 0$ and $r_1, r_2, \ldots, r_k \in (0,1)$, then the function $\sigma : (0,1) \to (0,\infty)$, defined by $\sigma(r) = u_p(1-r^2)/u_p(r^2)$, satisfies*

$$\sqrt[k]{\prod_{i=1}^{k} \sigma(r_i)} \leq \sigma\left(\sqrt[k]{\prod_{i=1}^{k} r_i}\right). \tag{3.12}$$

*In the previous inequality equality holds if and only if $r_1 = r_2 = \ldots = r_k$. In partic-
ular, for $k = 2$ we obtain the inequality*

$$\sqrt{\sigma(r_1)\sigma(r_2)} \leq \sigma(\sqrt{r_1 r_2}) \quad \text{for all } r_1, r_2 \in (0,1),$$

where equality holds if and only if $r_1 = r_2$.

Proof. By Lemma 3.3 the function $t \in (0,\infty) \mapsto \log Q(t) \in \mathbb{R}$ is strictly convex, where $Q : (0,\infty) \to (0,\infty)$ is defined by $Q(t) = u_p(e^{-t})/u_p(1 - e^{-t})$. Let
$t_1, t_2, \ldots, t_k > 0$ and $s = (t_1 + t_2 + \ldots + t_k)/k$. Due to Jensen's inequality (1.37) it
follows that the following inequality holds:

$$\log Q(s) \leq \frac{1}{k} \sum_{i=1}^{k} \log Q(t_i),$$

and that in this inequality equality holds if and only if $t_1 = t_2 = \ldots = t_k$. Since the
function log is strictly increasing, we conclude that

$$Q(s) \leq \left[\prod_{i=1}^{k} Q(t_i) \right]^{1/k}$$

or equivalently

$$\left[\prod_{i=1}^{k} \frac{u_p(1 - e^{-t_i})}{u_p(e^{-t_i})} \right]^{1/k} \leq \frac{u_p(1 - e^{-s})}{u_p(e^{-s})}$$

and that in the inequalities equality holds if and only if $t_1 = t_2 = \ldots = t_k$. If we put
in the last inequality $t_i = -2\log r_i$ for all $i \in \{1,2,\ldots,k\}$, we obtain the inequality
(3.12) and see that in (3.12) we have equality if and only if $r_1 = r_2 = \ldots = r_k$.   □

The next result is an analogue of Theorem 1.8 obtained by R. Balasubramanian et al.
[36] for Bessel functions.

Corollary 3.1. (Á. Baricz [41]) *If $c < 0$, $\kappa > 0$ and $r_1, r_2 \in (0,1)$, then the function
$\sigma : (0,1) \to (0,\infty)$, defined by $\sigma(r) = u_p(1 - r^2)/u_p(r^2)$, satisfies*

$$\frac{1}{\sigma(r_1)} + \frac{1}{\sigma(r_2)} \geq \frac{2}{\sigma(\sqrt{r_1 r_2})},$$

$$\sigma(r_1) + \sigma(r_2) \geq 2\sigma\left(\sqrt{1 - \sqrt{(1 - r_1^2)(1 - r_2^2)}} \right).$$

In the above inequalities equality holds if and only if $r_1 = r_2$.

Proof. By Theorem 3.1 we know that $\sqrt{\sigma(r_1)\sigma(r_2)} \leq \sigma(\sqrt{r_1 r_2})$. Using this fact and the geometric-harmonic mean inequality between the unweighted harmonic and geometric means of the non-negative values $\sigma(r_1)$, $\sigma(r_2)$ we obtain that

$$\frac{1}{\sigma(r_1)} + \frac{1}{\sigma(r_2)} \geq \frac{2}{\sqrt{\sigma(r_1)\sigma(r_2)}} \geq \frac{2}{\sigma(\sqrt{r_1 r_2})}.$$

On the other hand, let $Q : (0,\infty) \to (0,\infty)$ be the function defined in Lemma 3.3. Due to Lemma 3.3 and to the arithmetic-geometric mean inequality between the geometric and arithmetic means of the values $Q(t_1)$, $Q(t_2)$, we deduce

$$Q\left(\frac{t_1 + t_2}{2}\right) \leq \sqrt{Q(t_1)Q(t_2)} \leq \frac{Q(t_1) + Q(t_2)}{2},$$

for all $t_1, t_2 > 0$. If $e^{-t_1} = 1 - r_1^2$ and $e^{-t_2} = 1 - r_2^2$, then this relation becomes

$$\sigma\left(\sqrt{1 - \sqrt{(1 - r_1^2)(1 - r_2^2)}}\right) \leq \sqrt{\sigma(r_1)\sigma(r_2)} \leq \frac{\sigma(r_1) + \sigma(r_2)}{2}.$$

□

3.1.3 Inequalities Involving Ratios of Hypergeometric Functions

In this section we use Lemma 3.4 to obtain analogous results for the Gaussian hypergeometric function.

Theorem 3.2. (Á. Baricz [41]) *Let* $T(r) = F(a,b,a+b,r)$ *be the zero-balanced hypergeometric function. If* $a, b > 0$ *and* $r_1, r_2, \ldots, r_k \in (0,1)$, *then the function* $m : (0,1) \to (0,\infty)$, *defined by* $m(r) = T(1 - r^2)/T(r^2)$, *satisfies*

$$\sqrt[k]{\prod_{i=1}^{k} m(r_i)} \leq m\left(\sqrt[k]{\prod_{i=1}^{k} r_i}\right). \tag{3.13}$$

In the previous inequality equality holds if and only if $r_1 = r_2 = \ldots = r_k$. *In particular, for* $k = 2$ *we obtain the inequality*

$$\sqrt{m(r_1)m(r_2)} \leq m(\sqrt{r_1 r_2}) \text{ for all } r_1, r_2 \in (0,1),$$

where equality holds if and only if $r_1 = r_2$.

Proof. By Lemma 3.4 the function $t \in (0,\infty) \mapsto \log K(t) \in \mathbb{R}$ is strictly convex, where $K(t) = T(e^{-t})/T(1 - e^{-t})$. So by applying the same proof as in the case of Theorem 3.1, by replacing Q with K, we see that the theorem is true.   □

Remark 3.1. Our approach is simpler than the method used by R. Balasubramanian et al. [36]. We note that if we use the arithmetic-geometric mean inequality for the values $m(r_1)$ and $m(r_2)$, then [36, Theorem 1.5] actually implies the inequality

$$\sqrt{m(r_1)m(r_2)} \le m(\sqrt{r_1 r_2})$$

for $r_1, r_2 \in (0,1)$ but only for $a, b \in (0,1)$ and $a + b = 1$. The situation is similar in the case of the inequality (3.7), i.e. in the case of [195, Theorem 1.18] established by S.L. Qiu and M. Vuorinen. More precisely, using the arithmetic-geometric mean inequality for the values $\mu_a(r)$ and $\mu_a(t)$, we see that the inequality

$$\sqrt{\mu_a(r)\mu_a(t)} \le \mu_a(\sqrt{rt})$$

holds for $r, t \in (0,1)$ and $a \in (0, 1/2]$. This is another particular case of inequality

$$\sqrt{m(r)m(t)} \le m(\sqrt{rt}).$$

Corollary 3.2. (Á. Baricz [41]) *If $a, b > 0$, $r_1, r_2 \in (0,1)$, then the function $m : (0,1) \to (0,\infty)$, defined in Theorem 3.2, satisfies*

$$\frac{1}{m(r_1)} + \frac{1}{m(r_2)} \ge \frac{2}{m(\sqrt{r_1 r_2})}, \tag{3.14}$$

$$m(r_1) + m(r_2) \ge 2m\left(\sqrt{1 - \sqrt{(1-(r_1)^2)(1-(r_2)^2)}}\right). \tag{3.15}$$

In the above inequalities equality holds if and only if $r_1 = r_2$.

Proof. The proof of Corollary 3.1 also applies, mutatis mutandis, to the present corollary. □

3.1.4 Inequalities Involving Ratios of General Power Series

In this section we state a generalization of the Theorems 3.1 and 3.2 as well as of the Corollaries 3.1 and 3.2 for general power series.

Theorem 3.3. (Á. Baricz [41]) *Suppose that the power series $f(x) = \sum_{n \ge 0} A_n x^n$, where $A_n > 0$ for all $n \ge 0$, is convergent for all $x \in (0,1)$. Furthermore, suppose that the sequence $\{(n+1)A_{n+1}/A_n - n\}_{n \ge 0}$ is (strictly) decreasing. Let the function $m_f : (0,1) \to (0,\infty)$ be defined by $m_f(r) = f(1-r^2)/f(r^2)$. Then*

$$\sqrt[k]{\prod_{i=1}^{k} m_f(r_i)} \le m_f\left(\sqrt[k]{\prod_{i=1}^{k} r_i}\right) \tag{3.16}$$

for all $r_1, r_2, \ldots, r_k \in (0,1)$, where equality holds if and only if $r_1 = r_2 = \ldots = r_k$. In particular, for $k = 2$ the inequalities

$$\sqrt{m_f(r_1)m_f(r_2)} \leq m_f(\sqrt{r_1 r_2}),$$

$$\frac{1}{m_f(r_1)} + \frac{1}{m_f(r_2)} \geq \frac{2}{m_f(\sqrt{r_1 r_2})}, \tag{3.17}$$

$$m_f(r_1) + m_f(r_2) \geq 2m_f\left(\sqrt{1 - \sqrt{(1-(r_1)^2)(1-(r_2)^2)}}\right) \tag{3.18}$$

hold for all $r_1, r_2 \in (0,1)$ and in all these inequalities equality holds if and only if $r_1 = r_2$.

Proof. We use the same method as in the proofs of Theorems 3.1 and 3.2. Since $f(x)$ is convergent for all $x \in (0,1)$, we can deduce that

$$g(x) = (1-x)f'(x) = \sum_{n \geq 0} [(n+1)A_{n+1} - nA_n]x^n$$

is convergent too for all $x \in (0,1)$. By hypothesis we can prove that for $x \in (0,1)$ the following inequalities hold

$$[\log(xh_f(x))]' \geq 0$$

and

$$[\log(1-x)h_f(x)]' \leq 0,$$

where $h_f(x) = f'(x)/f(x)$. These inequalities imply that the functions $t \in (0,\infty) \mapsto f(e^{-t}) \in (0,\infty)$, $t \in (0,\infty) \mapsto 1/f(1-e^{-t}) \in (0,\infty)$ are log-convex. Consequently their product is also log-convex. Moreover, applying the same argument as in the proof of Lemma 3.3 we see that $t \in (0,\infty) \mapsto f(e^{-t})/f(1-e^{-t}) \in (0,\infty)$ is actually strictly log-convex. Thus applying Jensen's inequality (1.37) we obtain (3.16) after substitutions $e^{-t_i} = r_i$, for all $i \in \{1, 2, \ldots, k\}$.

For (3.17) and (3.18) we use inequality (3.16) for $k = 2$ and geometric-harmonic mean, respectively arithmetic-geometric mean inequalities for $m_f(r_1)$ and $m_f(r_2)$. $\qquad \square$

Remark 3.2. If we consider the function $q_f : \mathbb{N} \to \mathbb{R}_+^*$, defined by

$$q_f(n) = (n+1)A_{n+1}/A_n,$$

where A_n is the nth coefficient of the power series $f(x) = \sum_{n \geq 0} A_n x^n$ occurring in Theorem 3.3, then the condition concerning the sequence $\{(n+1)A_{n+1}/A_n - n\}_{n \geq 0}$ becomes $q_f(n) + 1 \geq q_f(n+1)$ for all $n \geq 0$. In particular in the case of hypergeometric functions we have that $q_G(n) = (a+n)(b+n)/(a+b+n)$, and for the Bessel functions $q_B(n) = -c/(4p + 2b + 4n + 2)$.

3.2 Functional Inequalities Involving Special Functions

This section is motivated by the open problem (see the previous section) of G.D. Anderson et al. [17]. We prove a chain of inequalities for hypergeometric functions, generalized and normalized Bessel functions, Kummer functions and for some general power series. Some particular cases and refinements are given. Note that all results of this section may be found in the paper of the author [45].

3.2.1 Inequalities Involving Gaussian Hypergeometric Functions

Let F be the Gaussian hypergeometric function defined by (3.1). For $a, b, c > 0$ let us consider the function $m_{a,b,c} : (0,1) \to (0,\infty)$, defined by

$$m_{a,b,c}(r) = \frac{F(a,b,c,1-r^2)}{F(a,b,c,r^2)}. \tag{3.19}$$

Since $a, b, c > 0$ and $r \in (0,1)$, it follows that $F(a,b,c,r^2) > 0$, thus $m_{a,b,c}$ is well defined, and in fact it is a decreasing homeomorphism when $c \leq a+b$. By using the properties (see the discussion after (3.1)) of the function $r \in (0,1) \mapsto F(a,b,c,r) \in (0,\infty)$, it is clear that the function $m_{a,b,c}$ maps $(0,1)$ into $(0,\infty)$ if $c \leq a+b$. Moreover, if $c > a+b$, then we obtain that $m_{a,b,c}$ maps $(0,1)$ into

$$\left(\frac{\Gamma(c-a)\Gamma(c-b)}{\Gamma(c)\Gamma(c-a-b)}, \frac{\Gamma(c)\Gamma(c-a-b)}{\Gamma(c-a)\Gamma(c-b)} \right) \subset (0,\infty).$$

Note that proceeding exactly as in the proof of Theorem 3.2 and Corollary 3.2 (see also the paper of Á. Baricz [41]) we can prove that the inequalities (3.14) and (3.15) hold for all $a, b, c > 0$, satisfying $c^2 - (a+b)c + ab > 0$. This result is the general form of Theorem 3.2 and Corollary 3.2.

Theorem 3.4. (Á. Baricz [45]) *If $a, b, c > 0$ satisfy $c^2 - (a+b)c + ab \geq 0$ and $r_1, \ldots, r_k \in (0,1)$, then the function $m_{a,b,c} : (0,1) \to (0,\infty)$, defined by (3.19), satisfies*

$$\sqrt[k]{\prod_{i=1}^{k} m_{a,b,c}(r_i)} \leq m_{a,b,c}\left(\sqrt[k]{\prod_{i=1}^{k} r_i} \right). \tag{3.20}$$

In the previous inequality equality holds if and only if $r_1 = r_2 = \ldots = r_k$. In particular, for $k = 2$ the inequalities

$$\sqrt{m_{a,b,c}(r_1)m_{a,b,c}(r_2)} \leq m_{a,b,c}(\sqrt{r_1 r_2}),$$

$$\frac{1}{m_{a,b,c}(r_1)} + \frac{1}{m_{a,b,c}(r_2)} \geq \frac{2}{m_{a,b,c}(\sqrt{r_1 r_2})}, \tag{3.21}$$

$$m_{a,b,c}(r_1) + m_{a,b,c}(r_2) \geq 2m_{a,b,c}\left(\sqrt{1 - \sqrt{(1-(r_1)^2)(1-(r_2)^2)}}\right) \qquad (3.22)$$

hold for all $r_1, r_2 \in (0,1)$ and in all these inequalities equality holds if and only if $r_1 = r_2$.

Proof. Consider the function $C : (0,\infty) \to (0,\infty)$, defined by $C(t) = T(e^{-t})/T(1 - e^{-t})$, where for $r \in (0,1)$ we use the notation $T(r) = F(a,b,c,r)$. Just like in Lemma 3.4 we can prove that C is strictly log-convex. We just need to prove that the sequence $\{A_n\}_{n\geq 0}$ with $A_n = [(n+1)d_{n+1} - nd_n]/d_n$ for all $n \geq 0$, is decreasing, where

$$d_n = \frac{(a)_n(b)_n}{(c)_n n!} \quad \text{for all } n \geq 0.$$

By the ascending factorial notation we have $A_n = [ab + (a+b-c)n]/(c+n)$ for all $n \geq 0$. Therefore the inequality $A_{n-1} \geq A_n$ for all $n \geq 1$, is equivalent to $c^2 - (a+b)c + ab \geq 0$, which holds by the hypothesis.

By using the proofs of Theorem 3.1 and Corollary 3.1 with Q replaced by C we see that the conclusion of the theorem is true. $\square$

The purpose of this section is to prove that the analogues of the inequalities (3.6), (3.7), (3.13), (3.14) and (3.15) hold for hypergeometric functions too (not only for their ratios). Similarly we determine some conditions concerning the parameters of generalized and normalized Bessel functions (for the results of their ratios see the previous section and the paper of the author [41]), Kummer functions and for some power series in order to have analogous inequalities. Theorems 3.6, 3.7, 3.9 and 3.10 are the main results of this section.

For the simplicity of notations in what follows, $H(r,s)$, $G(r,s)$, $A(r,s)$, and $A_2(r,s)$ will stand, respectively, for the unweighted harmonic, geometric, arithmetic, and the second-order power means of positive numbers r and s, i.e.,

$$H(r,s) = \frac{2rs}{r+s}, \quad G(r,s) = \sqrt{rs}, \quad A(r,s) = \frac{r+s}{2}, \quad A_2(r,s) = \sqrt{\frac{r^2+s^2}{2}}.$$

It is known that

$$H(r,s) \leq G(r,s) \leq A(r,s) \leq A_2(r,s) \quad \text{for all } r,s > 0.$$

We now state our main results for hypergeometric functions.

Theorem 3.5. (Á. Baricz [45]) *For $a,b \in (0,1]$ let $T(r) = F(a,b,a+b,r)$ be the zero-balanced Gaussian hypergeometric function. Then*

$$T(G(r,s)) \leq H(T(r), T(s)) \qquad (3.23)$$

for all $r, s \in (0, q)$, *where* $q = 0.7153318630\ldots$ *is the unique positive root of the equation*

$$2\log(1-x) + x/(1-x) = 0.$$

Furthermore, equality holds in (3.23) *if and only if* $r = s$.

Proof. We first assume that $a, b \in (0, 1)$. We prove that $\varphi : (-\log q, \infty) \mapsto \mathbb{R}$, defined by $\varphi(t) = 1/T(e^{-t})$, is strictly concave. The function φ is differentiable, so it suffices to show that φ' is strictly decreasing. Since $\varphi'(t) = e^{-t}T'(e^{-t})/T^2(e^{-t})$, we need to demonstrate that the function $\psi : (0, q) \mapsto \mathbb{R}$, defined by $\psi(x) = xT'(x)/T^2(x)$, is strictly increasing (because $t \mapsto e^{-t}$ is strictly decreasing and the composition of a strictly increasing and a strictly decreasing function is strictly decreasing). We prove this by representing ψ as a product of three positive and strictly increasing functions. Let us denote

$$T_0(x) = F(1, 1, 2, x) = \sum_{n \geq 0} \frac{(1)_n (1)_n}{(2)_n} \frac{x^n}{n!} = \sum_{n \geq 0} \frac{x^n}{n+1} \quad \text{for all } x \in (-1, 1).$$

From the Taylor series expansion of the function $\log(1-x)$ we can deduce easily that $T_0(x) = -x^{-1}\log(1-x)$ for all $x \in (0, 1)$. We have

$$\psi(x) = \left[\frac{x}{T_0(x)}\right] \cdot \left[\frac{T'(x)}{T(x)}\right] \cdot \left[\frac{T_0(x)}{T(x)}\right] \quad \text{for all } x \in (0, q). \qquad (3.24)$$

We claim that each factor of this product is strictly increasing and positive.

For the first factor we study the function $f_1 : (0, q) \to \mathbb{R}$, defined by $f_1(x) = x/T_0(x) = -x^2/\log(1-x)$. It is clear that $f_1'(x) > 0$ is equivalent to $2\log(1-x) + x/(1-x) < 0$. Now the function $h : [0, q] \to \mathbb{R}$, defined by $h(x) = 2\log(1-x) + x/(1-x)$, is convex and $h(0) = h(q) = 0$. These properties imply that $h(x) < 0$ for all $x \in (0, q)$, thus f_1 is strictly increasing on $(0, q)$.

To prove that the function $f_2 : (0, 1) \to (0, \infty)$, defined by $f_2(x) = T'(x)/T(x)$, is strictly increasing we denote

$$d_n = \frac{(a)_n (b)_n}{(a+b)_n} \cdot \frac{1}{n!} \quad \text{and} \quad A_n = \frac{(n+1)d_{n+1}}{d_n} \quad \text{for all } n \geq 0.$$

By the ascending factorial notation we obtain

$$A_n = (a+n)(b+n)/(a+b+n) = n + ab/(a+b+n) \quad \text{for all } n \geq 0.$$

So the inequality $A_{n+1} > A_n$ for all $n \geq 0$ is equivalent to

$$(a+b+n)(a+b+n+1) > ab \quad \text{for all } n \geq 0.$$

The last inequality is clearly true. Therefore the sequence $\{A_n\}_{n\geq 0}$ is strictly decreasing. Hence by Lemma 3.1 the function f_2 is strictly increasing[1] on $(0,1)$ and thus it is strictly increasing on $(0,q)$ too.

For the function $f_3 : (0,1) \to (0,\infty)$, defined by $f_3(x) = F_0(x)/F(x)$, we apply again Lemma 3.1. Set $B_n = [(n+1)d_n]^{-1}$ for all $n \geq 0$. From a simple computation we get that $B_{n+1} > B_n$ for all $n \geq 0$ is equivalent to

$$n(1-ab) + a + b - 2ab > 0 \quad \text{for all} \ \ n \geq 0.$$

The last inequality clearly holds because $1 - ab > 0$ and $(a+b)/2 \geq \sqrt{ab} > ab$. Consequently, the sequence $\{B_n\}_{n\geq 0}$ is strictly increasing. By Lemma 3.1 the function f_3 is strictly increasing[2] on $(0,1)$ and thus it is strictly increasing on $(0,q)$ too.

Summing up, all the functions f_1, f_2 and f_3 are strictly increasing on $(0,q)$. It is obvious that they are positive on $(0,q)$. Therefore ψ is strictly increasing.

We can observe that this reasoning remains true if $a = 1$ and $b \in (0,1)$ or $b = 1$ and $a \in (0,1)$. Finally, when $a = b = 1$, it is clear that the function ψ reduces to $f_1 f_2$, which by the above argument is strictly increasing.

Since ψ is strictly increasing, φ is strictly concave. Therefore

$$A(\varphi(t_1), \varphi(t_2)) \leq \varphi(A(t_1,t_2)) \quad \text{for all} \ \ t_1, t_2 > -\log q,$$

where equality holds if and only if $t_1 = t_2$. This result implies the inequality

$$A(\varphi(-\log r), \varphi(-\log s)) \leq \varphi(-\log G(r,s)) \quad \text{for all} \ \ r,s \in (0,q),$$

where equality holds if and only if $r = s$.   □

We note that recently G.D. Anderson et al. [21, Theorem 3.7] extended and improved elegantly Theorem 3.5 by showing that for all $a,b \in (0,1)$ and $r,s \in (0,1)$ the inequality

$$T(A(r,s)) \leq H(T(r),T(s))$$

is valid. Moreover, G.D. Anderson et al. [21, Theorem 1.4] extended this result to the Gaussian hypergeometric function $r \mapsto F(a,b,c,r)$ (see Theorem 3.11 for more details).

Theorem 3.6. (Á. Baricz [45]) *For $a,b > 0$ let $T(r) = F(a,b,a+b,r)$ be the zero-balanced Gaussian hypergeometric function. Then*

$$T(G(r,s)) \leq G(T(r),T(s)) \leq T(1-G(1-r,1-s)) \leq A(T(r),T(s))$$

for all $r,s \in (0,1)$. In each of these inequalities equality holds if and only if $r = s$.

---

[1] Note that this result was proved by R. Balasubramanian et al. in Lemma 2.1 [36, p. 260].

[2] We note that this result was proved by G.D. Anderson et al. [11, p. 1715].

Proof. (1) The function $\varphi_1 : (0,\infty) \to (1,\infty)$, defined by $\varphi_1(t) = T(e^{-t})$, is strictly log-convex. This is guaranteed by the proof of Lemma 3.4. Using this property of φ_1 we obtain

$$\varphi_1(A(t_1,t_2)) \leq G(\varphi_1(t_1), \varphi_1(t_2))$$

or equivalently

$$T\left(e^{-A(t_1,t_2)}\right) \leq G(T(e^{-t_1}), T(e^{-t_2}))$$

for all $t_1, t_2 > 0$, where equality holds if and only if $t_1 = t_2$. By choosing $t_1 = -\log r$ and $t_2 = -\log s$, it follows that $T(G(r,s)) \leq G(T(r), T(s))$ for all $r, s \in (0,1)$, where equality holds if and only if $r = s$.

(2) The second inequality follows from the fact that the function $\varphi_2 : (0,\infty) \to (0,1)$, defined by $\varphi_2(t) = 1/T(1 - e^{-t})$, is strictly log-convex, which was also proved in the proof of Lemma 3.4. Using this fact we obtain

$$\varphi_2(A(t_1,t_2)) \leq G(\varphi_2(t_1), \varphi_2(t_2))$$

or equivalently

$$T\left(1 - e^{-A(t_1,t_2)}\right) \geq G(T(1 - e^{-t_1}), T(1 - e^{-t_2}))$$

for all $t_1, t_2 > 0$, where equality holds if and only if $t_1 = t_2$. By choosing $t_1 = -\log(1 - r)$ and $t_2 = -\log(1 - s)$, it follows that

$$T(1 - G(1 - r, 1 - s)) \geq G(T(r), T(s)) \quad \text{for all} \ \ r, s \in (0,1),$$

where equality holds if and only if $r = s$.

(3) Let $\varphi_3 : (0,\infty) \to (1,\infty)$ be defined by $\varphi_3(t) = T(1 - e^{-t})$. We prove that φ_3 is strictly convex. For this it is enough to show that $t \mapsto \varphi_3'(t) = e^{-t}T'(1 - e^{-t})$ is a strictly increasing function on $(0,\infty)$. Let $\psi : (0,1) \to \mathbb{R}$ be defined by $\psi(x) = (1 - x)T'(x)$. Since this function is the quotient of the functions

$$x \in (0,1) \mapsto T'(x) = \sum_{n \geq 0} (n+1)d_{n+1}x^n$$

and

$$x \in (0,1) \mapsto \frac{1}{1-x} = \sum_{n \geq 0} x^n$$

and the sequence $\{(n+1)d_{n+1}\}_{n \geq 0}$ is strictly increasing,[3] it follows by Lemma 3.1 that ψ is strictly increasing. Since $\psi(1 - e^{-t}) = \varphi_3'(t)$ for all $t > 0$, it follows that φ_3' is strictly increasing. Thus φ_3 is strictly convex.

---

[3] See Theorem 6(4) established by G.D. Anderson et al. [17] and also the first part of Lemma 2.1 obtained by G.D. Anderson et al. [11, p. 1715].

By applying this property of φ_3 we obtain

$$\varphi_3\left(A(t_1,t_2)\right) \leq A(\varphi_3(t_1),\varphi_3(t_2))$$

or equivalently

$$T\left(1-e^{-A(t_1,t_2)}\right) \leq A(T(1-e^{-t_1}),T(1-e^{-t_2}))$$

for all $t_1,t_2 > 0$, where equality holds if and only if $t_1 = t_2$. By choosing $t_1 = -\log(1-r)$ and $t_2 = -\log(1-s)$, it follows that

$$T(1-G(1-r,1-s)) \leq A(T(r),T(s))$$

for all $r,s \in (0,1)$, where equality holds if and only if $r=s$.   □

We know that

$$F(1,1,2,x) = -\frac{1}{x}\log(1-x) \quad \text{for all} \quad x \in (0,1).$$

Consequently, by choosing $a=b=1$ in Theorems 3.5 and 3.6, we obtain the following result.

Corollary 3.3. (Á. Baricz [45]) *For $r,s \in (0,1)$ we have the following chain of inequalities:*

$$\log\left[(1-G(r,s))^{-1/G(r,s)}\right] \leq G\left(\log\left[(1-r)^{-1/r}\right],\log\left[(1-s)^{-1/s}\right]\right)$$

$$\leq \log\left[G(1-r,1-s)^{-1/(1-G(1-r,1-s))}\right]$$

$$\leq \log\left[G((1-r)^{-1/r},(1-s)^{-1/s})\right].$$

Moreover, if q denotes the unique positive root of the equation

$$2\log(1-x)+x/(1-x)=0,$$

then

$$\log\left[(1-G(r,s))^{-1/G(r,s)}\right] \leq H\left(\log\left[(1-r)^{-1/r}\right],\log\left[(1-s)^{-1/s}\right]\right)$$

for all $r,s \in (0,q)$. In each of these above inequalities equality holds if and only if $r=s$.

3.2.2 Inequalities Involving Generalized Bessel Functions

Using the same idea as in the proof of Theorem 3.6 we can prove the following results for Bessel functions.

Theorem 3.7. (Á. Baricz [45]) If $c < 0$ and $\kappa > \max\{0, -c/4 - 1\}$, then

$$u_p(G(r,s)) \le G(u_p(r), u_p(s)) \le A(u_p(r), u_p(s)) \le u_p(1 - G(1 - r, 1 - s)) \quad (3.25)$$

for all $r, s \in (0,1)$, where in each of these inequalities equality holds if and only if $r = s$.

Proof. (1) In the proof of Lemma 3.3 we have seen that the function $\varphi_4 : (0, \infty) \to (0, \infty)$, defined by $\varphi_4(t) = u_p(e^{-t})$, is strictly log-convex. Hence

$$\varphi_4(A(t_1, t_2)) \le G(\varphi_4(t_1), \varphi_4(t_2))$$

or equivalently

$$u_p\left(e^{-A(t_1, t_2)}\right) \le G(u_p(e^{-t_1}), u_p(e^{-t_2}))$$

for all $t_1, t_2 \in (0, \infty)$, where equality holds if and only if $t_1 = t_2$. By choosing $t_1 = -\log r$ and $t_2 = -\log s$, it follows that

$$u_p(G(r,s)) \le G(u_p(r), u_p(s)) \quad \text{for all } r, s \in (0,1),$$

where equality holds if and only if $r = s$.

(2) We have $u_p(x) > 0$, for all $x \in (0,1)$, so the arithmetic-geometric mean inequality implies the second inequality in (3.25).

(3) Let $\varphi_5 : (0, \infty) \to (0, \infty)$ be defined by $\varphi_5(t) = u_p(1 - e^{-t})$. We prove that this function is strictly concave. For this it is enough to show that the function $t \in (0, \infty) \mapsto \varphi_5'(t) = e^{-t} u_p'(1 - e^{-t})$ is strictly decreasing. Let $\psi : (0, 1) \to (0, \infty)$ be defined by $\psi(x) = (1 - x) u_p'(x)$. Notice that ψ is the quotient of the functions

$$x \in (0, 1) \mapsto u_p'(x) = \sum_{n \ge 0} (n+1) b_{n+1} x^n$$

and

$$x \in (0, 1) \mapsto \frac{1}{1-x} = \sum_{n \ge 0} x^n.$$

Put $\alpha_n = (n+1) b_{n+1}$ and $\beta_n = 1$ for all $n \ge 0$. We claim that the sequence $\{\alpha_n / \beta_n\}_{n \ge 0} = \{\alpha_n\}_{n \ge 0}$ is strictly decreasing. Fix any $n \ge 1$. Since

$$\alpha_n = (n+1) b_{n+1} = -\frac{c}{4(\kappa+n)n} \cdot n b_n = -\frac{c}{4(\kappa+n)n} \cdot \alpha_{n-1},$$

we have

$$\alpha_n - \alpha_{n-1} = -\frac{M(n)}{(\kappa+n)n} \alpha_{n-1},$$

where $M(n) = n^2 + \kappa n + c/4 > 0$. Taking into account that $n^2 \geq 2n - 1$, $\kappa > 0$ and $\kappa > -c/4 - 1$, we see that $M(n) \geq (\kappa + 2)n + c/4 - 1 \geq \kappa + c/4 + 1 > 0$. Therefore we have $\alpha_n < \alpha_{n-1}$ for all $n \geq 1$. Consequently, the sequence $\{\alpha_n\}_{n\geq0}$ is strictly decreasing. By Lemma 3.1 it follows that ψ is strictly decreasing. Since

$$\psi(1 - e^{-t}) = \varphi_5'(t) \text{ for all } t \in (0,\infty),$$

it follows that φ_5' is strictly decreasing. Thus φ_5 is strictly concave.

By applying this property of φ_5 we obtain

$$\varphi_5\left(A(t_1,t_2)\right) \geq A(\varphi_5(t_1), \varphi_5(t_2))$$

or equivalently

$$u_p\left(1 - e^{-A(t_1,t_2)}\right) \geq A(u_p(1 - e^{-t_1}), u_p(1 - e^{-t_2}))$$

for all $t_1, t_2 \in (0,\infty)$, where equality holds if and only if $t_1 = t_2$. By choosing $t_1 = -\log(1 - r)$ and $t_2 = -\log(1 - s)$, it follows that

$$u_p(1 - G(1 - r, 1 - s)) \geq A(u_p(r), u_p(s)) \text{ for all } r,s \in (0,1),$$

where equality holds if and only if $r = s$. $\qquad\square$

As a special case of (1.24) we obtain for $z = x \in (0,1)$ some known elementary functions as $\cosh, \sinh$. These are also special cases of the generalized and normalized Bessel functions (for $c = -1$ and $b = 1$ in (1.14)). So we have the following corollary.

Corollary 3.4. (Á. Baricz [45]) *For $r,s \in (0,1)$ we have the following inequalities*

$$\cosh G(r,s) \leq G(\cosh r, \cosh s) \leq A(\cosh r, \cosh s) \leq \cosh\sqrt{1 - G(1 - r^2, 1 - s^2)},$$

$$\frac{\sinh G(r,s)}{G(r,s)} \leq \frac{G(\sinh r, \sinh s)}{G(r,s)} \leq A\left(\frac{\sinh r}{r}, \frac{\sinh s}{s}\right) \leq \frac{\sinh\sqrt{1 - G(1 - r^2, 1 - s^2)}}{\sqrt{1 - G(1 - r^2, 1 - s^2)}},$$

where in each of these inequalities equality holds if and only if $r = s$.

Proof. Since
$$u_{-1/2}(x^2) = \sqrt{\pi/2} \cdot x^{1/2} I_{-1/2}(x) = \cosh x,$$

taking $c = -1, b = 1, p = -1/2$ in Theorem 3.7 and changing r with r^2, respectively s with s^2, (3.25) leads to the first chain of inequalities. Similarly, we have

$$u_{1/2}(x^2) = \sqrt{\pi/2} \cdot x^{-1/2} I_{1/2}(x) = \frac{\sinh x}{x}.$$

Therefore, applying Theorem 3.7 for $c = -1, b = 1, p = 1/2$ and changing r with r^2, respectively s with s^2, we get the second chain of inequalities. $\qquad\square$

3.2.3 Inequalities Involving Confluent Hypergeometric Functions

By a confluent hypergeometric function, also known as a Kummer function, we mean the function

$$\Phi(a,c,r) = \sum_{n\geq 0} \frac{(a)_n}{(c)_n} \frac{r^n}{n!} \quad \text{for all } r \in (-1,1) \tag{3.26}$$

defined for $a,c \in \mathbb{R}$ with $c \neq 0,-1,-2,\ldots$. Observe that this is another special case of the generalized hypergeometric function $_qF_r$ defined by (1.2). Indeed, when $q=1$, $r=1$, $a_1 = a$ and $b_1 = c$, then the series in (1.2) becomes the series (3.26).

Define the function $m_\Phi(r) : (0,1) \to (0,\infty)$ by

$$m_\Phi(r) = \frac{\Phi(a,c,1-r^2)}{\Phi(a,c,r^2)},$$

where $a,c > 0$. The next result deals with this function.

Theorem 3.8. (Á. Baricz [45]) *If $a,c > 0$ and $r_1, r_2, \ldots, r_k \in (0,1)$, then*

$$\sqrt[k]{\prod_{i=1}^{k} m_\Phi(r_i)} \leq m_\Phi\left(\sqrt[k]{\prod_{i=1}^{k} r_i}\right). \tag{3.27}$$

In the previous inequality equality holds if and only if $r_1 = r_2 = \ldots = r_k$. In particular, for $k=2$ the inequalities

$$\sqrt{m_\Phi(r_1)m_\Phi(r_2)} \leq m_\Phi(\sqrt{r_1 r_2}),$$

$$\frac{1}{m_\Phi(r_1)} + \frac{1}{m_\Phi(r_2)} \geq \frac{2}{m_\Phi(\sqrt{r_1 r_2})},$$

$$m_\Phi(r_1) + m_\Phi(r_2) \geq 2m_\Phi\left(\sqrt{1 - \sqrt{(1-(r_1)^2)(1-(r_2)^2)}}\right).$$

hold for all $r_1, r_2 \in (0,1)$ and in each of these inequalities equality holds if and only if $r_1 = r_2$.

Proof. Let us consider the function $V : (0,\infty) \to (0,\infty)$, defined by

$$V(t) = \Phi(a,c,e^{-t})/\Phi(a,c,1-e^{-t}).$$

Just like in Lemma 3.4 we can prove that V is strictly log-convex. We just need to prove that the sequence $\{A_n\}_{n\geq 0}$ with $A_n = [(n+1)e_{n+1} - ne_n]/e_n$ for all $n \geq 0$, is strictly decreasing, where

$$e_n = \frac{(a)_n}{(c)_n n!} \quad \text{for all } n \geq 0.$$

Since $(c+n)(n+1)e_{n+1} = (a+n)e_n$ for all $n \geq 0$, we have $A_n = (a+n)/(c+n) - n$ for all $n \geq 0$. Therefore to prove that $A_n < A_{n-1}$ for all $n \geq 1$, it is enough to show that $N(n) = n^2 + (2c-1)n + c^2 - 2c + a > 0$ for all $n \geq 1$. But $n^2 \geq 2n - 1$ and $2c + 1 > 0$, thus we have $N(n) \geq (2c+1)n + c^2 - 2c + a - 1 \geq c^2 + a > 0$.

By using the proofs of Theorem 3.1 and Corollary 3.1 with Q replaced by V we see that the conclusion of the theorem is true.   □

The next theorem is the analogue of Theorem 3.6 for Kummer's hypergeometric functions.

Theorem 3.9. (Á. Baricz [45]) *For $c > a > 0$ let $S(r) = \Phi(a,c,r)$ be the confluent hypergeometric function. Then*

$$S(G(r,s)) \leq G(S(r),S(s)) \leq S(1 - G(1 - r, 1 - s)) \leq A(S(r),S(s)) \qquad (3.28)$$

for all $r,s \in (0,1)$. In each of these inequalities equality holds if and only if $r = s$.

Proof. Consider the functions $\varphi_6 : (0,\infty) \rightarrow (0,\infty)$ and $\varphi_7 : (0,\infty) \rightarrow (0,\infty)$ defined by $\varphi_6(t) = S(e^{-t})$ and $\varphi_7(t) = S(1 - e^{-t})$. In view of the proof of Theorem 3.8, the first inequality in (3.28) follows from the fact that φ_6 is strictly log-convex, the second inequality in (3.28) is also true, because $1/\varphi_7$ is strictly log-convex too. The third inequality in (3.28) follows from the fact that φ_7 is strictly concave. For this we prove that $t \in (0,\infty) \mapsto e^{-t}\Phi'(1 - e^{-t})$ is strictly decreasing, or equivalently $x \in (0,1) \mapsto (1-x)\Phi'(x)$ is strictly decreasing. Therefore by Lemma 3.1 it is enough to show that the sequence $\{A_n\}_{n\geq 0}$, defined by $A_n = (n+1)e_{n+1}$ for all $n \geq 0$, is strictly decreasing, where the sequence $\{e_n\}_{n\geq 0}$ is defined in the proof of Theorem 3.8. From the definition of e_n we obtain

$$A_n = (n+1)e_{n+1} = \frac{a+n}{n(c+n)} \cdot ne_n = \frac{a+n}{n(c+n)}A_{n-1} \quad \text{for all } n \geq 1.$$

Therefore we need to prove that $P(n) = n^2 + (c-1)n - a > 0$ for all $n \geq 1$. Since $n^2 \geq 2n - 1$, we have $P(n) \geq (c+1)n - a - 1 \geq c - a \geq 0$. This implies that the function φ_7 is strictly concave, i.e. we have

$$\varphi_7(A(t_1,t_2)) \geq A(\varphi_7(t_1), \varphi_7(t_2)) \quad \text{for all } t_1, t_2 > 0,$$

and equality holds if and only if $t_1 = t_2$. Taking $e^{-t_1} = 1 - r$ and $e^{-t_2} = 1 - s$, we deduce the third inequality.   □

3.2.4 Inequalities Involving General Power Series and Concluding Remarks

In this section we formulate a generalization of Theorems 3.6, 3.7 and 3.9. Moreover, we give some concluding remarks on the results of Sects. 3.1 and 3.2 by presenting some improvements of G.D. Anderson et al. [21] and pointing out some applications of the results.

Theorem 3.10. (Á. Baricz [45]) *Let us consider the power series* $f(x) = \sum_{n \geq 0} A_n x^n$, *where* $A_n > 0$ *for all* $n \geq 0$, *and* $f(x)$ *is convergent for all* $x \in (0,1)$. *Let us denote* $B_{n+1} = (n+1)A_{n+1}$ *for all* $n \geq 0$. *Then the following assertions are true:*

(a) *If the sequence* $\{B_{n+1}\}_{n \geq 0}$ *is decreasing, then*

$$f(G(r,s)) \leq G(f(r),f(s)) \leq A(f(r),f(s)) \leq f(1 - G(1-r, 1-s))$$

whenever $r, s \in (0,1)$.

(b) *If the sequence* $\{B_{n+1}\}_{n \geq 0}$ *is increasing and the sequence* $\{(B_{n+1} - B_n)/A_n\}_{n \geq 0}$ *is decreasing, then*

$$f(G(r,s)) \leq G(f(r),f(s)) \leq f(1 - G(1-r, 1-s)) \leq A(f(r),f(s))$$

whenever $r, s \in (0,1)$.

In each of these inequalities equality holds if and only if $r = s$.

Proof. (a) The assumption on the sequence $\{B_n\}_{n \geq 0}$ guarantees that the function $t \in (0, \infty) \mapsto f(e^{-t}) \in (0, \infty)$ is strictly log-convex (see Theorem 3.3) and the function $t \in (0, \infty) \mapsto f(1 - e^{-t}) \in (0, \infty)$ is strictly convex. These imply the first and the third inequality. The second inequality is obvious, so the proof of the first part is complete.

(b) In the same manner we can prove that the functions $t \in (0, \infty) \mapsto f(e^{-t}) \in (0, \infty)$ and $t \in (0, \infty) \mapsto 1/f(1 - e^{-t}) \in (0, \infty)$ are strictly log-convex and the function $t \in (0, \infty) \mapsto f(1 - e^{-t}) \in (0, \infty)$ is strictly concave. These properties imply the first, second and the third inequality, therefore the proof is complete. □

Remark 3.3. The first inequality in the Theorems 3.6, 3.7 and 3.10 can be proved by using a much simpler argument. Let us consider the convergent power series $f(x) = \sum_{n \geq 0} A_n x^n$, where $A_n > 0$ for all $n \geq 0$ and $x \in (0,1)$. Let us denote $P_n(x) = \sum_{k=0}^{n} A_k x^k$. Then by the Cauchy–Bunyakovsky–Schwarz inequality (1.35) we have

$$\left[\sum_{k=0}^{n} \left(\sqrt{A_k} r^{k/2} \right) \left(\sqrt{A_k} s^{k/2} \right) \right]^2 \leq \left(\sum_{k=0}^{n} A_k r^k \right) \left(\sum_{k=0}^{n} A_k s^k \right).$$

But this inequality is equivalent to $P_n(\sqrt{rs}) \leq \sqrt{P_n(r)P_n(s)}$, so using the fact that $\lim_{n \to \infty} P_n(x) = f(x)$, we obtain immediately the inequality $f(G(r,s)) \leq G(f(r),f(s))$. Moreover this shows that the above inequality remains true for all $r, s \in (0, \infty)$ if the corresponding power series $f(x) = \sum_{n \geq 0} A_n x^n$ is convergent for all $x \in (0, \infty)$. Thus the first inequality in Theorem 3.7 in fact holds for all $r, s \in (0, \infty)$ because the series $u_p(x)$ is convergent for all $x \in \mathbb{R}$. We note that this simple idea to use the Cauchy–Bunyakovsky–Schwarz inequality (1.35) in order to deduce the inequality

$$f(G(r,s)) \leq G(f(r),f(s))$$

may be found in Theorem 177 of the book of G.H. Hardy et al. [112] (for further results and interesting examples see the paper of C.P. Niculescu [163]).

Remark 3.4. Furthermore, if in addition $\{B_{n+1}/A_n\}_{n\geq 0} = \{(n+1)A_{n+1}/A_n\}_{n\geq 0}$ is strictly increasing, i.e. in view of Lemma 3.1 the function $x \mapsto f'(x)/f(x)$ is strictly increasing for $x \in (0,1)$, then f is log-convex. This implies that $f(A(r,s)) \leq G(f(r),f(s))$ holds for all $r,s \in (0,1)$. Thus it is clear that the above inequality improves the first inequality, i.e.

$$f(G(r,s)) \leq G(f(r),f(s))$$

in Theorem 3.10. It is known that the zero-balanced hypergeometric function $x \mapsto T(x) = F(a,b,a+b,x)$ is log-convex on $(0,1)$ for all $a,b > 0$, because by Lemma 2.1 established by R. Balasubramanian et al. [36, p. 260] the function $x \mapsto T'(x)/T(x)$ is strictly increasing on $(0,1)$. Therefore the first inequality in Theorem 3.6 and Corollary 3.3 can be improved as follows for all $r,s \in (0,1)$

$$T(G(r,s)) \leq T(A(r,s)) \leq G(T(r),T(s)),$$

$$\log\left[(1-G(r,s))^{-1/G(r,s)}\right] \leq \log\left[(1-A(r,s))^{-1/A(r,s)}\right]$$
$$\leq G\left(\log\left[(1-r)^{-1/r}\right],\log\left[(1-s)^{-1/s}\right]\right).$$

The situation is the same for the confluent hypergeometric function, i.e. the first inequality in Theorem 3.9 can be improved in the following way:

$$S(G(r,s)) \leq S(A(r,s)) \leq G(S(r),S(s)), \quad r,s \in (0,1).$$

Here we have used the fact that the sequence $C_n = (n+1)e_{n+1}/e_n = (a+n)/(c+n)$ is strictly increasing when $c > a > 0$ and consequently by Lemma 3.1 the function $x \in (0,1) \mapsto S'(x)/S(x)$ is strictly increasing too.

Remark 3.5. Even if the function $x \in (0,1) \mapsto u_p(x) \in (0,\infty)$ is log-concave on $(0,1)$ (see Lemma 3.2), the function $x \in (0,\infty) \mapsto u_{-1/2}(x^2) = \cosh x \in (0,\infty)$ is log-convex and increasing, thus the first inequality (in the first chain of inequalities) in Corollary 3.4 can be improved as follows:

$$\cosh G(r,s) \leq \cosh A(r,s) \leq G(\cosh r, \cosh s)$$

for all $r,s \in (0,\infty)$. Further the first inequality (in the second chain of inequalities) in Corollary 3.4 can be improved also as follows:

$$\frac{\sinh G(r,s)}{G(r,s)} \leq \frac{\sinh A(r,s)}{A(r,s)} \leq \frac{G(\sinh r, \sinh s)}{G(r,s)}$$

for all $r,s \in (0,\infty)$, since the function $x \in (0,\infty) \mapsto \mathscr{I}_{1/2}(x) = (\sinh x)/x$ is log-convex and increasing. This is justified by the fact that the function, defined by (1.24), i.e. $x \in \mathbb{R} \mapsto \mathscr{I}_p(x) \in [1,\infty)$, is log-convex for all $p \geq -1/2$. This result was obtained by E. Neuman [159] (see also Theorem 3.23). Thus actually the first inequality in (3.25) can be improved too, as follows:

$$u_p(G(r,s)) \leq u_p(A(r,s)) \leq G(u_p(r),u_p(s))$$

for all $r,s \in (0,\infty)$ and all $c < 0$, $\kappa \geq 1/2$. Here we used the fact that $\mathscr{I}_p$ is log-convex for all $p \geq -1/2$, which in view of relation $\mathscr{I}_{\kappa-1}(x\sqrt{-c}) = \lambda_p(x) = u_p(x^2)$ implies that λ_p is log-convex too for all $c < 0$, $\kappa \geq 1/2$ on $\mathbb{R}$. Hence $\lambda_p(A(r,s)) \leq G(\lambda_p(r),\lambda_p(s))$ holds for all $r,s \in (0,\infty)$. Changing r with $\sqrt{r}$ and s with $\sqrt{s}$, the asserted result follows.

During the course of writing the thesis [54] we have found the work of G.D. Anderson et al. [21] in which the authors improved some results obtained by Á. Baricz in [45]. The following results, which we state here without proof, improve and extend the results of Theorems 3.5 and 3.6.

Theorem 3.11. (G.D. Anderson, M.K. Vamanamurthy and M. Vuorinen [21]) *If $a,b,c > 0$ and $F(r) = F(a,b,c,r)$, then the following assertions are true:*

(a) *If $ab/(a+b+1) < c$, then $x \in (0,1) \mapsto \log F(x) \in (0,\infty)$ is convex. In particular,*

$$F(A(r,s)) \leq G(F(r),F(s))$$

for all $r,s \in (0,1)$, with equality if and only if $r = s$.
(b) *If $(a-c)(b-c) > 0$, then $t \in (0,\infty) \mapsto \log F(1 - e^{-t}) \in (0,\infty)$ is concave. In particular,*

$$G(F(r),F(s)) \leq F(1 - G(1-r,1-s))$$

for all $r,s \in (0,1)$, with equality if and only if $r = s$.
(c) *If $a+b \geq c$, then $t \in (0,\infty) \mapsto F(1 - e^{-t}) \in (0,\infty)$ is convex. In particular,*

$$F(1 - G(1-r,1-s)) \leq A(F(r),F(s))$$

for all $r,s \in (0,1)$, with equality if and only if $r = s$.
(d) *If $a+b \geq c > 2ab$ and $c \geq a+b-1/2$, then $x \in (0,1) \mapsto 1/F(x) \in (0,1)$ is concave. In particular,*

$$F(A(r,s)) \leq H(F(r),F(s))$$

for all $r,s \in (0,1)$, with equality if and only if $r = s$.

In addition, G.D. Anderson et al. [21] improved Lemma 3.2 and Theorem 3.7 as follows. These results are related to Remark 3.5. See also Theorem 3.24 for more details.

Theorem 3.12. (G.D. Anderson, M.K. Vamanamurthy and M. Vuorinen [21]) *If* $\kappa, R > 0$ *and* $c < 0$, *then the following assertions are true:*

(a) *The function* $t \in (0, \infty) \mapsto \log u_p(Re^{-t}) \in (0, \infty)$ *is convex.*
(b) *The function* $x \in (0, \infty) \mapsto \log u_p(x) \in (0, \infty)$ *is concave.*
(c) *The function* $x \in (0, \infty) \mapsto u_p(x) \in (0, \infty)$ *is convex.*

In particular, for all $c < 0$ *and* $\kappa > 0$ *the following chain of inequalities holds for all* $r, s > 0$

$$u_p(G(r,s)) \le G(u_p(r), u_p(s)) \le u_p(A(r,s)) \le A(u_p(r), u_p(s)),$$

where in each of these inequalities equality holds if and only if $r = s$.

(d) *If in addition* $\kappa > -1 - cR/4$, *then the function* $t \in (0, \infty) \mapsto u_p(R(1 - e^{-t})) \in (0, \infty)$ *is concave, and thus the inequality*

$$A(u_p(r), u_p(s)) \le u_p(R - G(R - r, R - s))$$

holds for all $r, s \in (0, R)$ *with equality if and only if* $r = s$.

We note that Theorems 3.3 and 3.10 has been also complemented and improved elegantly by G.D. Anderson et al. [21, Theorems 1.5 and 3.1]. Moreover, the results of Theorem 3.9 has been extended to the case of generalized hypergeometric functions [21, Theorem 3.18]. Motivated by the results of Á. Baricz [41, 45] and G.D. Anderson et al. [21], recently Y. Chu et al. [85] presented also some interesting functional inequalities for Gaussian hypergeometric functions and for some other special functions arising in the theory of quasiconformal maps and Ramanujan's modular equations. Although the functional inequalities derived in Sects. 3.1 and 3.2 are interesting in their own right, they are interesting from many aspects. Firstly, because these inequalities yield new inequalities for many familiar elementary functions. Secondly, because the monotonicity and convexity properties deduced here in order to obtain the functional inequalities for the Gaussian hypergeometric functions and generalized Bessel functions can be useful to solve other interesting problems in quasiconformal analysis, economic theory and engineering sciences. For example, recently G.D. Anderson et al. [14] used among other some monotonicity and convexity properties of the function $t \mapsto m(\sqrt{r})$, where $r = 1/(1 + e^t)$, in order to deduce sharp estimates for the hyperbolic metric and the induced distance function on a plane domain. Recently, R. Boucekkine and J.R. Ruiz-Tamarit [76] pointed out that the Gaussian hypergeometric functions are also useful in economic dynamics, notably in the assessment of the transition dynamics of endogenous economic growth models. They focused on the famous Lucas-Uzawa model, which solved by the means of Gaussian hypergeometric functions and showed that the use of Gaussian hypergeometric functions allows for explicit representation of the equilibrium dynamics of all variables in level. One of the crucial facts in the study of R. Boucekkine and J.R. Ruiz-Tamarit [76, Lemma 1] it was the monotonicity property of the function

$$r \mapsto \frac{F(a, b, a+1, r)}{F(a-1, b, a+1, r)}.$$

Finally, we note that the study of complete elliptic integrals, Gaussian hypergeometric functions and related functions, like the modulus of the Grötzsch ring, the generalized modulus, the (a,b,c)-modular function, the Legendre $\mathscr{M}$-function, and certain combinations of the generalized elliptic integrals is in active progress. We refer to the recent papers of G.D. Anderson et al. [12, 13, 20, 22], V. Heikkala et al. [113, 114] and to the references therein.

3.3 Landen-Type Inequality for Bessel Functions

Some of the most important properties of the complete elliptic integral of the first kind $\mathscr{K}$ are the Landen identities proved in 1771 (see for example G. Almkvist and B.C. Berndt [7]):

$$\mathscr{K}\left(\frac{2\sqrt{r}}{1+r}\right) = (1+r)\mathscr{K}(r), \quad \mathscr{K}\left(\frac{1-r}{1+r}\right) = \frac{1+r}{2}\mathscr{K}\left(\sqrt{1-r^2}\right), \qquad (3.29)$$

where $r \in (0,1)$.

Recently G.D. Anderson et al. [17, p. 79] posed the following problem:

Find an analog of Landen's transformation formulas in (3.29) for $F(a,b,a+b,r)$. In particular, if $h(r) = F(a,b,a+b,r^2)$ and $a,b \in (0,1)$, is it true that

$$h\left(\frac{2\sqrt{r}}{1+r}\right) \le \rho h(r) \qquad (3.30)$$

for some constant ρ and all $r \in (0,1)$?

Since $2\sqrt{r}/(1+r) > r$ for $r \in (0,1)$, ρ must be greater than 1. S.L. Qiu and M. Vuorinen [194, Theorem 1.2] found the answer to the above problem and proved that 2 is the best value of ρ, for which (3.30) holds.

Motivated by this problem we prove in this section an analogous inequality for the generalized and normalized Bessel functions. Theorems 3.13 and 3.14 are the main results of this section. Before we state our main results in this section we need the following lemma.

Lemma 3.5. (Á. Baricz [39]) *Let $\delta > 1$, and let $q : (0,1) \to (0,\infty)$ be defined by $q(x) = \delta^{x^2}$. Then there exists a constant ρ such that*

$$q\left(\frac{2\sqrt{x}}{1+x}\right) \le \rho q(x) \quad \text{for all } x \in (0,1).$$

The best constant ρ is $\delta^{4\sqrt{2}-5}$.

Proof. The inequality $q(2\sqrt{x}/(1+x)) \le \rho q(x)$ is equivalent to $\delta^{4x/(x+1)^2} \le \rho \delta^{x^2}$. Since $4x/(x+1)^2 > x^2$ for all $x \in (0,1)$, ρ must be greater than 1. Therefore if we take the logarithm of both sides of inequality $\delta^{4x/(x+1)^2} \le \rho \delta^{x^2}$, then it is enough to find the maximum value (which will be $(\log \rho)/(\log \delta)$) of the function $f : (0,1) \rightarrow (0,\infty)$, defined by $f(x) = 4x/(x+1)^2 - x^2$. For this function we have that $f(0) = f(1) = 0$,

$$f'(x) = 2\left[\frac{2(1-x)}{(1+x)^3} - x\right] \text{ and } f''(x) = 2\left[\frac{4(x-2)}{(x+1)^4} - 1\right].$$

Since $(x+1)^3 f'(x) = -2(x^2 + x + 2)(x^2 + 2x - 1)$, we get that $x_0 = \sqrt{2} - 1$ is the positive root of the equation $f'(x) = 0$. It is also easy to see that $f''(x) < 0$ for all $x \in (0,1)$. Thus the function f is concave on $(0,1)$ and its maximum value is $f(x_0) = 4\sqrt{2} - 5$. This implies that the best value of ρ for which the inequality $\delta^{4x/(x+1)^2} \leq \rho \delta^{x^2}$ holds is $e^{f(x_0)\log A} = \delta^{4\sqrt{2}-5}$.   □

3.3.1 Landen-Type Inequality for Generalized Bessel Functions

By Lemma 1.2 we know that the function u_p satisfies

$$4\kappa u'_p(z) = -cu_{p+1}(z) \tag{3.31}$$

for all $z \in \mathbb{C}$. Clearly this relation remains true for $z = x \in (0,1)$. Now if we suppose that $\kappa > 0$ and $c < 0$, it follows from the representation (1.20) that $u_p(x) > 0$ for all $x \in (0,1)$. Consequently, in this case (3.31) implies that the function u_p is strictly increasing on $(0,1)$ and that the function $\lambda_p : (0,1) \rightarrow \mathbb{R}$, defined by $\lambda_p(x) = u_p(x^2)$, is strictly increasing too. Since $2\sqrt{r}/(1+r) > r$ for all $r \in (0,1)$ it follows that

$$\lambda_p\left(\frac{2\sqrt{r}}{1+r}\right) > \lambda_p(r) \text{ for all } r \in (0,1).$$

The following result contains a sufficient condition for the parameters b, c, p for the reverse inequality.

Theorem 3.13. (Á. Baricz [39]) *Let $\delta > 1$ be a fixed real number. If $4\kappa \log \delta \geq -c > 0$, then*

$$\lambda_p\left(\frac{2\sqrt{r}}{1+r}\right) < \rho \cdot \lambda_p(r)$$

holds for all $r \in (0,1)$ and $\rho \geq \delta^{4\sqrt{2}-5}$.

Proof. Let us consider the function $\varphi : (0,1) \rightarrow (0,\infty)$, defined by $\varphi(x) = u_p(x)/\delta^x$, where $\delta > 1$. From the series representation of the exponential function we have

$$\delta^x = e^{x\log \delta} = \sum_{n \geq 0} \frac{(x\log \delta)^n}{n!},$$

where the right-hand side is convergent for all $x \in (0,1)$. We want to prove that function φ is decreasing on $(0,1)$ for all $\kappa \geq -c/(4\log \delta) > 0$. If we consider $\alpha_n = b_n$ and $\beta_n = (\log \delta)^n/n!$ for all $n \geq 0$, where b_n is defined by (1.18), then by Lemma 3.1 it is enough to show that the sequence $A_n = n! \cdot b_n/(\log \delta)^n$ is decreasing for $n \geq 0$. By the ascending factorial notation

$$4A_n \cdot (\kappa + n - 1)\log \delta = (-c)A_{n-1} \text{ for all } n \geq 1.$$

By the hypothesis we obtain that $A_n > 0$ for all $n \geq 0$, therefore the inequality $A_n \leq A_{n-1}$ for all $n \geq 1$ is equivalent to the inequality $\kappa + n - 1 \geq -c/(4\log \delta)$, which clearly holds.

Now because $1 > 4r/(r+1)^2 > r^2 > 0$ for $r \in (0,1)$ and φ is decreasing we obtain that (using the fact that $u_p(x) > 0$ for all $x \in (0,1)$)

$$u_p\left(4r/(1+r)^2\right) \cdot \delta^{r^2} < u_p(r^2) \cdot \delta^{4r/(1+r)^2}.$$

But this is equivalent to

$$\lambda_p\left(\frac{2\sqrt{r}}{1+r}\right) \cdot q(r) < \lambda_p(r) \cdot q\left(\frac{2\sqrt{r}}{1+r}\right),$$

where $q(r) = \delta^{r^2}$, therefore using Lemma 3.5 the proof is complete.   □

We know that if we take $c = -1$ and $b = 1$ in (1.14), then the generalized Bessel function reduces to the modified Bessel function defined by (1.7). Now if we consider the modified Bessel functions of the first kind of real order which can be expressed with elementary functions as cosh and sinh, we may obtain (by putting in Theorem 3.13 $c = -1$, $b = 1$ and $p = -1/2$, $p = 1/2$ respectively) the following interesting examples for the functions which satisfy the Landen-type inequality.

Corollary 3.5. (Á. Baricz [39]) *For $\mathscr{I}_{1/2}(r) = (\sinh r)/r$ the Landen-type inequality, i.e.*

$$\mathscr{I}_{1/2}\left(\frac{2\sqrt{r}}{1+r}\right) < \rho \cdot \mathscr{I}_{1/2}(r)$$

holds for all $r \in (0,1)$, $\rho \geq \delta^{4\sqrt{2}-5}$ and $\delta > e^{1/6} \simeq 1.181360413....$

Corollary 3.6. (Á. Baricz [39]) *For $\mathscr{I}_{-1/2}(r) = \cosh r$ the Landen-type inequality, i.e.*

$$\mathscr{I}_{-1/2}\left(\frac{2\sqrt{r}}{1+r}\right) < \rho \cdot \mathscr{I}_{-1/2}(r)$$

holds for all $r \in (0,1)$, $\rho \geq \delta^{4\sqrt{2}-5}$ and $\delta > e^{1/2} \simeq 1.648721271....$

3.3.2 Landen-Type Inequality for General Power Series

In this section we formulate a generalization of Theorem 3.13.

Theorem 3.14. (Á. Baricz [39]) *Suppose that the power series* $f(x) = \sum_{n \geq 0} A_n x^n$ *is convergent for all* $x \in (0,1)$, *where* $A_n > 0$ *for all* $n \geq 0$. *Furthermore suppose that for a given* $\delta > 1$ *the sequence* $\{n! A_n / (\log \delta)^n\}_{n \geq 0}$ *is decreasing. If we define* $\lambda_f(x) = f(x^2)$, *then*

$$\lambda_f \left(\frac{2\sqrt{r}}{1+r} \right) < \rho \cdot \lambda_f(r)$$

holds for all $r \in (0,1)$ *and* $\rho \geq \delta^{4\sqrt{2}-5}$.

Proof. Let us consider the function $\varphi_f : (0,1) \rightarrow (0,\infty)$, defined by $\varphi_f(x) = f(x)/\delta^x$. By the hypothesis and Lemma 3.1 we get that the function φ_f is decreasing. Since for $r \in (0,1)$ we have $4r/(1+r)^2 > r^2$, we obtain that $\varphi_f(4r/(1+r)^2) < \varphi_f(r^2)$, which actually is equivalent to

$$\lambda_f \left(\frac{2\sqrt{r}}{1+r} \right) \cdot q(r) < \lambda_f(r) \cdot q\left(\frac{2\sqrt{r}}{1+r} \right),$$

where $q(r) = \delta^{r^2}$. Using Lemma 3.5 the proof is complete.   □

Now if we use Theorem 3.14 for Kummer functions, i.e. for confluent hypergeometric functions defined by (3.26), then we obtain the following result.

Corollary 3.7. (Á. Baricz [39]) *Let* $\delta \geq e$. *If* $c \geq a/(\log \delta) > 0$, *then* $\Phi(a,c,r)$ *satisfies the Landen-type inequality, i.e.*

$$\lambda_\Phi \left(\frac{2\sqrt{r}}{1+r} \right) < \rho \lambda_\Phi(r)$$

for all $r \in (0,1)$ *and all* $\rho \geq \delta^{4\sqrt{2}-5}$, *where* $\lambda_\Phi(r) = \Phi(a,c,r^2)$.

Proof. By Theorem 3.14 it is enough to show that the sequence

$$B_n = \frac{(a)_n}{(c)_n} \cdot \frac{1}{(\log \delta)^n}$$

is decreasing for $n \geq 0$. By the ascending factorial notation we obtain that

$$(c+n-1)(\log \delta)B_n = (a+n-1)B_{n-1},$$

therefore $B_n \leq B_{n-1}$, for all $n \geq 1$ is equivalent to $c \geq a/(\log \delta)$. Thus the proof is complete.   □

Remark 3.6. We note that if we take $\delta = e$ in Lemma 3.5, then $e^{4x/(x+1)^2} \leq \rho e^{x^2}$ holds for $x \in (0,1)$ and the best constant ρ is $e^{4\sqrt{2}-5} = 1.928715521\ldots$. Moreover, Corollary 3.7 for $\delta = e$ becomes: if $c \geq a > 0$, then $\Phi(a,c,r)$ satisfies the Landen-type inequality, i.e.

$$\lambda_\Phi \left(\frac{2\sqrt{r}}{1+r} \right) < \rho \lambda_\Phi(r)$$

for all $r \in (0,1)$ and all $\rho \geq e^{4\sqrt{2}-5}$, where $\lambda_\Phi(r) = \Phi(a,c,r^2)$. Actually this result for the confluent hypergeometric function generalizes the inequality $e^{4x/(x+1)^2} < \rho e^{x^2}$, because $\Phi(a,a,x) = e^x$.

3.4 Convexity of Hypergeometric Functions with Respect to Hölder Means

3.4.1 Introduction and Preliminaries

Let be given a nonempty interval $\Lambda \subseteq (0,\infty)$ and a function $f : \Lambda \to (0,\infty)$. The function f is said to be multiplicatively convex if for all $r,s \in \Lambda$ and all $\lambda \in [0,1]$ the inequality

$$f(r^\lambda s^{1-\lambda}) \leq [f(r)]^\lambda [f(s)]^{1-\lambda} \tag{3.32}$$

holds. The function f is called multiplicatively concave if

$$f(r^\lambda s^{1-\lambda}) \geq [f(r)]^\lambda [f(s)]^{1-\lambda} \tag{3.33}$$

for all $r,s \in \Lambda$ and all $\lambda \in (0,1)$. If for $r \neq s$ and $\lambda \in (0,1)$ the inequality (3.32) (respectively (3.33)) is strict, then f is said to be strictly multiplicatively convex (respectively strictly multiplicatively concave).

It can be proved (see for example C.P. Niculescu [163, Theorem 2.3]) that a continuous function $f : \Lambda \to (0,\infty)$ is multiplicatively convex (respectively strictly multiplicatively convex) if and only if

$$f(\sqrt{rs}) \leq \sqrt{f(r)f(s)} \quad \text{(respectively } f(\sqrt{rs}) < \sqrt{f(r)f(s)})$$

for all $r,s \in \Lambda$ with $r \neq s$.

A similar characterization of the continuous (strictly) multiplicatively concave functions holds too.

In Theorem 3.6 (see also Á. Baricz [45, Theorem 1.10]) we proved for $a,b > 0$ that the function $T : (0,1) \to (0,\infty)$, defined by $T(r) = F(a,b,a+b,r)$, satisfies the following chain of inequalities

$$T(G(r,s)) \leq G(T(r),T(s)) \leq T(1 - G(1-r,1-s)) \leq A(T(r),T(s)) \tag{3.34}$$

for all $r, s \in (0, 1)$. We note that since the function $r \mapsto T'(r)/T(r)$ is strictly increasing (see R. Balasubramanian et al. [36, Lemma 2.1]) for $a, b > 0$ and $r \in (0, 1)$, we have that T is strictly log-convex on $(0, 1)$, i.e. the following inequality

$$T(A(r, s)) \leq G(T(r), T(s)) \tag{3.35}$$

holds for all $r, s \in (0, 1)$. So because T is strictly increasing on $(0, 1)$, combining (3.34) with (3.35) we obtain

$$T(G(r, s)) \leq T(A(r, s)) \leq G(T(r), T(s)) \leq A(T(r), T(s)) \tag{3.36}$$

for all $a, b > 0$ and $r, s \in (0, 1)$. On the other hand we proved in Theorem 3.5 (see also Á. Baricz [45, Theorem 1.10]) that for $a, b \in (0, 1]$ the inequality (3.23) holds for all $r, s \in (0, q)$, where q is the unique positive root of the equation $2\log(1 - x) + x/(1 - x) = 0$. Recall that recently G.D. Anderson et al. [21, Theorem 3.7] improved Theorem 3.5 by showing that for all $a, b \in (0, 1)$ and $r, s \in (0, 1)$ the inequality

$$T(A(r, s)) \leq H(T(r), T(s))$$

and hence (3.23) is valid. Using (3.23), the harmonic-geometric mean inequality imply

$$T(H(r, s)) \leq T(G(r, s)) \leq H(T(r), T(s)), \tag{3.37}$$

where $a, b \in (0, 1]$ and $r, s \in (0, 1)$.

In fact using (3.36) and (3.37) we see that the inequality

$$T(M(r, s)) \leq M(T(r), T(s)) \quad \text{for all } r, s \in (0, 1) \tag{3.38}$$

holds under certain conditions on a, b and for M being the unweighted harmonic, geometric and arithmetic mean.

Taking into account the inequalities (3.36) and (3.37), it is natural to ask for which means the inequality (3.38) remains true. In this section our aim is to answer this question partially for Hölder means.

3.4.2 Convexity of Hypergeometric Functions with Respect to Hölder Means

Let $\Lambda \subseteq \mathbb{R}$ be a non-degenerate interval and let $\varphi : \Lambda \to \mathbb{R}$ be a strictly monotonic continuous function. The function $M_\varphi : \Lambda^2 \to \Lambda$, defined by

$$M_\varphi(r, s) = \varphi^{-1}\left(A(\varphi(r), \varphi(s))\right),$$

is called the quasi-arithmetic mean associated with φ, while the function φ is said to be the generating function of the quasi-arithmetic mean M_φ. For details see J. Aczél [3], Z. Daróczy [86] and J. Matkowski [144].

A function $f : \Lambda \to \mathbb{R}$ is said to be convex with respect to the mean M_φ (or M_φ−convex) if for all $r, s \in \Lambda$ and all $\lambda \in (0, 1)$ the inequality

$$f(M_\varphi^{(\lambda)}(r, s)) \leq M_\varphi^{(\lambda)}(f(r), f(s)) \tag{3.39}$$

holds, where $M_\varphi^{(\lambda)}(r, s) = \varphi^{-1}(\lambda \varphi(r) + (1 - \lambda)\varphi(s))$ is the weighted version of M_φ. If for $r \neq s$ the inequality (3.39) is strict, then f is said to be strictly convex with respect to M_φ. For details see D. Borwein et al. [75], J. Matkowski and J. Rätz [146, 147].

It can be proved (see [75]) that $f : \Lambda \to \mathbb{R}$ is (strictly) convex with respect to M_φ if and only if $\varphi \circ f \circ \varphi^{-1}$ is (strictly) convex in the usual sense on $\varphi(\Lambda)$.

Among the quasi-arithmetic means the Hölder means (or power means) are of special interest. They are associated with the function $\varphi_p : (0, \infty) \to \mathbb{R}$, defined by

$$\varphi_p(r) = \begin{cases} r^p, & \text{if } p \neq 0 \\ \log r, & \text{if } p = 0. \end{cases}$$

Thus the Hölder mean $H_p : (0, \infty) \times (0, \infty) \to (0, \infty)$ is defined by

$$H_p(r, s) = \begin{cases} [A(r^p, s^p)]^{1/p}, & \text{if } p \neq 0 \\ G(r, s), & \text{if } p = 0. \end{cases}$$

Theorem 3.15. (Á. Baricz [48]) *For all $a, b > 0$ and $p \in [0, 1]$ the zero-balanced hypergeometric function $r \in (0, 1) \mapsto T(r) = F(a, b, a + b, r)$ is convex with respect to the Hölder means H_p.*

Proof. Using the definition of convexity with respect to Hölder means it is enough to show that for all $\lambda, r, s \in (0, 1)$, $a, b > 0$ and $p \in (0, 1]$, $p \neq 0$

$$T([\lambda r^p + (1 - \lambda)s^p]^{1/p}) \leq [\lambda [T(r)]^p + (1 - \lambda)[T(s)]^p]^{1/p}. \tag{3.40}$$

Moreover we need to prove that for all $\lambda, r, s \in (0, 1)$ and $a, b > 0$

$$T(r^\lambda s^{1-\lambda}) \leq [T(r)]^\lambda [T(s)]^{1-\lambda} \tag{3.41}$$

holds, i.e. the zero-balanced hypergeometric function is multiplicatively convex on $(0, 1)$.

First assume that $p = 0$. Then we need to prove that (3.41) holds. Using the first inequality in (3.34) and Theorem 2.3 due to C.P. Niculescu [163] the desired result follows. Note that in fact (3.41) can be proved using the Hölder-Rogers inequality (1.38). For this let us denote $P_n(r) = \sum_{k=0}^{n} d_k r^k$, where

$$d_k = \frac{(a)_k (b)_k}{(a + b)_k} \cdot \frac{1}{k!} \quad \text{for all } k \in \{0, 1, \ldots, n\}.$$

Then by the Hölder-Rogers inequality (1.38) we have

$$\sum_{k=0}^{n}(d_k{}^{\lambda}r^{\lambda k})(d_k{}^{1-\lambda}s^{(1-\lambda)k}) \leq \left(\sum_{k=0}^{n}d_k r^k\right)^{\lambda}\left(\sum_{k=0}^{n}d_k s^k\right)^{1-\lambda}. \tag{3.42}$$

But this is equivalent to $P_n(r^{\lambda}s^{1-\lambda}) \leq [P_n(r)]^{\lambda}[P_n(s)]^{1-\lambda}$, so using the fact that

$$\lim_{n\to\infty}P_n(r) = T(r),$$

we obtain immediately (3.41).

Now assume that $p \neq 0$. In order to establish the convexity of T with respect to H_p in this case we need to show that the function $\varphi_p \circ T \circ \varphi_p^{-1}$ is convex in the usual sense. Let us denote $f_G(r) = (\varphi_p \circ T \circ \varphi_p^{-1})(r) = [T(r^{1/p})]^p$. Setting $q = 1/p \geq 1$ we have $f_G(r) = [T(r^q)]^{1/q}$, thus a simple computation shows that

$$f_G'(r) = r^{q-1}\frac{T'(r^q)}{T(r^q)}[T(r^q)]^{1/q} = \frac{1}{q}f_G(r)\frac{d(\log T(r^q))}{dr} > 0. \tag{3.43}$$

We know by Lemma 2.1 due to R. Balasubramanian et al. [36] that the function T is log-convex on $(0,1)$. On the other hand the function $r \mapsto r^q$ is convex on $(0,1)$. Thus by monotonicity of T for all $\lambda, r, s \in (0,1)$ we obtain

$$T([\lambda r + (1-\lambda)s]^q) \leq T(\lambda r^q + (1-\lambda)s^q) \leq [T(r^q)]^{\lambda}[T(s^q)]^{1-\lambda}.$$

This shows that the function $r \mapsto T(r^q)$ is log-convex and consequently $r \mapsto d(\log T(r^q))/dr$ is increasing. From (3.43) we obtain that f_G is increasing, therefore f_G' is increasing too as a product of two positive and increasing functions.   □

Taking into account the above proof we note that Theorem 3.15 may be generalized easily in the following way. The proof of the next theorem is similar, so we omit the details.

Theorem 3.16. (Á. Baricz [48]) *For all $a,b > 0$ and $p \in [0,m]$, where $m \in \{1,2,\dots\}$, the zero-balanced hypergeometric function $r \in (0,1) \mapsto F(a,b,a+b,r^m)$ is convex with respect to Hölder means H_p. In particular the complete elliptic integral of the first kind $\mathscr{K}$ is convex on $(0,1)$ with respect to means H_p where $p \in [0,2]$, i.e. for all $\lambda, r, s \in (0,1)$ and $p \in (0,2]$ one has*

$$\mathscr{K}([\lambda r^p + (1-\lambda)s^p]^{1/p}) \leq [\lambda[\mathscr{K}(r)]^p + (1-\lambda)[\mathscr{K}(s)]^p]^{1/p}.$$

Moreover for all $\lambda, r, s \in (0,1)$

$$\mathscr{K}(r^{\lambda}s^{1-\lambda}) \leq [\mathscr{K}(r)]^{\lambda}[\mathscr{K}(s)]^{1-\lambda}$$

holds, i.e. the complete elliptic integral $\mathscr{K}$ is multiplicatively convex on $(0,1)$.

Consider the function $r \in (0,1) \mapsto T(r) = F(a,b,a+b,r) \in (0,\infty)$. Due to G.D. Anderson et al. [21, Theorem 3.7] we know that the function $r \mapsto 1/T(r)$ is strictly concave on $(0,1)$ for all $a,b \in (0,1]$. This implies that we have

$$T(H_{-1}^{(\lambda)}(r,s)) \leq T(\lambda r + (1-\lambda)s) \leq H_{-1}^{(\lambda)}(T(r),T(s)), \qquad (3.44)$$

where $\lambda, r, s \in (0,1)$ and $a,b \in (0,1]$. Here $H_{-1}^{(\lambda)}(r,s) = [\lambda/r + (1-\lambda)/s]^{-1}$ is the weighted harmonic mean and we used the harmonic-arithmetic mean inequality between the weighted harmonic and arithmetic means of r and s. We note that in fact (3.44) shows that the zero-balanced hypergeometric function T is convex on $(0,1)$ for all $a,b \in (0,1]$ with respect to the Hölder mean H_{-1}.

The following result is similar to Theorem 3.16.

Theorem 3.17. (Á. Baricz [48]) *If $a,b,p > 0$ and $m \in \{1,2,3,\dots\}$, then the function $r \in (0,1) \mapsto f_m(r) = F(a,b,a+b,r^m) - 1$ is strictly convex with respect to the Hölder means H_p. In particular, for $m = 1$ and $m = 2$ the functions $f_1(r) = F(a,b,a+b,r) - 1$ and $f_2(r) = 2\mathcal{K}(r)/\pi - 1$ are strictly convex on $(0,1)$ with respect to means H_p, where $p > 0$. Consequently, by using the notation $T(r) = F(a,b,a+b,r)$, for all $\lambda, r, s \in (0,1)$ and $p > 0$ one has*

$$T([\lambda r^p + (1-\lambda)s^p]^{1/p}) < 1 + [\lambda[T(r)-1]^p + (1-\lambda)[T(s)-1]^p]^{1/p},$$

$$\frac{2}{\pi}\mathcal{K}([\lambda r^p + (1-\lambda)s^p]^{1/p}) < 1 + \left[\lambda\left(\frac{2}{\pi}\mathcal{K}(r)-1\right)^p + (1-\lambda)\left(\frac{2}{\pi}\mathcal{K}(s)-1\right)^p\right]^{1/p}.$$

Proof. We just need to show that $r \in (0,1) \mapsto [f_m(r^{1/p})]^p$ is strictly convex. Let us denote $\gamma(r) = [f_m(r^{1/p})]^p = [T(r^{m/p}) - 1]^p$. Setting $q = 1/p > 0$ we get $\gamma(r) = [T(r^{qm}) - 1]^{1/q}$. A simple computation shows that

$$\gamma'(r) = m\left[\frac{r^{mq}T'(r^{mq})}{T(r^{mq})-1}\right] \cdot \left[\frac{T(r^{mq})-1}{r^q}\right]^{1/q}.$$

Now taking $r^{mq} = x \in (0,1)$ we need only to prove that the function

$$x \in (0,1) \mapsto g_m(x) = m\left[\frac{xT'(x)}{T(x)-1}\right] \cdot \left[\frac{T(x)-1}{x^{1/m}}\right]^{1/q}$$

is strictly increasing.

To prove the asserted result we show that the each term on the right-hand side of the above relation is strictly increasing. Thus g_m will be strictly increasing as a product of two positive and strictly increasing functions.

Using the notation

$$d_n = \frac{(a)_n(b)_n}{(a+b)_n} \cdot \frac{1}{n!} \quad \text{for all } n \geq 0,$$

by Lemma 3.1 it is clear that $x \in (0,1) \mapsto xT'(x)/(T(x)-1)$ is strictly increasing because

$$\frac{xT'(x)}{T(x)-1} = \frac{\sum\limits_{n \geq 1} nd_n x^n}{\sum\limits_{n \geq 1} d_n x^n} = \frac{\sum\limits_{n \geq 0} (n+1)d_{n+1} x^n}{\sum\limits_{n \geq 0} d_{n+1} x^n}$$

and clearly the sequence $\{(n+1)d_{n+1}/d_{n+1}\}_{n \geq 0} = \{n+1\}_{n \geq 0}$ is strictly increasing.

Now, since $1/q > 0$, to prove that $x \in (0,1) \mapsto [(T(x)-1)/\sqrt[m]{x}]^{1/q}$ is strictly increasing it is enough to show that $x \in (0,1) \mapsto (T(x)-1)/\sqrt[m]{x}$ is strictly increasing. We have that

$$x^{1+1/m} \frac{d}{dx}\left[\frac{T(x)-1}{x^{1/m}}\right] = xT'(x) - \frac{T(x)-1}{m} = \sum_{n \geq 1} nd_n x^n - \frac{1}{m}\sum_{n \geq 1} d_n x^n$$

which is clearly positive, because by assumption $1/m \leq 1 \leq n$ and $d_n > 0$. Thus the proof is complete.   □

3.4.3 Convexity of General Power Series with Respect to Hölder Means

Let us consider the power series

$$f(r) = \sum_{n \geq 0} A_n r^n \quad (\text{where } A_n > 0 \text{ for all } n \geq 0) \tag{3.45}$$

and suppose that it is convergent for all $r \in (0,1)$. In this section our aim is to find conditions for the convexity of f with respect to the Hölder means. From the proof of Theorem 3.15 it is clear that the fact that T is log-convex was sufficient for T to be convex with respect to H_p for $p \in (0,1]$. Moreover, taking into account the proof of Theorem 3.17 we observe that the statement of this theorem holds for an arbitrary power series. Thus our main result in this section is the following:

Theorem 3.18. (Á. Baricz [48]) *Let f be defined by (3.45), $m \in \{1,2,\dots\}$, and for all $n \geq 0$ let us denote $B_n = (n+1)A_{n+1}/A_n$. Then the following assertions are true:*

(a) *If the sequence $\{B_n\}_{n \geq 0}$ is increasing, then $r \in (0,1) \mapsto f(r^m)$ is convex with respect to H_p for $p \in [0,m]$.*
(b) *If the sequence $\{B_n - n\}_{n \geq 0}$ is increasing, then $r \in (0,1) \mapsto f(r^m)$ is convex with respect to H_p for $p \in [0,\infty)$.*
(c) *The function $r \in (0,1) \mapsto f(r^m) - 1$ is strictly convex with respect to H_p for $p \in (0,\infty)$.*

Proof. (a) First assume that $p = 0$. Then a simple application of the Hölder-Rogers inequality (1.38) gives the multiplicatively convexity of f. Now let $p \neq 0$. Then by Lemma 3.1 it is clear that $r \in (0,1) \mapsto f'(r)/f(r)$ is increasing. Let us denote $\phi(r) = (\phi_p \circ f \circ \phi_p^{-1})(r) = [f(r^{m/p})]^p$. Setting $q = m/p \geq 1$ we have $\phi(r) = [f(r^q)]^{m/q}$, thus a simple computation shows that

$$\phi'(r) = mr^{q-1}\frac{f'(r^q)}{f(r^q)}[f(r^q)]^{m/q} = \frac{m}{q}\phi(r)\frac{d(\log f(r^q))}{dr} \geq 0. \qquad (3.46)$$

On the other hand the function $r \in (0,1) \mapsto r^q \in (0,1)$ is convex. Therefore because f is strictly increasing and log-convex, one has for all $\lambda, r, s \in (0,1)$

$$f([\lambda r + (1-\lambda)s]^q) \leq f(\lambda r^q + (1-\lambda)s^q) \leq [f(r^q)]^\lambda [f(s^q)]^{1-\lambda}.$$

This shows that the function $r \in (0,1) \mapsto f(r^q)$ is log-convex too and consequently $r \in (0,1) \mapsto d(\log f(r^q))/dr$ is increasing. From (3.46) we obtain that ϕ is increasing, therefore ϕ' is increasing too as a product of two positive and increasing functions.

(b) Let us denote $Q(r) = d(\log f(r))/dr = f'(r)/f(r)$. Using again Lemma 3.1 from the fact that the sequence $\{B_n - n\}_{n \geq 0}$ is increasing we get that the function

$$r \mapsto (1-r)Q(r) = (1-r)\frac{f'(r)}{f(r)} = \frac{\sum\limits_{n \geq 0}[(n+1)A_{n+1} - nA_n]r^n}{\sum\limits_{n \geq 0}A_n r^n}$$

is increasing too. Thus the function $r \mapsto \log[(1-r)Q(r)]$ will be also increasing, i.e. $d\log[(1-r)Q(r)]/dr \geq 0$ for all $r \in (0,1)$. This in turn implies that

$$\frac{Q'(r)}{Q(r)} \geq \frac{1}{1-r} \quad \text{for all } r \in (0,1). \qquad (3.47)$$

Taking into account (3.46) for $q = m/p > 0$ we just need to show that

$$r \mapsto \phi'(r) = \frac{m}{q}\phi(r)\frac{d(\log f(r^q))}{dr} = \frac{m}{q}\phi(r)Q(r^q) \geq 0$$

is increasing. Now using (3.47) we get that

$$\phi''(r) = \frac{m}{q}\phi(r)Q(r^q)\left[\frac{m}{q}Q(r^q) + \frac{d\log(Q(r^q))}{dr}\right]$$

$$\geq \frac{m}{q}\phi(r)Q(r^q)\left[\frac{m}{q}Q(r^q) + \frac{1}{1-r^q}\right] \geq 0,$$

which completes the proof of this part.

(c) We just need to show that $r \mapsto [f(r^{m/p}) - 1]^p$ is strictly convex on $(0,1)$. Let us denote $\sigma(r) = [f(r^{m/p}) - 1]^p$. Setting $q = 1/p > 0$ we get $\sigma(r) = [f(r^{qm}) - 1]^{1/q}$. Now by using the proof of Theorem 3.17 with T replaced by f we see that the conclusion of this part is true.   □

3.4.4 Concluding Remarks

As we have seen in Theorem 3.18 the log-convexity of the power series was crucial to prove convexity properties with respect to Hölder means. The following theorem contains sufficient conditions for a differentiable log-convex function to be convex with respect to Hölder means.

Theorem 3.19. (Á. Baricz [48]) *Let $f : \Lambda \subseteq (0,\infty) \to (0,\infty)$ be a differentiable function. The following assertions are true:*

(a) *If the function f is (strictly) increasing and (strictly) log-convex, then f is (strictly) convex with respect to the Hölder means H_p for $p \in [0,1]$.*

(b) *If the function f is (strictly) decreasing and (strictly) log-convex, then f is (strictly) convex with respect to the Hölder means H_p for $p \in [1,\infty)$. Moreover, if f is decreasing then f is (strictly) multiplicatively convex if and only if it is (strictly) convex with respect to the Hölder means H_p for $p \in [0,\infty)$.*

Proof. (a) Suppose that $p = 0$. Then using the arithmetic-geometric mean inequality, monotonicity of f and the log-convexity property, one has

$$f(r^\lambda s^{1-\lambda}) \leq f(\lambda r + (1-\lambda)s) \leq [f(r)]^\lambda [f(s)]^{1-\lambda}$$

for all $r, s \in \Lambda$ and $\lambda \in (0,1)$. Now assume that $p \neq 0$. Let us denote $g(r) = [f(r^{1/p})]^p$ and $q = 1/p \geq 1$. Then $g(r) = [f(r^q)]^{1/q}$ and

$$g'(r) = \frac{1}{q} g(r) \frac{d[\log f(r^q)]}{dr} > 0. \tag{3.48}$$

In this case $r \mapsto r^q$ is convex, thus

$$f([\lambda r + (1-\lambda)s]^q) \leq f(\lambda r^q + (1-\lambda)s^q) \leq [f(r^q)]^\lambda [f(s^q)]^{1-\lambda} \tag{3.49}$$

holds for all $r, s \in \Lambda$ and $\lambda \in (0,1)$, which means that $r \mapsto f(r^q)$ is log-convex too. Thus by (3.48) g' is increasing as a product of two increasing functions.

(b) Using the same notation as in part (a) $q = 1/p \in (0,1]$ and consequently $r \mapsto r^q$ is concave. But f is decreasing, thus (3.49) holds again. Now suppose that f is multiplicatively convex and decreasing. For $p \in (0,1]$ we have $q = 1/p \geq 1$ and $r \mapsto r^q$ is log-concave. Thus

$$f([\lambda r + (1-\lambda)s]^q) \leq f((r^q)^\lambda (s^q)^{1-\lambda}) \leq [f(r^q)]^\lambda [f(s^q)]^{1-\lambda} \qquad (3.50)$$

holds for all $r,s \in \Lambda$ and $\lambda \in (0,1)$. When $p \geq 1$, then $q = 1/p \in (0,1]$ and $r \mapsto r^q$ is concave. Thus using the fact that f is decreasing, one has

$$f([\lambda r + (1-\lambda)s]^q) \leq f(\lambda r^q + (1-\lambda)s^q) \leq f((r^q)^\lambda (s^q)^{1-\lambda}) \leq [f(r^q)]^\lambda [f(s^q)]^{1-\lambda} \qquad (3.51)$$

for all $r,s \in \Lambda$ and $\lambda \in (0,1)$. So (3.50) and (3.51) imply that $r \mapsto f(r^q)$ is log-convex and consequently g is convex. Finally it is clear that the convexity of f with respect to Hölder means H_p, $p \in [0,\infty)$ implies the convexity of f with respect to H_0 and this is the multiplicatively convexity. Thus the proof is complete.   □

The decreasing homeomorphism $m : (0,1) \to (0,\infty)$ defined by

$$m(r) = F(a,b,a+b,1-r^2)/F(a,b,a+b,r^2)$$

and other various forms of this function were studied by R. Balasubramanian et al. [36] and also by S.L. Qiu and M. Vuorinen [195] (see also the references therein). In [36] the authors proved that for $a \in (0,2)$ and $b \in (0,2-a]$ the inequality

$$m(G(r,s)) \geq H(m(r),m(s)) \qquad (3.52)$$

holds for all $r,s \in (0,1)$. In Corollary 3.2 we proved that in fact (3.52) holds for all $a,b > 0$ and $r,s \in (0,1)$. Our aim in what follows is to generalize (3.52). Recall that in [36] in order to prove (3.52) the authors proved that the function $K : (0,\infty) \to (0,\infty)$ defined by $K(t) = F(e^{-t})/F(1-e^{-t})$ is strictly convex. However, in Lemma 3.4 we proved that K is strictly log-convex. In order to present another generalization of (3.52) we prove that in fact K is strictly convex with respect to Hölder means H_p, where $p \in [1,\infty)$.

Corollary 3.8. (Á. Baricz [48]) *For all $a,b > 0$ and $p \geq 1$ the function K is strictly convex on $(0,\infty)$ with respect to Hölder means H_p, i.e. for all $\lambda, r, s \in (0,1)$ and $a,b > 0$, $p \geq 1$ one has*

$$\frac{\lambda}{[m(r)]^p} + \frac{1-\lambda}{[m(s)]^p} \geq \frac{1}{[m(\alpha(r,s))]^p}$$

or equivalently

$$H_p^{(\lambda)} \left(\frac{1}{m(r)}, \frac{1}{m(s)} \right) \geq \frac{1}{m(\alpha(r,s))},$$

where $\log \alpha(r,s) = -H_p^{(\lambda)}(\log(1/r), \log(1/s))$ *and*

$$H_p^{(\lambda)}(r,s) = \begin{cases} [\lambda r^p + (1-\lambda)s^p)]^{1/p}, & \text{if } p \neq 0 \\ r^\lambda s^{1-\lambda}, & \text{if } p = 0 \end{cases}$$

is the weighted version of H_p.

Proof. By Lemma 3.4 we know that K is strictly decreasing and strictly log-convex. Thus by part (b) of Theorem 3.19 we get that K is strictly convex on $(0,\infty)$ with respect to Hölder means H_p for $p \in [1,\infty)$. This means that

$$K(H_p^{(\lambda)}(t_1,t_2)) \leq H_p^{(\lambda)}(K(t_1), K(t_2))$$

holds for all $t_1, t_2 > 0$, $\lambda \in (0,1)$ and $a, b > 0$. Now, let $e^{-t_1} = r^2 \in (0,1)$ and $e^{-t_2} = s^2 \in (0,1)$. Then we get $K(t_1) = 1/m(r)$, $K(t_2) = 1/m(s)$ and $K(H_p^{(\lambda)}(t_1,t_2)) = 1/m(\alpha(r,s))$. Thus the proof is complete.   □

Remark 3.7. When $\lambda = 1/2$ and $p = 1$ we get that $\alpha(r,s) = G(r,s)$, thus the inequality in Corollary 3.8 reduces to (3.52).

For the multiplicatively convexity (concavity) and convexity with respect to Hölder means of Euler's gamma function we refer to the papers of D. Gronau and J. Matkowski [103, 104], C.P. Niculescu [163] and T. Trif [225].

Finally, we note that very recently X. Zhang et al. [237] extended some of the main results of this section to the case of convexity with respect to two Hölder means.

3.5 Askey's and Grünbaum's Inequality for Generalized Bessel Functions

In 1973 F.A. Grünbaum [108] established the following inequality:

$$1 + J_0(z) \geq J_0(x) + J_0(y), \tag{3.53}$$

where $x, y \geq 0$ and $z^2 = x^2 + y^2$. In fact this result on the Bessel function J_0 arose first in the context of a problem involving the Boltzmann equation (see the papers of F.A. Grünbaum [106, 107]). To solve this problem there was sufficient to establish an inequality for the Legendre polynomials, which was proved in [105] also by F.A. Grünbaum. The inequality (3.53), as we mentioned above, was proved in [108] by F.A. Grünbaum by using the well known fact (see the book of G. Szegő [222]) that the spherical functions on a sphere (i.e. the Legendre polynomials) approach the spherical functions on the plane (the Bessel functions) as the radius approaches infinity.

In the same year R. Askey [28] extended the above Grünbaum inequality for the function $\mathscr{J}_p$, defined by (1.23), i.e.

$$\mathscr{J}_p(x) = 2^p \Gamma(p+1) x^{-p} J_p(x),$$

by showing that

$$\mathscr{J}_p(x) + \mathscr{J}_p(y) \le 1 + \mathscr{J}_p(z) \tag{3.54}$$

holds for all $x, y, z, p \ge 0$ such that $z^2 = x^2 + y^2$. It is worth mentioning here that in 1975 and 1977 A.Mcd. Mercer [148, 149] deduced and extended too the above inequality using a different approach. Recently E. Neuman [161] obtained a different upper bound for the sum $\mathscr{J}_p(x) + \mathscr{J}_p(y)$. Moreover, he found lower and upper bounds for the function $\mathscr{J}_p$ using Gegenbauer polynomials.

In this section we prove that the inequality (3.54) holds also for the generalized and normalized Bessel functions. Note that the results of this section may be found in the paper of Á. Baricz and E. Neuman [58]. Using Gegenbauer polynomials too in this section we find lower and upper bounds for the generalized and normalized Bessel functions. Namely, we prove that the function $\lambda_p(x) = u_p(x^2)$, $x \ge 0$ satisfies the Askey-type inequality

$$1 + \lambda_p(z) \ge \lambda_p(x) + \lambda_p(y),$$

where $x, y \ge 0$, $z^2 = x^2 + y^2$ and u_p is the generalized Bessel function. A part of the results (when $c > 0$) are simple extensions of the results of R. Askey [28], E. Neuman [161], while for the results when $c < 0$ we give different proofs.

3.5.1 Askey's and Grünbaum's Inequality for Generalized Bessel Functions

Our first main result in this section reads as follows.

Theorem 3.20. (Á. Baricz and E. Neuman [58]) *If $p, b, c \in \mathbb{R}$ are such that $\kappa > 1$, then the function $\lambda_p : [0, \infty) \to \mathbb{R}$, defined by $\lambda_p(x) = u_p(x^2)$, satisfies*

$$1 + \lambda_p(z) \ge \lambda_p(x) + \lambda_p(y), \tag{3.55}$$

whenever $x, y, z \ge 0$ and $z^2 = x^2 + y^2$.

Proof. If $c = 0$, then we have $\lambda_p(x) \equiv 1$, so there is nothing to prove. Assume that $c > 0$. We know that if $c = 1$ and $b = 1$ then $\lambda_p(x) = \mathscr{J}_p(x)$. Due to the result of R. Askey [28, Theorem 2], i.e. inequality (3.54), we know that $\mathscr{J}_p(x) + \mathscr{J}_p(y) \le 1 + \mathscr{J}_p(z)$ holds, when $x, y, z, p \ge 0$ and $z^2 = x^2 + y^2$. Now in (3.54) changing x with $x\sqrt{c}$, y with $y\sqrt{c}$ and p with $\kappa - 1$ (in other words $\mathscr{J}_{\kappa-1}(x\sqrt{c})$ becomes $\lambda_p(x)$), we get immediately that $\lambda_p(x) + \lambda_p(y) \leq 1 + \lambda_p(z)$ holds. For $c < 0$ the situation is a little different. By the definition of λ_p, inequality (3.55) is equivalent to

$$1 + u_p(x^2 + y^2) \geq u_p(x^2) + u_p(y^2). \tag{3.56}$$

But for this it is enough to show that

$$1 + u_p(x + y) \geq u_p(x) + u_p(y) \tag{3.57}$$

and after changing x with x^2 and y with y^2, inequality (3.56) follows. Observe that for inequality (3.57) it is enough to show that the function $f(x) = u_p(x) - 1$ is super-additive, i.e. $f(x+y) \geq f(x) + f(y)$ for all $x, y \geq 0$. Now we prove that if the function $g(x) = f(x)/x$ is increasing, then $f(x)$ is super-additive. We have

$$f(x+y) = x\frac{f(x+y)}{x+y} + y\frac{f(x+y)}{x+y} \geq x\frac{f(x)}{x} + y\frac{f(y)}{y} = f(x) + f(y).$$

Thus, we need to prove that $h : (0, \infty) \to \mathbb{R}$, defined by $h(x) = (u_p(x) - 1)/x$, is increasing and the proof is complete. On the other hand

$$h'(x) = (xu'_p(x) - (u_p(x) - 1))/x^2$$

and this is clearly positive if and only if $xu'_p(x) \geq u_p(x) - 1$ holds. By definition $u_p(x) = \sum_{n \geq 0} b_n x^n$ and thus the above inequality becomes $\sum_{n \geq 1} (n-1)b_n x^n \geq 0$. This clearly holds because $c < 0$ and $\kappa > 1$ implies that $b_n > 0$ for all $n \geq 0$.   □

Remark 3.8. We know that for $c = -1$ and $b = 1$ the function w_p reduces to the modified Bessel function I_p, defined by (1.7). Therefore, if we denote λ_p in this case ($c = -1$ and $b = 1$) with $\mathscr{I}_p$ defined by (1.24), i.e. $\mathscr{I}_p(x) = 2^p \Gamma(p+1)x^{-p}I_p(x)$, then we obtain an Askey type inequality for modified Bessel functions:

$$\mathscr{I}_p(x) + \mathscr{I}_p(y) \leq 1 + \mathscr{I}_p(z), \tag{3.58}$$

which holds for all $p > 0$ and $x, y, z \geq 0$ such that $z^2 = x^2 + y^2$. To show that for $\mathscr{I}_0 = I_0$ Grünbaum's inequality holds we may proceed as in the proof of Theorem 3.20. We need to prove that the function $\varphi(x) = I_0(\sqrt{x}) - 1$ is super-additive, and this is guaranteed by the fact that $x \mapsto \varphi(x)/\sqrt{x}$ is increasing. Moreover, there is another way to prove that the functions $f(x) = u_p(x) - 1$ and φ are super-additive. Namely, Petrović's inequality ([170], see also the book of D.S. Mitrinović [156, Theorem 1, p. 22]) states that if the function $\phi : [0, \infty) \to \mathbb{R}$ is convex, then

$$\phi(x_1) + \phi(x_2) + \ldots + \phi(x_n) \leq \phi(x_1 + x_2 + \ldots + x_n) + (n-1)\phi(0)$$

holds for all $x_1, x_2, \ldots, x_n \geq 0$. For $n = 2$ we get that if $\phi(0) = 0$ and ϕ is convex, then ϕ is super-additive. Observe that $f(0) = 0$, $\varphi(0) = 0$ and both functions are convex. Therefore we obtain Grünbaum's inequality for modified Bessel function of zero order

$$1 + I_0(z) \geq I_0(x) + I_0(y),$$

which holds for all $x, y, z \geq 0$ such that $z^2 = x^2 + y^2$. Further, it is worth mentioning that from the proof of Theorem 3.20 inequality (3.55) also holds when $c < 0$ and $\kappa > 0$, more precisely the function $f(x) = u_p(x) - 1$ is super-additive too in this case. This implies that in fact (3.58) holds true for all $p > -1$. This was pointed out also in [44] by the author. In particular, choosing $p = -1/2$, from (3.58) we obtain the following interesting inequality

$$1 + \cosh z \geq \cosh x + \cosh y,$$

whenever $x, y, z \geq 0$ and $z^2 = x^2 + y^2$. It is important to note here that the analogous of the above Grünbaum type inequality for the cosine function does not hold. Moreover, it can be shown that the condition $p > 0$ for the Askey type inequality (3.54) cannot be changed to $p > -1$.

3.5.2 Lower and Upper Bounds for Generalized Bessel Functions

In this section we give a different upper bound to those given in (3.55) for the sum $\lambda_p(x) + \lambda_p(y)$.

Theorem 3.21. (Á. Baricz and E. Neuman [58]) *For the parameters* $p, b, c \in \mathbb{R}$ *with* $\kappa \geq 1/2$, *we have*

$$[\lambda_p(x) + \lambda_p(y)]^2 \leq [1 + \lambda_p(x+y)][1 + \lambda_p(x-y)] \quad \text{for all } x, y \in \mathbb{R}. \qquad (3.59)$$

Proof. First we prove that (3.59) holds for $\kappa > 1/2$. For $c = 0$ we have equality. Assume that $c > 0$. By Theorem 2.1, established by E. Neuman [161], we know that (3.59) holds for $c = 1$ and $b = 1$, i.e. for the function $\mathcal{J}_p$ (see the inequality (3.60)). Using (3.60) and changing x with $x\sqrt{c}$, y with $y\sqrt{c}$ and p with $\kappa - 1$, we get immediately (3.59). Now suppose that $c < 0$. By Lemma 1.3 we know that

$$\lambda_p(x) = \int_0^1 \cosh(tx\sqrt{-c})d\mu_{p,b}(t),$$

where $p + b/2 > 0$ and $\mu_{p,b}$ is a probability measure on the interval $[0, 1]$.
    Using the identities

$$\cosh\alpha + \cosh\beta = 2\cosh\left(\frac{\alpha+\beta}{2}\right)\cosh\left(\frac{\alpha-\beta}{2}\right), \ 2\cosh^2\left(\frac{\alpha}{2}\right) = 1 + \cosh\alpha,$$

as well as the Cauchy–Bunyakovsky–Schwarz inequality for integrals (1.36), we obtain

$$|\lambda_p(x) + \lambda_p(y)|$$

$$\leq \int_0^1 |\cosh(xt\sqrt{-c}) + \cosh(yt\sqrt{-c})| d\mu_{p,b}(t)$$

$$= 2\int_0^1 \left|\cosh\frac{(x+y)t\sqrt{-c}}{2}\cosh\frac{(x-y)t\sqrt{-c}}{2}\right| d\mu_{p,b}(t)$$

$$\leq 2\left(\int_0^1 \cosh^2\frac{(x+y)t\sqrt{-c}}{2} d\mu_{p,b}(t)\right)^{1/2}\left(\int_0^1 \cosh^2\frac{(x-y)t\sqrt{-c}}{2} d\mu_{p,b}(t)\right)^{1/2}$$

$$= \left[\int_0^1 [1 + \cosh((x+y)t\sqrt{-c})] d\mu_{p,b}(t) \int_0^1 [1 + \cosh((x-y)t\sqrt{-c})] d\mu_{p,b}(t)\right]^{1/2}$$

$$= [1 + \lambda_p(x+y)]^{1/2}[1 + \lambda_p(x-y)]^{1/2}.$$

From (3.59) we have that for all $x, y \in \mathbb{R}$ and $p > -1/2$ one has

$$[\mathscr{J}_p(x) + \mathscr{J}_p(y)]^2 \leq [1 + \mathscr{J}_p(x+y)][1 + \mathscr{J}_p(x-y)], \qquad (3.60)$$

$$[\mathscr{I}_p(x) + \mathscr{I}_p(y)]^2 \leq [1 + \mathscr{I}_p(x+y)][1 + \mathscr{I}_p(x-y)] \qquad (3.61)$$

respectively. Using the identities

$$[\cos x + \cos y]^2 = [1 + \cos(x+y)][1 + \cos(x-y)],$$

$$[\cosh x + \cosh y]^2 = [1 + \cosh(x+y)][1 + \cosh(x-y)],$$

we obtain that (3.60) and (3.61) hold for all $x, y \in \mathbb{R}$ and all $p \geq -1/2$. Here we used the fact that $\mathscr{J}_{-1/2}(x) = \cos x$ and $\mathscr{I}_{-1/2}(x) = \cosh x$. Thus in view of the relations $\mathscr{J}_{\kappa-1}(x\sqrt{c}) = \lambda_p(x)$ (where $c \geq 0$) and $\mathscr{I}_{\kappa-1}(x\sqrt{-c}) = \lambda_p(x)$ (where $c \leq 0$) it follows that (3.59) holds for all $\kappa \geq -1/2$, and with this the proof is complete. $\qquad \square$

Remark 3.9. When $x = y$, the inequality (3.59) reduces to $2\lambda_p^2(x) \leq 1 + \lambda_p(2x)$ which bears resemblance (in fact is the generalization) of the double-angle formula for the functions cos and cosh, i.e. $2\cos^2 x = 1 + \cos(2x)$ and $2\cosh^2 x = 1 + \cosh(2x)$.

Our next goal is to establish computable lower and upper bounds for the function λ_p. We recall some basic facts about the Gegenbauer polynomials G_m^p ($p > -1/2, m \in \mathbb{N}$) and the Gauss-Gegenbauer quadrature formula. These polynomials are orthogonal on $[-1, 1]$ with the weight function $\tau(t) = (1-t^2)^{p-1/2}$. The explicit formula for G_m^p (see the book of M. Abramowitz and I.A. Stegun [1, Eq. 22.3.4]) is:

$$G_m^p(t) = \sum_{n=0}^{[m/2]} (-1)^n \frac{\Gamma(p+m-n)}{\Gamma(p)n!(m-2n)!}(2t)^{m-2n}.$$

In particular we have

$$G_2^p(t) = 2p(p+1)t^2 - p. \tag{3.62}$$

The classical Gauss-Gegenbauer quadrature formula with the remainder (see the book of K.E. Atkinson [29]) is

$$\int_{-1}^{1} (1-t^2)^{p-1/2} g(t)\, \mathrm{dt} = \sum_{i=1}^{m} \tau_i f(t_i) + \gamma_m f^{(2m)}(\alpha), \tag{3.63}$$

where $f \in C^{2m}([-1,1])$, γ_m is a positive number and does not depend on f, α is an intermediate point in $(-1,1)$. The nodes t_i ($i \in \{1,2,\dots,m\}$) are the roots of G_m^p and the weights τ_i are given explicitly by (see (15.3.2) in the book of G. Szegő [222])

$$\tau_i = \pi \left(\frac{2^{1-p}}{\Gamma(p)}\right)^2 \cdot \frac{\Gamma(2p+m)}{m!(1-t_i^2)} \cdot [(G_m^p)'(t_i)]^{-2} \quad \text{for all} \;\; i \in \{1,2,\dots,m\}. \tag{3.64}$$

We are now in a position to prove the following result.

Theorem 3.22. (Á. Baricz and E. Neuman [58]) *Let $p,b \in \mathbb{R}$ be such that $\kappa \geq 1/2$. Then the following assertions are true:*

(a) *If $c \in [0,1]$ and $x \in [-\pi/2, \pi/2]$, then*

$$\cos\left(\sqrt{\frac{c}{2\kappa}}x\right) \leq \lambda_p(x) \leq \frac{1}{3\kappa}\left[2\kappa - 1 + (\kappa+1)\cos\left(\sqrt{\frac{3c}{2(\kappa+1)}}x\right)\right]. \tag{3.65}$$

(b) *If $c \leq 0$ and $x \in \mathbb{R}$, then*

$$\cosh\left(\sqrt{\frac{-c}{2\kappa}}x\right) \leq \lambda_p(x). \tag{3.66}$$

Equality holds in (3.65) and in (3.66) if $x = 0$ or $c = 0$ or $\kappa = 1/2$.

Proof. (a) We know by Theorem 2.2, obtained by E. Neuman [161], that (3.65) holds for $c = 1$, $b = 1$ and for all $x \in [-\pi/2, \pi/2]$, $p > -1/2$, i.e.

$$\cos\left(\frac{x}{\sqrt{2(p+1)}}\right) \leq \mathscr{J}_p(x) \leq \frac{1}{3(p+1)}\left[2p+1+(p+2)\cos\left(\sqrt{\frac{3}{2(p+2)}}x\right)\right]. \tag{3.67}$$

Observe that if $p = -1/2$, then we have equality in (3.67), thus we may conclude that (3.67) holds for all $p \geq -1/2$. Now changing in (3.67) x with $x\sqrt{c}$ and p with $\kappa - 1$, we obtain (3.65).

(b) First assume that $\kappa > 1/2$. In order to establish the lower bound in (3.66) we use the Gauss-Gegenbauer quadrature formula (3.63) with the function $f(t) = \cosh(xt\sqrt{-c})$ and $m = 2$. Since we have $f^{(4)}(t) = x^4 c^2 \cosh(xt\sqrt{-c}) \geq 0$ for $t \in [-1,1], x \in \mathbb{R}$ and $c \leq 0$, we obtain that

$$\tau_1 f(t_1) + \tau_2 f(t_2) \leq \int_{-1}^{1} (1-t^2)^{\kappa-3/2} \cosh(tx\sqrt{-c})\,dt. \qquad (3.68)$$

After changing p with $\kappa - 1$ in (3.62), (3.64) we get that

$$G_2^{\kappa-1}(t) = 2(\kappa-1)\kappa t^2 - (\kappa-1)$$

with roots $t_1 = -t_2 = 1/\sqrt{2\kappa}$ and

$$\tau_1 = \tau_2 = [2B(\kappa-1/2,1/2)]^{-1}.$$

Therefore using (3.68) we get that

$$B\left(\kappa-\frac{1}{2},\frac{1}{2}\right)\cosh\left(\sqrt{\frac{-c}{2\kappa}}x\right) \leq \int_{-1}^{1}(1-t^2)^{\kappa-3/2}\cosh(tx\sqrt{-c})\,dt.$$

Application of Lemma 1.3 gives the asserted inequality (3.66). Now taking in (3.66) $c = -1$ and $b = 1$, one has

$$\cosh\left(\frac{x}{\sqrt{2(p+1)}}\right) \leq \mathscr{I}_p(x) \qquad (3.69)$$

for all $p > -1/2$ and $x \in \mathbb{R}$. Clearly in (3.69) we have equality when $p = -1/2$, thus (3.69) holds for all $p \geq -1/2$ and $x \in \mathbb{R}$. From this it follows that (3.66) holds for all $\kappa \geq 1/2$ (we used the relation $\mathscr{I}_{\kappa-1}(x\sqrt{-c}) = \lambda_p(x)$). $\qquad \square$

Sharper lower and upper bounds for λ_p can be obtained using higher order quadrature formulas (3.63) with even and odd numbers of knots, respectively.

3.6 Inequalities Involving Modified Bessel Functions

In this section we present the intrinsic properties, including logarithmic convexity (concavity), of the modified Bessel functions of the first kind and some other related functions. Moreover, we establish some inequalities involving the functions under discussion. Recall that for $c = -1$ and $b = 1$ the function u_p, defined by (1.20), reduces to

$$\gamma_p(x) = \mathscr{I}_p(\sqrt{x}) = 2^p \Gamma(p+1)x^{-p/2}I_p(\sqrt{x}) = \sum_{n \geq 0} \frac{(1/4)^n}{(p+1)_n n!}x^n, \qquad (3.70)$$

where $p \neq -1, -2, \ldots$ and $x \in \mathbb{R}$. The following function

$$v_p(x) = x \frac{I_p(x)}{I_{p+1}(x)} \tag{3.71}$$

is of special interest in finite elasticity. H.C. Simpson and S.J. Spector [214] used the function v_0 to prove that a nonlinearly elastic cylinder eventually becomes unstable in uniaxial compression. Moreover, they proved that for any $p > 0$ the function v_p has application in the buckling and necking of such cylinders. For more details the interested reader is referred to the paper of H.C. Simpson and S.J. Spector [213] and to the references therein. For later use, let us record a formula

$$v_p(x) = 2(p+1) \frac{\gamma_p(x^2)}{\gamma_{p+1}(x^2)} \tag{3.72}$$

which is an immediate consequence of (3.71) and (3.70).

The results of this section may be found in [59], which is a joint work of the author with E. Neuman and is a continuation of an earlier work of E. Neuman [159]. The main results of this section provide either enhancements of some results derived in [159] (see Theorem 3.23) or they contain complementary results (see Theorem 3.24) to those included in Theorem 3.23. Moreover, as an application of the log-concavity results on modified Bessel functions of the first kind we extend van der Corput's inequality to modified Bessel functions.

Before we state and prove the main results of this section, let us recall a slight modification of Lemma 3.1 which will be used in the sequel. For the convenience we give a simple proof of this version of Lemma 3.1 to see why we can replace the condition $x \in (0,1)$ occurring in Lemma 3.1 with $x \in (0,\infty)$. For an alternative proof of Lemma 3.6 see also the paper of V. Heikkala et al. [114].

Lemma 3.6. (M. Biernacki and J. Krzyż [74], S. Ponnusamy and M. Vuorinen [186]) *Suppose that the power series $f(x) = \sum_{n \geq 0} \alpha_n x^n$ and $g(x) = \sum_{n \geq 0} \beta_n x^n$, where $\alpha_n \in \mathbb{R}$ and $\beta_n > 0$ for all $n \geq 0$, both converge for all $x \in \mathbb{R}$. Then the function f/g is (strictly) increasing (decreasing) on $(0,\infty)$ if the sequence $\{\alpha_n/\beta_n\}_{n \geq 0}$ is (strictly) increasing (decreasing).*

Proof. (Á. Baricz [46]) We consider only the case when the sequence $\{\alpha_n/\beta_n\}_{n \geq 0}$ is strictly increasing and prove that this assumption implies that f/g is strictly increasing. The other cases are similar, so we omit the details.

If $x > y > 0$, then inequality $f(x)/g(x) > f(y)/g(y)$ is equivalent to

$$\sum_{n \geq 0} \alpha_n x^n \cdot \sum_{n \geq 0} \beta_n y^n > \sum_{n \geq 0} \alpha_n y^n \cdot \sum_{n \geq 0} \beta_n x^n,$$

and this is true if

$$\alpha_i \beta_j x^i y^j + \alpha_j \beta_i x^j y^i > \alpha_j \beta_i x^i y^j + \alpha_i \beta_j x^j y^i \tag{3.73}$$

for all $i, j \in \mathbb{N}$. This can be verified by adding up the corresponding parts of these inequalities for i and then summing for j. But (3.73) can be transformed to the inequality $(\alpha_i \beta_j - \alpha_j \beta_i)(x^i y^j - x^j y^i) > 0$, which clearly holds because by assumptions we have that if $i > j$ ($i < j$ respectively) then $\alpha_i / \beta_i > \alpha_j / \beta_j$ ($\alpha_i / \beta_i < \alpha_j / \beta_j$ respectively) and $x^i y^j - x^j y^i = (xy)^j (x^{i-j} - y^{i-j}) > 0$ ($x^i y^j - x^j y^i = (xy)^j (x^{i-j} - y^{i-j}) < 0$ respectively). $\qquad\square$

It is worth mentioning that this lemma was used, among other things, to prove many interesting inequalities for the zero-balanced Gauss hypergeometric function (see the works of R. Balasubramanian et al. [36], S. Ponnusamy and M. Vuorinen [186]) and also for the generalized Bessel functions (see the first three sections of this chapter or see the papers of Á. Baricz [39, 41, 45] for more details).

Theorem 3.23. (Á. Baricz and E. Neuman [59]) *Let the numbers a and b be such that $a > 1$ and $b = 1/(4 \log a)$. Then the following assertions are true:*

(a) *If $p \geq b - 1$, then the function $x \in (0, 2b] \mapsto \mathscr{I}_p(x) \in (1, \infty)$ is strictly log-convex.*

(b) *If $a \geq e^{1/4}$ and $p \in [b - 1, 0]$, then the function $x \in (0, 2b] \mapsto I_p(x) \in (0, \infty)$ is strictly log-convex.*

(c) *If $p \geq b - 1$, $\alpha \in (0, 1)$, $x, y \in (0, 2b]$ and $x \neq y$, then the following inequalities are valid:*

$$I_p(\alpha x) < \alpha^p I_p(x) [\mathscr{I}_p(x)]^{\alpha - 1}, \tag{3.74}$$

$$\left[x I'_p(x) \right]^2 < p I_p^2(x) + x^2 I_p(x) I''_p(x), \tag{3.75}$$

$$\frac{I_p(A(x,y))}{\sqrt{I_p(x) I_p(y)}} < \left(\frac{A(x,y)}{G(x,y)} \right)^p, \tag{3.76}$$

$$v_{p-1}^2(x) - (2p - 1) v_{p-1}(x) - x^2 < 2p. \tag{3.77}$$

Proof. (a) In order to prove that the function $\mathscr{I}_p$ is strictly log-convex it suffices to show that its logarithmic derivative $\left[\log \mathscr{I}_p(x) \right]' = \mathscr{I}'_p(x) / \mathscr{I}_p(x)$ is a strictly increasing function. Using (3.70) we obtain that $\mathscr{I}'_p(x) / \mathscr{I}_p(x) = 2x \gamma'_p(x^2) / \gamma_p(x^2)$. It is clear that the last function is strictly increasing if the function $\phi(x) = \sqrt{x} \gamma'_p(x) / \gamma_p(x)$ is strictly increasing. To prove this let us represent the function ϕ as the product of three functions:

$$\phi(x) = \frac{\sqrt{x}}{a^x} \cdot \frac{a^x}{\gamma_p(x)} \cdot \gamma'_p(x),$$

where $a > 1$ and $x > 0$. One can verify that the function $x \in (0, 2b] \mapsto \sqrt{x}/a^x \in (0, \infty)$ is an increasing function. It follows from (3.70) that $x \in (0, \infty) \mapsto \gamma'_p(x) \in (0, \infty)$ is strictly increasing on its domain. To complete the proof of monotonicity of the function ϕ it suffices to show that the function $x \mapsto a^x/\gamma_p(x)$ is increasing for $x > 0$. Using the series expansion for a^x and (3.70) we obtain

$$\frac{a^x}{\gamma_p(x)} = \frac{\sum\limits_{n \geq 0} \alpha_n x^n}{\sum\limits_{n \geq 0} \beta_n x^n},$$

where $\alpha_n = (\log a)^n/n!$ and $\beta_n = (1/4)^n/[(p+1)_n n!]$ for $n \geq 0$. Clearly $\beta_n > 0$ for all n. Using Lemma 3.6, we see that the sequence $\{A_n\}_{n\geq0}$, where $A_n = \alpha_n/\beta_n$, is increasing, because

$$A_{n+1}/A_n = (4\log a)(p+n+1) = (p+n+1)/b \geq b/b = 1.$$

(b) Using (3.70), we obtain $I_p(x) = (x/2)^p \mathscr{I}_p(x)/\Gamma(p+1)$. The function $x \mapsto x^p$ is log-convex for $p \leq 0$, while the function $x \mapsto \mathscr{I}_p(x)$ is strictly log-convex for $b - 1 \leq p \leq 0$ by part (a). Thus the product in question is a strictly log-convex function in x for all $p \in [b-1, 0]$. The assumption $a \geq e^{1/4}$ implies that the interval $[b-1, 0]$ is not empty.

(c) Let us note that for $p > -1/2$ and $x > 0$, Theorem 3.23 has been established by E. Neuman [159]. Since the proofs of the inequalities (3.74)–(3.77) go along the lines introduced in [159], they are not included in this section.   □

Theorem 3.24. (Á. Baricz and E. Neuman [59]) *The following assertions hold:*

(a) *If $p > -1$, then the function $x \in (0, \infty) \mapsto \gamma_p(x) \in (0, \infty)$ is strictly log-concave.*
(b) *If $p > 0$, then the function $x \in (0, \infty) \mapsto I_p(\sqrt{x}) \in (0, \infty)$ is strictly log-concave.*
(c) *If $p > 0$, then the function $x \in (0, \sqrt{3}p] \mapsto I_p(x) \in (0, \infty)$ is strictly log-concave.*
(d) *If $p > -1$, then the function $x \in (0, \infty) \mapsto v_{p-1}(x) \in (0, \infty)$ is strictly increasing.*
(e) *If $p > -1$, $\alpha \in (0, 1)$, $x, y > 0$ and $x \neq y$, then the following inequalities are true:*

$$I_p(\alpha x) > \alpha^p I_p(x) \left[\mathscr{I}_p(x)\right]^{\alpha^2-1}, \tag{3.78}$$

$$\left[xI_p'(x)\right]^2 > 2pI_p^2(x) - xI_p(x)I_p'(x) + x^2 I_p(x)I_p''(x), \tag{3.79}$$

$$\frac{I_p(A_2(x,y))}{\sqrt{I_p(x)I_p(y)}} > \left(\frac{A_2(x,y)}{G(x,y)}\right)^p, \tag{3.80}$$

$$1 + xI_{p-1}'(x)/I_{p-1}(x) > xI_p'(x)/I_p(x), \tag{3.81}$$

$$v_{p-1}^2(x) - 2(p-1)v_{p-1}(x) - x^2 > 4p. \tag{3.82}$$

Proof. (a) In view of Lemma 3.6, the result follows immediately following the proof of Lemma 3.2 and by choosing $c = -1$ and $b = 1$. However, for the sake of completeness we give the proof here. In order to establish the assertion it suffices to prove that the logarithmic derivative of γ_p is strictly decreasing. For this aim we will employ Lemma 3.6. Making use of (3.70) we obtain

$$\frac{\gamma_p'(x)}{\gamma_p(x)} = \sum_{n\geq 0} \frac{(1/4)^{n+1}}{(p+1)_{n+1}n!} x^n \Big/ \sum_{n\geq 0} \frac{(1/4)^n}{(p+1)_n n!} x^n.$$

Taking into consideration the notation used in Lemma 3.6 we obtain

$$A_n = \alpha_n/\beta_n = \frac{1}{4(p+n+1)} \quad \text{for all } n \geq 0.$$

This in turn implies that $A_n > A_{n+1}$ for all $n \geq 0$, because $p > -1$.

(b) It follows from (3.70) that

$$I_p(\sqrt{x}) = \frac{x^{p/2}\gamma_p(x)}{2^p\Gamma(p+1)}.$$

Thus the function $x \mapsto I_p(\sqrt{x})$ is strictly log-concave as the product of the log-concave function $x \mapsto x^{p/2}$ and the strictly log-concave function γ_p. Hence the assertion follows.

(c) In order to prove that the modified Bessel function of the first kind I_p is strictly log-concave when $p > 0$ and $0 < x \leq \sqrt{3}p$, it suffices to show that its logarithmic derivative $I_p'(x)/I_p(x)$ is strictly decreasing, i.e., that $I_p(x)I_p''(x) - [I_p'(x)]^2 < 0$ holds for all $x \in (0, \sqrt{3}p]$ and $p > 0$. Making use of the inequality (3.79) we obtain

$$I_p(x)I_p''(x) - [I_p'(x)]^2 < \frac{1}{x}I_p(x)\left[I_p'(x) - 2p\frac{1}{x}I_p(x)\right].$$

In order to complete the proof of part (c) it suffices to show that

$$xI_p'(x)/I_p(x) < 2p. \tag{3.83}$$

Using Amos' inequality [10, p. 243]

$$I_{p-1}(x)I_{p+1}(x) - I_p^2(x) < 0, \tag{3.84}$$

which holds for all $x > 0$ and $p \geq 0$, together with formulas

$$I_{p-1}(x) = (p/x)I_p(x) + I_p'(x)$$

and

$$I_{p+1}(x) = I_p'(x) - (p/x)I_p(x)$$

the inequality (3.84) can be written as $xI_p'(x)/I_p(x) < \sqrt{x^2+p^2}$. Since $0 < x \leq \sqrt{3}p$, we have $xI_p'(x)/I_p(x) < 2p$ which completes the proof of (3.83).

(d) Making use of (3.72), we obtain $v_p(\sqrt{x}) = 2(p+1)\gamma_p(x)/\gamma_{p+1}(x)$. We need to use Lemma 3.6 again with

$$\alpha_n = \frac{2(p+1)(1/4)^n}{(p+1)_n n!} \quad \text{and} \quad \beta_n = \frac{(1/4)^n}{(p+2)_n n!} \quad \text{for all } n \geq 0$$

to conclude that the sequence $\{\alpha_n/\beta_n\}_{n\geq 0}$, where $\alpha_n/\beta_n = 2(p+n+1)$ is strictly increasing, hence the function $x \mapsto v_{p-1}(\sqrt{x})$ is strictly increasing. This in turn implies that the assertion (d) holds true.

(e) Since γ_p is strictly log-concave, the inequality

$$\gamma_p(\alpha x + (1-\alpha)y) > [\gamma_p(x)]^\alpha [\gamma_p(y)]^{1-\alpha} \tag{3.85}$$

holds true for all $\alpha \in (0,1)$, $p > -1$, $x,y \geq 0$ and $x \neq y$. Letting $y = 0$ and taking into account that $\gamma_p(0) = 1$ we obtain $\gamma_p(\alpha x) > [\gamma_p(x)]^\alpha$. This in conjunction with (3.70) gives

$$I_p(\sqrt{\alpha x}) > (\sqrt{\alpha})^p I_p [\gamma_p(x)]^{\alpha-1}.$$

Replacing x by x^2, α by α^2, and utilizing (3.70) we obtain the inequality (3.78). For the proof of the inequality (3.79) we appeal again to the logarithmic concavity of the function γ_p. This in turn implies that the function $x \mapsto \gamma_p'(x)/\gamma_p(x)$ is decreasing. Utilizing (3.70), one obtains

$$\frac{\gamma_p'(x)}{\gamma_p(x)} = \frac{I_p'(\sqrt{x})}{2\sqrt{x}I_p(\sqrt{x})} - \frac{p}{2x}.$$

Hence

$$\left[\frac{\gamma_p'(x)}{\gamma_p(x)}\right]' = \frac{I_p''(\sqrt{x})}{4xI_p(\sqrt{x})} - \frac{I_p'(\sqrt{x})}{4x\sqrt{x}I_p(\sqrt{x})} - \frac{1}{4x}\left[\frac{\gamma_p'(x)}{\gamma_p(x)}\right]^2 + \frac{p}{2x^2} < 0.$$

Replacing x by x^2 completes the proof of (3.79). In order to establish the inequality (3.80) we let $\alpha = 1/2$ in (3.85) and next apply (3.70) to obtain

$$I_p^2\left(\sqrt{\frac{x+y}{2}}\right) > \left(\frac{x+y}{2\sqrt{xy}}\right)^p I_p(\sqrt{x})I_p(\sqrt{y}).$$

Replacing x by x^2 and y by y^2 we obtain the assertion (3.80). Inequality (3.81) is an immediate consequence of the property of v_{p-1} established in part (d). This in conjunction with (3.71) gives $[xI_{p-1}(x)/I_p(x)]' > 0$. The assertion (3.81) now follows. For the proof of (3.82) we divide both sides of (3.79) by $I_p(x)$ to obtain

$$[xI_p'(x)]^2/I_p(x) + xI_p'(x) - 2pI_p(x) > x^2 I_p''(x),$$

which holds for all $x > 0$ and $p > -1$. Combining this with the differential equation (1.6) we obtain

$$h_p^2(x) + h_p(x) - 2p > x^2 + p^2 - h_p(x), \qquad (3.86)$$

where $h_p(x) = xI_p'(x)/I_p(x)$. Taking into account that $xI_p'(x) = xI_{p-1} - pI_p(x)$, we obtain $h_p(x) = v_{p-1}(x) - p$. Substituting this into (3.86) gives the desired result. This completes the proof. $\qquad\square$

Corollary 3.9. (Á. Baricz and E. Neuman [59]) *Let the numbers a and b be such that $1 < a < e^{1/4}$ and $b = 1/(4\log a)$. If $x, y \in (0, 2b]$ and $p \geq b - 1$, then the following inequalities are valid:*

$$\frac{I_p(A(x,y))}{\sqrt{I_p(x)I_p(y)}} \leq \left(\frac{A(x,y)}{G(x,y)}\right)^p \leq \left(\frac{A_2(x,y)}{G(x,y)}\right)^p \leq \frac{I_p(A_2(x,y))}{\sqrt{I_p(x)I_p(y)}}. \qquad (3.87)$$

In each of these inequalities equality holds if and only if $x = y$.

Proof. From $a < e^{1/4}$ it follows that $p \geq b - 1 > 0$. Making use of (3.76) and (3.80), we obtain (3.87). $\qquad\square$

Before we state the next corollary, let us introduce a two-parameter function

$$v_{p,q} : (0, \infty) \to (0, \infty) \quad \text{by} \quad v_{p,q}(x) = xI_p(x)/I_q(x).$$

Let us note that $v_{p,p+1} = v_p$. For this function we have the following.

Corollary 3.10. (Á. Baricz and E. Neuman [59]) *Let $a > e^{1/4}$ and $b = 1/(4\log a)$. If $p \geq 2b/\sqrt{3}$ and $b - 1 \leq q \leq 0$, then the function $x \in (0, 2b] \mapsto v_{p,q}(x) \in (0, \infty)$ is strictly log-concave.*

Proof. We establish the logarithmic concavity of the function $x \mapsto v_{p,q}(x)$ by showing that each of the following functions x, $x \mapsto I_p(x)$, and $x \mapsto 1/I_q(x)$ is log-concave. Clearly the identity function $x \mapsto x$ is log-concave for all $x > 0$. Using part (c) of Theorem 3.24 we see that the function $x \mapsto I_p(x)$ is strictly log-concave for all $0 < x \leq p\sqrt{3}$ provided $p > 0$. Finally, it follows from part (b) of Theorem 3.23 that the function $x \mapsto 1/I_q(x)$ is strictly log-concave if $0 < x \leq 2b$ and $b - 1 \leq q \leq 0$. Since $2b \leq p\sqrt{3}$, the assertion follows. $\qquad\square$

Remark 3.10. During the course of writing this monograph we have found the Ph.D. thesis of D. Hammarwall [110] in which the author used part (a) (for $p = 0$) and part (b) of Theorem 3.24 to prove that the minimum mean squared-error estimation is in fact an increasing function. For more details we refer the reader to [110, Theorem 6.4].

Recall that in 1992 E. Neuman [159] proved that the function $\mathscr{I}_p : \mathbb{R} \to [1,\infty)$, defined by

$$\mathscr{I}_p(x) = {}_0F_1(p+1,x^2/4) = 2^p\Gamma(p+1)x^{-p}I_p(x) = \sum_{n\geq 0}\frac{(1/4)^n}{(p+1)_n n!}\cdot x^{2n}, \quad (3.88)$$

is strictly log-convex, when $p > -1/2$. Observe that in particular we obtain

$$\mathscr{I}_{-1/2}(x) = \sqrt{\pi/2}\cdot x^{1/2}I_{-1/2}(x) = \cosh x,$$

thus actually $\mathscr{I}_p$ is strictly log-convex for all $p \geq -1/2$.

In 1996 C. Giordano et al. [100] using Hölder-Rogers's inequality for integrals claimed that all derivatives of the function $x \in (0,\infty) \mapsto \mathscr{I}_p(x) \in [1,\infty)$ are strictly log-convex for all $p > -1/2$. In 1997 M.E.H. Ismail [119] proved that in fact just the derivatives of even order of the function $\mathscr{I}_p$ will be strictly log-convex. Recall that $\mathscr{I}_{-1/2}(x) = \cosh x$, thus in this case it is clear that for all $k \in \mathbb{N}$ the function

$$x \mapsto \mathscr{I}_{-1/2}^{(2k)}(x) = \cosh x$$

is strictly log-convex on $\mathbb{R}$ and

$$x \mapsto \mathscr{I}_{-1/2}^{(2k+1)}(x) = \sinh x$$

is strictly log-concave on $(0,\infty)$. The above results suggest the following.

Theorem 3.25. (Á. Baricz [54]) *If $p > -1$, then the following assertions are true:*

(a) *The function $x \mapsto \mathscr{I}_p(x)$ is strictly log-convex for $0 < x \leq \sqrt{2(p+1)(p+2)}$.*
(b) *The function $x \mapsto \mathscr{I}_p'(x)$ is strictly log-concave for $0 < x \leq \sqrt{2(p+2)}$.*

Proof. (a) In what follows we prove that under the stated assumptions the function $x \mapsto \mathscr{I}_p'(x)/\mathscr{I}_p(x)$ is strictly increasing. From the theory of Bessel functions the recurrence formula (see part (g) of Remark 1.1)

$$xI_{p+1}(x) = xI_p'(x) - pI_p(x) \tag{3.89}$$

is well-known. Thus (3.88) yields

$$\mathscr{I}_p'(x) = 2^p\Gamma(p+1)\left[-px^{-p-1}I_p(x) + x^{-p}I_p'(x)\right] = 2^p\Gamma(p+1)x^{-p}I_{p+1}(x). \tag{3.90}$$

Using again (3.88) it suffices to show that $[I_{p+1}(x)/I_p(x)]' > 0$, i.e. in view of (3.89) the following inequality holds

$$[I_{p+1}(x)]^2 - I_p(x)I_{p+2}(x) < \frac{1}{x}I_p(x)I_{p+1}(x). \tag{3.91}$$

Recall the following recurrence formula from the book of G.N. Watson [227, p. 79]:

$$2(p+1)I_{p+1}(x) = x[I_p(x) - I_{p+2}(x)].$$

Together with (3.89) it yields the inequality (see C.M. Joshi and S.K. Bissu [126])

$$\frac{I_{p+1}(x)}{I_p(x)} = \frac{x}{2(p+1)}\left[1 - \frac{I_{p+2}(x)}{I_p(x)}\right] < \frac{x}{2(p+1)}, \tag{3.92}$$

since for $p > -1$ and $x > 0$ the modified Bessel function is positive. On the other hand, it is known that for all $p > -1$ and $x > 0$ the Turán-type inequality[4]

$$[\mathscr{I}_{p+1}(x)]^2 \le \mathscr{I}_p(x)\mathscr{I}_{p+2}(x) \tag{3.93}$$

or equivalently

$$[I_{p+1}(x)]^2 - I_p(x)I_{p+2}(x) < \frac{[I_{p+1}(x)]^2}{p+2}$$

holds. Using (3.92) and (3.93) the required inequality (3.91) follows.

(b) First observe that for $p > -1$ the power series

$$\mathscr{I}_p'(x) = [2^p\Gamma(p+1)x^{-p}I_p(x)]' = \sum_{n\ge1}\frac{(1/4)^{n-1}}{2(p+1)_n(n-1)!}\cdot x^{2n-1}$$

has only positive coefficients, therefore it makes sense to study the log-concavity of $x \mapsto \mathscr{I}_p'(x)$. Moreover, it is enough to show that $x \mapsto \mathscr{I}_p''(x)/\mathscr{I}_p'(x)$ is a strictly decreasing function on $(0, \sqrt{2(p+2)}]$. Application of (3.89) and (3.90) yields

$$\frac{\mathscr{I}_p''(x)}{2^p\Gamma(p+1)} = x^{-p}\left[I_{p+2}(x) + \frac{1}{x}I_{p+1}(x)\right]. \tag{3.94}$$

Hence from (3.90) and (3.94) we just need to prove that $f_p : (0, \sqrt{2(p+2)}] \to (0,\infty)$, defined by

$$f_p(x) = 1/x + I_{p+2}(x)/I_{p+1}(x),$$

is strictly decreasing for all $p > -1$. For this observe that from the following recurrence relation (see the book of G.N. Watson [227, p. 79])

$$xI_{p-1}(x) = xI_p'(x) + pI_p(x)$$

we have that

$$x^2[I_{p+1}(x)]^2 f_p'(x) = x^2 I_{p+1}(x)I_{p+2}'(x) - x^2 I_{p+2}I_{p+1}'(x) - [I_{p+1}(x)]^2 < 0,$$

---

[4] This Turán type inequality was established by V.K. Thiruvenkatachar and T.S. Nanjundiah [224], and was also proved by C.M. Joshi and S.K. Bissu [125]. Recently it was rediscovered and generalized by the author in [51, 55].

and this is equivalent to the inequality

$$\frac{xI'_{p+2}(x)}{I_{p+2}(x)} < \frac{xI'_{p+1}(x)}{I_{p+1}(x)} + \frac{I_{p+1}(x)}{xI_{p+2}(x)}. \tag{3.95}$$

Using inequality (3.81) it is clear that

$$\frac{xI'_{p+2}(x)}{I_{p+2}(x)} < 1 + \frac{xI'_{p+1}(x)}{I_{p+1}(x)}$$

holds for all $p > -3$ and $x > 0$. Finally, in view of (3.95) it remains to prove that $xI_{p+2}(x) < I_{p+1}(x)$ holds for all $p > -1$ and $x > 0$. Using the inequality (3.92) for $p+1$, i.e.

$$2(p+2)I_{p+2}(x) < xI_{p+1}(x),$$

where $p > -2$ and $x > 0$, the asserted result follows. $\qquad\square$

The following inequality involving hyperbolic sine and cosine functions of J.G. van der Corput (see D.S. Mitrinović [156, p. 270]) is of special interest in this section:

$$|\cosh a - \cosh b| \geq |a - b|\sqrt{\sinh a \sinh b} \quad \text{for all} \ \ a, b \geq 0. \tag{3.96}$$

Taking into account the relation $\mathscr{I}_{-1/2}(x) = \cosh x$, the inequality (3.96) may be written as

$$|\mathscr{I}_{-1/2}(a) - \mathscr{I}_{-1/2}(b)| \geq |a - b|\sqrt{\mathscr{I}'_{-1/2}(a)\mathscr{I}'_{-1/2}(b)} \quad \text{for all} \ \ a, b \geq 0.$$

The following result is an extension of van der Corput's inequality (3.96).

Theorem 3.26. (Á. Baricz [54]) *If* $a, b \in [0, \sqrt{2(p+2)}]$ *and* $p > -1$, *then*

$$|\mathscr{I}_p(a) - \mathscr{I}_p(b)| \geq |a - b|\sqrt{\mathscr{I}'_p(a)\mathscr{I}'_p(b)}. \tag{3.97}$$

Proof. Without loss of generality it is enough to show that inequality (3.97) holds for $a < b$. Observe that the function $x \mapsto \mathscr{I}_p(x)$ is convex on $[a, b]$, since as power series has only positive coefficients. Now taking into account that in virtue of Theorem 3.25 the function $x \mapsto \mathscr{I}'_p(x)$ is log-concave on $[0, \sqrt{2(p+2)}]$, from the classical Hermite-Hadamard inequality (1.39) applied to the function $x \mapsto \mathscr{I}'_p(x)$, we have

$$\frac{\mathscr{I}_p(b) - \mathscr{I}_p(a)}{b - a} = \frac{1}{b - a}\int_a^b \mathscr{I}'_p(x)\,dx \geq \mathscr{I}'_p\left(\frac{a+b}{2}\right) \geq \sqrt{\mathscr{I}'_p(a)\mathscr{I}'_p(b)}.$$

$\qquad\square$

Remark 3.11. Since for all $n \geq 1$ and $p \neq -1, -2, \ldots$ one has $(p+1)(p+2)_{n-1} = (p+1)_n$, from (3.88) we have the following differentiation formula for the function $\mathscr{I}_p$, namely

$$2(p+1)\mathscr{I}'_p(x) = x\mathscr{I}_{p+1}(x), \tag{3.98}$$

where $x \in \mathbb{R}$ and $p \neq -1, -2, \ldots$. Now taking into account the relations $\mathscr{I}_{-1/2}(x) = \cosh x$ and $\mathscr{I}_{1/2}(x) = \sqrt{\pi/2} \cdot x^{-1/2} I_{1/2}(x) = (\sinh x)/x$, the inequality (3.96) may be written as

$$|\mathscr{I}_{-1/2}(a) - \mathscr{I}_{-1/2}(b)| \geq |a-b|\sqrt{ab\mathscr{I}_{1/2}(a)\mathscr{I}_{1/2}(b)},$$

and from (3.98) it follows that the inequality (3.97) is equivalent to

$$2(p+1)|\mathscr{I}_p(a) - \mathscr{I}_p(b)| \geq |a-b|\sqrt{ab\mathscr{I}_{p+1}(a)\mathscr{I}_{p+1}(b)}. \tag{3.99}$$

Remark 3.12. First note that when $p = -1/2$, the inequality (3.97) or (3.99) reduces to the inequality (3.96) with the interval of validity $[0, \sqrt{3}]$ for a and b. Secondly, we note that in particular we have (see (2.30) for $z = x$)

$$\mathscr{I}_{3/2}(x) = 3\sqrt{\pi/2} \cdot x^{-3/2} I_{3/2}(x) = -3\left(\frac{\sinh x}{x^3} - \frac{\cosh x}{x^2}\right),$$

thus we obtain from (3.99) for $p = 1/2$ the following inequality:

$$|b\sinh a - a\sinh b| \geq |a-b|\sqrt{(a\cosh a - \sinh a)(b\cosh b - \sinh b)},$$

where $a, b \in [0, \sqrt{5}]$.

3.7 Miscellaneous Inequalities Involving the Generalized Bessel Functions

The sine and cosine functions are particular cases of normalized Bessel functions, while the hyperbolic sine and hyperbolic cosine functions are particular cases of normalized modified Bessel functions. Thus it is natural to generalize some formulas and inequalities involving these elementary functions to normalized Bessel functions and normalized modified Bessel functions, respectively. In this section our aim is to extend some classical inequalities, like Mitrinović's inequality, Mahajan's inequality, Jordan's inequality and Redheffer's inequality, for the function λ_p, using an adequate integral representation of this function.

This section is organized as follows: in the first subsection we extend Mahajan's inequality using a result of L. Lorch and M.E. Muldoon [141]. On the other hand we offer the hyperbolic counterpart of Mitrinović's inequality [156, p. 240]. In the second subsection we use Redheffer's inequality to obtain a new lower bound for the function λ_p, while in the third subsection we extend Cusa's inequality to generalized and normalized Bessel functions. In the fourth subsection we obtain new bounds for the function λ_p by extending three improvements of the well-known Jordan's inequality. Finally, in the last subsection we prove that the integral

$$\varsigma_p(x) = \int_0^x \lambda_p(t)\,dt$$

is sub-additive (super-additive) under certain conditions on parameters b, c, p. This result extends the recent result of S. Koumandos [137], who among other things proved that the sine integral is sub-additive on $[0, \infty)$. We note that all results of this section may be found in the recent paper of the author [49].

3.7.1 Mitrinović's Inequality and Mahajan's Inequality

In 1979 A. Mahajan [143] extended a result of D.S. Mitrinović [156, p. 240] by proving that if $p > -1$, then

$$(x+1)\mathscr{J}_p\left(\frac{\pi}{x+1}\right) - x\mathscr{J}_p\left(\frac{\pi}{x}\right) > 1 \text{ for all } x > \pi\frac{\pi + \sqrt{\pi^2 + 32(p+2)}}{16(p+2)}. \quad (3.100)$$

D.S. Mitrinović has proved (3.100) in the case when $p = -1/2$, i.e.

$$(x+1)\cos\left(\frac{\pi}{x+1}\right) - x\cos\left(\frac{\pi}{x}\right) > 1, \quad (3.101)$$

but just for $x \geq \sqrt{3} \simeq 1.732050808\ldots$, while A. Mahajan's generalization yields a better interval of the validity for (3.101), namely $x > 1.407014637\ldots$. In 1987 L. Lorch and M.E. Muldoon [141] proved that the largest interval of validity for (3.101) is $(1, \infty)$ and that for (3.100) it is (x_1, ∞), where x_1 is the largest root of the equation $\varphi_1(x+1) = \varphi_1(x)$ with

$$\varphi_1(x) = x[\mathscr{J}_p(\pi/x) - 1].$$

In the next theorem our aim is to extend inequality (3.100) to the function λ_p. In the case of $c > 0$ we use the method of L. Lorch and M.E. Muldoon [141] mutatis mutandis, while for $c < 0$ we give a different proof. As we can see the analogous of inequality (3.101) for the hyperbolic cosine function holds just if $x \in (-1, 0)$, otherwise the inequality is reversed.

Theorem 3.27. (Á. Baricz [49]) *If $\kappa, c > 0$, then*

$$(x+1)\lambda_p\left(\frac{\pi}{x+1}\right) - x\lambda_p\left(\frac{\pi}{x}\right) > 1 \quad (3.102)$$

holds for all $x < -x_2 - 1$ or $x > x_2$, where x_2 is the largest root of the equation $\varphi_2(x+1) = \varphi_2(x)$ with $\varphi_2(x) = x[\lambda_p(\pi/x) - 1]$. Moreover, if $\kappa > 0$ and $c < 0$, then (3.102) holds for all $x \in (-1,0)$, while for $x \in (-\infty, -1) \cup (0,\infty)$, the inequality (3.102) is reversed.

Proof. Let us consider the function $\varphi_3 : \mathbb{R} \setminus \{-1,0\} \to \mathbb{R}$ defined by

$$\varphi_3(x) = (x+1)\lambda_p\left(\frac{\pi}{x+1}\right) - x\lambda_p\left(\frac{\pi}{x}\right) - 1.$$

It is easy to verify that $\varphi_3(x - 1/2) = \varphi_3(-x - 1/2)$ for all $x \neq -1, 0$. Thus clearly the graph of the function φ_3 is symmetric with respect to the straight line $x = -1/2$. Now let us distinguish the cases when $c > 0$ and $c < 0$.

Suppose that $c > 0$. It is known from part (d) of Lemma 1.1 that

$$[x^{-p}w_p(x)]' = -c \cdot x^{-p}w_{p+1}(x) \tag{3.103}$$

for all $\kappa > 0$ and $c, x \in \mathbb{R}$. Observe that using (1.19), (1.20) and (1.25) we obtain $\lambda_p(x) = 2^p \Gamma(\kappa) x^{-p} w_p(x)$. Thus from (3.103) one has

$$\lambda_p'(x) = -c \cdot 2^p \Gamma(\kappa) x^{-p} w_{p+1}(x),$$

$$\lambda_p''(x) = -c \cdot 2^p \Gamma(\kappa) x^{-p-1}[w_{p+1}(x) - cxw_{p+2}(x)],$$

so λ_p is a (decreasing and) concave function for all sufficiently small positive x, say $x \in (0, \alpha)$. Due to L. Lorch and M.E. Muldoon [141] we know that if a function f is concave on $(0, \beta]$, then

$$\frac{f(r)}{r} - \frac{f(s)}{s} > \left(\frac{1}{r} - \frac{1}{s}\right) f(0) \quad \text{for all } 0 < r < s \leq \beta. \tag{3.104}$$

Moreover, if f is continuous, then this inequality remains true for certain r and s, one or both possibly greater than β, provided that for every $s > \beta$ we restrict our attention to those values of r less than the smallest value of r for which (3.104) becomes an equality. Thus from (3.104) we obtain that

$$\frac{\lambda_p(r)}{r} - \frac{\lambda_p(s)}{s} > \left(\frac{1}{r} - \frac{1}{s}\right) \lambda_p(0) \quad \text{for all } 0 < r < s < \alpha.$$

Putting $r = \pi/(x+1)$ and $s = \pi/x$, we obtain $\varphi_3(x) > 0$ for all $x > x_2$. Clearly by the symmetry of the function φ_3 we have $\varphi_3(x) > 0$ for all $x < -x_2 - 1$.

Now assume that $c < 0$. If $x \in (-1,0)$, then

$$\varphi_3(x) = (x+1) \cdot \sum_{n \geq 0} b_n \left(\frac{\pi}{x+1}\right)^{2n} - x \cdot \sum_{n \geq 0} b_n \left(\frac{\pi}{x}\right)^{2n} - 1$$

$$= (x+1) \cdot \sum_{n\geq 1} b_n \left(\frac{\pi}{x+1}\right)^{2n} - x \cdot \sum_{n\geq 1} b_n \left(\frac{\pi}{x}\right)^{2n}$$

$$= \sum_{n\geq 1} b_n \pi^{2n} \left[\frac{1}{(x+1)^{2n-1}} - \frac{1}{x^{2n-1}}\right] > 0.$$

From the symmetry of the function φ_3 it is enough to show that $\varphi_3(x) < 0$ for all $x > 0$. By Lemma 1.2 we know that

$$(4\kappa)u_p'(x) = (-c)u_{p+1}(x), \tag{3.105}$$

for all $x \in \mathbb{R}$ and $\kappa \neq 0, -1, -2, \ldots$. From the l'Hospital rule and (3.105) it is easy to verify that

$$\lim_{x\to\infty} \varphi_3(x) = \lim_{x\to\infty} x \left[\lambda_p\left(\frac{\pi}{x+1}\right) - \lambda_p\left(\frac{\pi}{x}\right)\right]$$

$$= \lim_{x\to\infty} \pi \left[\frac{x^2}{(x+1)^2}\lambda_p'\left(\frac{\pi}{x+1}\right) - \lambda_p'\left(\frac{\pi}{x}\right)\right]$$

$$= \lim_{x\to\infty} 2\pi \left(-\frac{c}{4\kappa}\right) \left[\frac{x^2}{(x+1)^3}\lambda_{p+1}\left(\frac{\pi}{x+1}\right) - \frac{1}{x}\lambda_{p+1}\left(\frac{\pi}{x}\right)\right] = 0.$$

In what follows we prove that the function φ_3 is strictly concave on $(0,\infty)$. Thus using the above mentioned limit (the graph of the function φ_3 is tangent to the x-axis at infinity) we may conclude that φ_3 is negative, which completes the proof. For this let us consider the function $\varphi_4(x) = x^3 \lambda_p''(x)$. Since (3.105) implies that

$$\varphi_4'(x) = 2b_1(p)x^2$$
$$\times \left[3\lambda_{p+1}(x) + 12b_1(p+1)x^2\lambda_{p+2}(x) + 4b_1(p+1)b_1(p+2)x^4\lambda_{p+3}(x)\right],$$

so $\varphi_4'(x) > 0$ for all $x > 0$, it follows that φ_4 is strictly increasing (here $b_1(p) = (-c)/(4\kappa)$). Finally

$$\varphi_3''(x) = \frac{\pi^2}{(x+1)^3}\lambda_p''\left(\frac{\pi}{x+1}\right) - \frac{\pi^2}{x^3}\lambda_p''\left(\frac{\pi}{x}\right),$$

hence the required result follows. $\qquad\square$

Remark 3.13. Taking $b = 1$ and $c = -1$ in Theorem 3.27 we get the hyperbolic counterpart of the inequalities (3.100) and (3.101), namely

$$(x+1)\mathscr{I}_p\left(\frac{\pi}{x+1}\right) - x\mathscr{I}_p\left(\frac{\pi}{x}\right) > 1 \quad \text{for all } x \in (-1,0), \, p > -1.$$

In particular, for $p = -1/2$ we have

$$(x+1)\cosh\left(\frac{\pi}{x+1}\right) - x\cosh\left(\frac{\pi}{x}\right) > 1, \quad \text{whenever} \quad x \in (-1,0).$$

When $x < -1$ or $x > 0$ both inequalities are reversed.

3.7.2 Redheffer's Inequality

In 1969 R. Redheffer [198] established the following inequality:

$$\mathcal{J}_{1/2}(x) = \frac{\sin x}{x} \geq \frac{\pi^2 - x^2}{\pi^2 + x^2} \quad \text{for all} \quad x \in \mathbb{R}. \tag{3.106}$$

Throughout this section (as in the first chapter), it should be understood that functions such as $(\sin x)/x$, which have removable singularities at $x = 0$, have had these singularities removed in statements like (3.106). Recall that in 2004 E. Neuman [159, Theorem 2.2] proved, using Gauss-Gegenbauer quadrature formula, that if $p > -1/2$, then for all $x \in [-\pi/2, \pi/2]$

$$\frac{1}{3(p+1)}\left[2p+1+(p+2)\cos\left(\sqrt{\frac{3}{2(p+2)}}x\right)\right] \geq \mathcal{J}_p(x) \geq \cos\left(\frac{x}{\sqrt{2(p+1)}}\right). \tag{3.107}$$

Note that we have equality in (3.107) when $p = -1/2$. Observe that

$$\mathcal{J}_{\kappa-1}(x\sqrt{c}) = \lambda_p(x),$$

thus changing in the previous inequality p with $\kappa - 1$ and x with $x\sqrt{c}$, we deduce that (3.65) holds, i.e. if $c \in [0,1]$ and $\kappa \geq 1/2$, then for all $x \in [-\pi/2, \pi/2]$ we have

$$\frac{1}{3\kappa}\left[2\kappa-1+(\kappa+1)\cos\left(\sqrt{\frac{3c}{2(\kappa+1)}}x\right)\right] \geq \lambda_p(x) \geq \cos\left(\sqrt{\frac{c}{2\kappa}}x\right). \tag{3.108}$$

Using (3.106) we can find another lower bound for the function λ_p, which is valid for all real numbers. The key tool will be the Sonine integral formula (see G.N. Watson [227, p. 373]), which expresses any Bessel function in terms of an integral involving a Bessel function of lower order.

Theorem 3.28. (Á. Baricz [49]) *If $c \geq 0$ and $\kappa \geq 3/2$, then*

$$\lambda_p(x) \geq \frac{\pi^2 - cx^2}{\pi^2 + cx^2} \quad \text{for all} \quad x \in \mathbb{R}.$$

Proof. From the Sonine integral formula [227, p. 373] for Bessel functions

$$J_{q+p+1}(x) = \frac{x^{p+1}}{2^p \Gamma(p+1)} \int_0^{\pi/2} J_q(x\sin\theta)\sin^{q+1}\theta\cos^{2p+1}\theta d\theta,$$

which holds for all $p,q > -1$ and all $x \in \mathbb{R}$, we obtain immediately the following formula which will be useful in the sequel:

$$\mathscr{I}_{q+p+1}(x) = \frac{2}{B(p+1,q+1)} \int_0^{\pi/2} \mathscr{I}_q(x\sin\theta)\sin^{2q+1}\theta\cos^{2p+1}\theta d\theta, \quad (3.109)$$

where $p,q > -1$, $x \in \mathbb{R}$ and $B(p,q) = \Gamma(p)\Gamma(q)/\Gamma(p+q)$ is Euler's beta function. Changing in (3.109) p with $p - 1/2$ and taking $q = 1/2$ one has for all $p > -1/2$ and all $x \in \mathbb{R}$ that

$$\mathscr{I}_{p+1}(x) = \frac{2}{B\left(p+\frac{1}{2},\frac{3}{2}\right)} \int_0^{\pi/2} \mathscr{I}_{1/2}(x\sin\theta)\sin^2\theta\cos^{2p}\theta d\theta. \quad (3.110)$$

Using (3.106) it follows that for all $\theta \in [0,\pi/2]$ and $x \in \mathbb{R}$

$$\mathscr{I}_{1/2}(x\sin\theta)\sin^2\theta\cos^{2p}\theta \geq \frac{\pi^2 - x^2\sin^2\theta}{\pi^2 + x^2\sin^2\theta}\sin^2\theta\cos^{2p}\theta \geq \frac{\pi^2 - x^2}{\pi^2 + x^2}\sin^2\theta\cos^{2p}\theta.$$

Thus using (3.110), and (3.106) again we obtain

$$\mathscr{I}_{p+1}(x) \geq \frac{\pi^2 - x^2}{\pi^2 + x^2} \quad \text{for all } x \in \mathbb{R} \text{ and } p \geq -1/2.$$

Finally using again $\mathscr{I}_{\kappa-1}(x\sqrt{c}) = \lambda_p(x)$ and changing in the previous inequality p with $\kappa - 2$ and x with $x\sqrt{c}$, we get the required result.   □

Remark 3.14. Recently C.P. Chen et al. [81] established the following Redheffer-type inequality by using mathematical induction and infinite product representation of cosine function:

$$\mathscr{I}_{-1/2}(x) = \cos x \geq \frac{\pi^2 - 4x^2}{\pi^2 + 4x^2} \quad \text{for all } x \in \left[-\frac{\pi}{2},\frac{\pi}{2}\right]. \quad (3.111)$$

Observe that if we use (3.109) again, by changing p with $p - 1/2$ and taking $q = -1/2$ one has for all $p > -1/2$ and all $x \in \mathbb{R}$ that

$$\mathscr{I}_p(x) = \frac{2}{B\left(p+\frac{1}{2},\frac{1}{2}\right)} \int_0^{\pi/2} \mathscr{I}_{-1/2}(x\sin\theta)\cos^{2p}\theta d\theta. \quad (3.112)$$

Following the proof of Theorem 3.28 it is easy to verify that from (3.111) and (3.112) we have

$$\mathscr{J}_p(x) = \cos x \geq \frac{\pi^2 - 4x^2}{\pi^2 + 4x^2} \quad \text{for all } p \geq -\frac{1}{2} \text{ and all } x \in \left[-\frac{\pi}{2}, \frac{\pi}{2}\right].$$

Thus if we change p with $\kappa - 1$ and x with $x\sqrt{c}$, we get another lower bound for the function λ_p, namely if $c \geq 0$ and $\kappa \geq 1/2$, then for all $x \in [-\pi/2, \pi/2]$ one has

$$\lambda_p(x) \geq \frac{\pi^2 - 4cx^2}{\pi^2 + 4cx^2}.$$

It is worth mentioning that for $c \in [0,1]$ and $\kappa \geq 1/2$, the lower bound in (3.108) is better than the above lower bound, since direct application of (3.111) yields

$$\cos\left(\sqrt{\frac{c}{2\kappa}}x\right) \geq \frac{\pi^2 - 2cx^2/\kappa}{\pi^2 + 2cx^2/\kappa} \geq \frac{\pi^2 - 4cx^2}{\pi^2 + 4cx^2}.$$

It is worth also mentioning here that using the idea of the proof Theorem 3.28 we can prove the following results:

Theorem 3.29. (Á. Baricz [54]) *Suppose that $c \in [0,1]$ and $\kappa \geq 1/2$. Then the function λ_p is increasing on $[-\pi, 0]$ and is decreasing on $[0, \pi]$. Moreover, λ_p is concave, and consequently it is log-concave on $[-\pi/2, \pi/2]$. In addition, the inequality $\lambda_{p+1}(x) \geq \lambda_p(x)$ holds, provided that $x \in [-\pi, \pi]$.*

Proof. Since $\mathscr{J}_{\kappa-1}(x\sqrt{c}) = \lambda_p(x)$, clearly it is enough to show that if $p \geq -1/2$, then the function $\mathscr{J}_p$ is increasing on $[-\pi, 0]$, decreasing on $[0, \pi]$, concave on $[-\pi/2, \pi/2]$, and in addition the inequality $\mathscr{J}_{p+1}(x) \geq \mathscr{J}_p(x)$ holds for all $x \in [-\pi, \pi]$. The concavity of $\mathscr{J}_p$ was proved by the author in [51, Theorem 6], however, for the sake of completeness we sketch its proof.

The cosine function is increasing on $[-\pi, 0]$ and decreasing on $[0, \pi]$. From (3.112) clearly $\mathscr{J}_p$ is also increasing on $[-\pi, 0]$ and decreasing on $[0, \pi]$. Namely, if $x, y \in [0, \pi]$ such that $x \geq y$, then we have $\mathscr{J}_{-1/2}(x) \leq \mathscr{J}_{-1/2}(y)$. Changing in this inequality x with $x \sin \theta$ and y with $y \sin \theta$, where $\theta \in [0, \pi/2]$, multiplying both sides with $\cos^{2p} \theta$, and integrating, in view of (3.112) we obtain that $\mathscr{J}_p(x) \leq \mathscr{J}_p(y)$ for all $x, y \in [0, \pi], x \geq y$ and $p \geq -1/2$. Similarly, it can be shown that the function $\mathscr{J}_p$ is increasing on $[-\pi, 0]$ for all $p \geq -1/2$. On the other hand, since the cosine function is concave on $[-\pi/2, \pi/2]$, it follows that the integrand in (3.112) is concave in x. More precisely, since the cosine function is concave on $[-\pi/2, \pi/2]$, one has

$$\mathscr{J}_{-1/2}(\alpha x + (1-\alpha)y) \geq \alpha \mathscr{J}_{-1/2}(x) + (1-\alpha)\mathscr{J}_{-1/2}(y),$$

where $\alpha \in [0,1]$ and $x, y \in [-\pi/2, \pi/2]$. Changing x with $x \sin \theta$, y with $y \sin \theta$, multiplying both sides of the above inequality with $\cos^{2p} \theta$, and integrating, from (3.112) it follows that

$$\mathscr{J}_p(\alpha x + (1-\alpha)y) \geq \alpha \mathscr{J}_p(x) + (1-\alpha)\mathscr{J}_p(y)$$

holds for all $p \geq -1/2$, $\alpha \in [0,1]$ and $x,y \in [-\pi/2, \pi/2]$. Consequently, $\mathscr{J}_p$ is indeed concave on $[-\pi/2, \pi/2]$.

It remains to prove that $\mathscr{J}_{p+1}(x) \geq \mathscr{J}_p(x)$ for all $x \in [-\pi, \pi]$. Since $x \mapsto \mathscr{J}_{1/2}(x) = (\sin x)/x$ is increasing on $[-\pi, 0]$ and decreasing on $[0, \pi]$, we obtain the inequality:

$$x \cos x \leq \sin x \quad \text{for all} \quad x \in [-\pi, \pi].$$

Thus, we have

$$\mathscr{J}_{-1/2}(x) \leq \mathscr{J}_{1/2}(x) \quad \text{for all} \quad x \in [-\pi, \pi].$$

Changing x with $x \sin\theta$ and multiplying both sides of the above inequality with $\sin^2\theta \cos^{2p}\theta$, from (3.112) and (3.110) one has

$$B\left(p+\frac{1}{2},\frac{1}{2}\right)\mathscr{J}_p(x) - B\left(p+\frac{3}{2},\frac{1}{2}\right)\mathscr{J}_{p+1}(x) \leq B\left(p+\frac{1}{2},\frac{3}{2}\right)\mathscr{J}_{p+1}(x),$$

which holds for all $p \geq -1/2$ and $x \in [-\pi, \pi]$. Finally, the relation

$$B\left(p+\frac{1}{2},\frac{1}{2}\right) = B\left(p+\frac{3}{2},\frac{1}{2}\right) + B\left(p+\frac{1}{2},\frac{3}{2}\right)$$

yields $\mathscr{J}_p(x) \leq \mathscr{J}_{p+1}(x)$. Thus the proof is complete.   □

3.7.3 Cusa's Inequality and Related Inequalities

By using geometrical constructions Nicolaus da Cusa (1401–1464) discovered the inequality

$$\frac{\sin x}{x} \leq \frac{2+\cos x}{3} \Leftrightarrow \mathscr{J}_{1/2}(x) \leq \frac{2+\mathscr{J}_{-1/2}(x)}{3} \quad \text{for all} \quad x \in \left[-\frac{\pi}{2},\frac{\pi}{2}\right]. \quad (3.113)$$

Here the comment about removable singularities applies just as in (3.106). Willebrod Snellius (1581-1626) gave a proof for (3.113) in his book entitled "Cyclometicus," but his proof was quite obscure (for further details see the book of J. Sándor [203]). The first scientist who found an acceptable (geometrical) proof for (3.113) was Christiaan Huygens (Huyghens) (1629-1695). In his book "De circuli magnitudine inventa" Huygens used (3.113) in order to approximate π (for the history of this see the papers of J. Sándor and M. Bencze [205], A.P. Iuskevici [122]). In 1999 F. Qi et al. [190, p. 521] proved that

$$\frac{\sin x}{x} \geq \frac{1+\cos x}{2} \quad \text{for all} \quad x \in \left[-\frac{\pi}{2},\frac{\pi}{2}\right] \quad (3.114)$$

by using Chebyshev's integral inequality (1.40). Recently, there has been a considerable interest in Missouri Journal of Mathematical Sciences, regarding inequalities involving $\mathscr{J}_{1/2}(x) = (\sin x)/x$ (see for example the paper of J. Sándor [203] and the references therein). We note that the inequalities (3.113) and (3.114) may be used for a simple proof of the known fact that

$$\lim_{x \to 0} \frac{\sin x}{x} = 1.$$

On the other hand, it is known that (see D.S. Mitrinović [156, p. 238]) if $a \in (0, 1/2]$, then

$$\frac{2(1 + a\cos x)}{\pi} \leq \frac{\sin x}{x} \leq \frac{1 + a\cos x}{a+1} \quad \text{for all } x \in \left[-\frac{\pi}{2}, \frac{\pi}{2}\right]. \tag{3.115}$$

This trigonometric inequality represents a partial answer to the problem E 1277 proposed by A. Oppenheim. It was proved by W.B. Carver in American Mathematical Monthly 65, 206-209 (1958). As a generalization of this inequality, using the same idea as in the proof of Theorem 3.28, we recently proved in [51, Theorem 5] that if $0 < a \leq 1/2$, $p \geq -1/2$ and $x \in [-\pi/2, \pi/2]$, then the following inequality holds:

$$\frac{1 + 2a(p+1)\mathscr{J}_p(x)}{a(2p+1) + \pi/2} \leq \mathscr{J}_{p+1}(x) \leq \frac{1 + 2a(p+1)\mathscr{J}_p(x)}{a(2p+1) + (a+1)}. \tag{3.116}$$

Observe that if we change in (3.116) p with $\kappa - 1$ and x with $x\sqrt{c}$, then we obtain that for all $a \in (0, 1/2]$, $c \geq 0$, $\kappa \geq 1/2$ and $x\sqrt{c} \in [-\pi/2, \pi/2]$ the inequality

$$\frac{1 + 2a\kappa\lambda_p(x)}{a(2\kappa - 1) + \pi/2} \leq \lambda_{p+1}(x) \leq \frac{1 + 2a\kappa\lambda_p(x)}{a(2\kappa - 1) + (a+1)} \tag{3.117}$$

holds. In what follows our aim is to improve (3.117) using (3.113) and (3.114). First note that Cusa's inequality (3.113) is better than the right-hand side of (3.115). For this observe that

$$\frac{d}{da}\left(\frac{1 + a\cos x}{1+a}\right) = \frac{\cos x - 1}{(1+a)^2} \leq 0 \quad \text{for all } x \in \left[-\frac{\pi}{2}, \frac{\pi}{2}\right] \quad \text{and } a \in \left(0, \frac{1}{2}\right].$$

Thus the function

$$a \mapsto \frac{1 + a\cos x}{1+a}$$

is decreasing on $(0, 1/2]$, and hence the asserted result follows. Secondly, observe that when $x \in [-x_3, x_3]$, where $x_3 \simeq 0.7197987821\ldots$, then the inequality (3.114) is better than the left-hand side of (3.115) for all $a \in (0, 1/2]$. This is justified by the following inequality

$$\frac{1 + \cos x}{2} - \frac{2(1 + a\cos x)}{\pi} \geq \frac{1 + \cos x}{2} - \frac{2}{\pi}\left(1 + \frac{\cos x}{2}\right) \geq 0,$$

where $x \in [-x_3, x_3]$ and x_3 is the largest root of the equation $4 - \pi = (\pi - 2)\cos x$. Thus, a slight improvement of (3.117) is the following result. Since the proofs of these inequalities go along the lines in [51], they are not included in this section.

Theorem 3.30. (Á. Baricz [49]) *If $c \geq 0$ and $\kappa \geq 1/2$, then we have*

$$\frac{1 + 2\kappa\lambda_p(x)}{2\kappa + 1} \leq \lambda_{p+1}(x) \leq \frac{1 + \kappa\lambda_p(x)}{\kappa + 1} \quad \text{for all } x\sqrt{c} \in \left[-\frac{\pi}{2}, \frac{\pi}{2}\right]. \tag{3.118}$$

In what follows let us discuss the hyperbolic counterpart of the inequality (3.114). As we can see the Chebyshev integral inequality (1.40) is also useful here. More precisely, by using the same idea as in the proof of the inequality (3.114), putting $p(t) = 1$, $f(t) = \sinh t$, $g(t) = t$, $t \in [a,b] = [0,x]$, $x \in [0,\infty)$ in (1.40), we have

$$\int_0^x \sinh t\, dt \int_0^x t\, dt \leq \int_0^x dt \int_0^x t \sinh t\, dt.$$

A direct calculation yields

$$\frac{\sinh x}{x} \leq \frac{1 + \cosh x}{2} \Leftrightarrow \mathscr{I}_{1/2}(x) \leq \frac{1 + \mathscr{I}_{-1/2}(x)}{2} \quad \text{for all } x \in \mathbb{R}. \tag{3.119}$$

Looking for a generalization of (3.119) we obtain the following result.

Theorem 3.31. (Á. Baricz [49]) *If $c \leq 0$ and $\kappa \geq 1/2$, then*

$$\lambda_{p+1}(x) \leq \frac{1 + 2\kappa\lambda_p(x)}{2\kappa + 1} \quad \text{for all } x \in \mathbb{R}. \tag{3.120}$$

Proof. Recall that as a generalization of (1.27), recently S. András and Á. Baricz [24] proved the formula (1.33), i.e. in particular, if $c, x \in \mathbb{R}$ and $2p > 2q > -(b+1)$, then

$$\lambda_p(x) = \int_0^1 \lambda_q(tx) \frac{2t^{2q+b}(1 - t^2)^{p-q-1}}{B\left(q + \frac{b+1}{2}, p - q\right)}\, dt. \tag{3.121}$$

From this it follows that if $p > q > -1$, then

$$\mathscr{I}_p(x) = \frac{2}{B(q+1, p-q)} \int_0^1 \mathscr{I}_q(tx) t^{2q+1}(1 - t^2)^{p-q-1}\, dt, \tag{3.122}$$

and in particular taking $q = -1/2$ (changing p with $p+1$ and taking $q = 1/2$ respectively), we get that for all $p > -1/2$ and $x \in \mathbb{R}$

$$\mathscr{I}_p(x) = \frac{2}{B\left(p + \frac{1}{2}, \frac{1}{2}\right)} \int_0^{\pi/2} \mathscr{I}_{-1/2}(x\sin\theta)\cos^{2p}\theta\, d\theta, \tag{3.123}$$

$$\mathscr{I}_{p+1}(x) = \frac{2}{B\left(p + \frac{1}{2}, \frac{3}{2}\right)} \int_0^{\pi/2} \mathscr{I}_{1/2}(x\sin\theta)\sin^2\theta\cos^{2p}\theta\, d\theta. \tag{3.124}$$

Now changing in (3.119) x with $x\sin\theta$ and multiplying both sides of (3.119) with $\sin^2\theta\cos^{2p}\theta$, after integration we obtain

$$\mathscr{I}_{p+1}(x) \leq \frac{1+2(p+1)\mathscr{I}_p(x)}{2p+3}$$

for all $p \geq -1/2$ and $x \in \mathbb{R}$. Finally, observe that $\mathscr{I}_{\kappa-1}(x\sqrt{-c}) = \lambda_p(x)$, thus the proof is complete.   □

Remark 3.15. Recall that in Theorem 3.29 we proved that if $c \in [0,1]$ and $\kappa \geq 1/2$, then $\lambda_{p+1}(x) \geq \lambda_p(x)$ for all $x \in [-\pi,\pi]$. When $c \in [0,1]$, $\kappa \geq 1/2$ and $x \in [-\pi/2,\pi/2]$, then the left-hand side of inequality (3.118) provides an improvement of the above mentioned inequality, since under the same assumptions

$$\frac{1+2\kappa\lambda_p(x)}{2\kappa+1} \geq \lambda_p(x).$$

Here we used the fact that $\lambda_p(x) \leq 1$ under the stated hypothesis. For this let us recall the integral representation formula (1.27) from Lemma 1.3 (see also the paper of Á. Baricz and E. Neuman [58, Lemma 2.1])

$$\lambda_p(x) = \begin{cases} \dfrac{2}{B\left(p+\frac{b}{2},\frac{1}{2}\right)} \displaystyle\int_0^1 (1-t^2)^{p+\frac{b-2}{2}}\cos(tx\sqrt{c})\,\mathrm{d}t, & c \geq 0 \\[2em] \dfrac{2}{B\left(p+\frac{b}{2},\frac{1}{2}\right)} \displaystyle\int_0^1 (1-t^2)^{p+\frac{b-2}{2}}\cosh(tx\sqrt{-c})\,\mathrm{d}t, & c \leq 0, \end{cases}$$

which is valid for all $x \in \mathbb{R}$ and $\kappa > 1/2$. Since

$$2\int_0^1 (1-t^2)^{p+\frac{b-2}{2}}\,\mathrm{d}t = B\left(p+\frac{b}{2},\frac{1}{2}\right)$$

it follows that $\lambda_p(x) \leq 1$ when $c \geq 0$, $\kappa > 1/2$ and $x \in \mathbb{R}$. Because $\mathscr{I}_{-1/2}(x) = \cos x \leq 1$ for all $x \in \mathbb{R}$, from (3.112) it follows that $\mathscr{I}_p(x) \leq 1$ for all $p \geq -1/2$. Now using again the formula $\mathscr{I}_{\kappa-1}(x\sqrt{c}) = \lambda_p(x)$, we obtain $\lambda_p(x) \leq 1$ for all $c \geq 0$, $\kappa \geq 1/2$ and $x \in \mathbb{R}$.

Following the above argument it is clear that $\lambda_p(x) \geq 1$ for all $c \leq 0$, $\kappa \geq 1/2$ and $x \in \mathbb{R}$. This means that the inequality (3.120) improves the inequality $\lambda_{p+1}(x) \leq \lambda_p(x)$, where $c \leq 0$, $\kappa > 0$ and $x \in \mathbb{R}$, which can be verified easily in view of the infinite series representation of the function λ_p.

Another immediate application of the integral representation formula (1.27) is the following result.

Theorem 3.32. (Á. Baricz [49]) *If $c \leq 0$ and $\kappa \geq 1$, then $\lambda_{p+1}(x) \geq \sqrt{\lambda_p(x)}$ for all $x \in \mathbb{R}$. Moreover, if $c \in [0,1]$ and $\kappa \in [1/2,1]$, then $\lambda_{p+1}(x) \geq \sqrt{\lambda_p(x)}$ also holds for all $x \in [-\pi/2, \pi/2]$.*

Proof. Let us consider the function $\varphi_5(x) = \lambda_{p+1}^2(x) - \lambda_p(x)$. Since this function is even, it is enough to show the required inequality for $x \geq 0$ and $x \in [0, \pi/2]$, respectively. Applying (3.105) for p and $p+1$, we obtain

$$2\varphi_5'(x) = (-c)x\lambda_{p+1}(x)\left[\frac{2\lambda_{p+2}(x)}{\kappa+1} - \frac{1}{\kappa}\right].$$

Now if $c \leq 0$ and $\kappa \geq 1$, then in view of Remark 3.15 clearly

$$\lambda_{p+2}(x) \geq 1 \geq (\kappa+1)/(2\kappa),$$

and thus φ_5 is increasing. Hence $\varphi_5(x) \geq \varphi_5(0) = 0$. Finally, suppose that $c \in [0,1]$ and $\kappa \in [1/2,1]$. From (3.108) it follows that $\lambda_{p+1}(x) \geq 0$, provided $x \in [-\pi/2, \pi/2]$. Now using again Remark 3.15 one has

$$\lambda_{p+2}(x) \leq 1 \leq (\kappa+1)/(2\kappa),$$

and thus φ_5 is increasing. This completes the proof. □

Remark 3.16. Taking $b = 1$, $c = 1$ and $p = -1/2$ in Theorem 3.32 we obtain for all $x \in [-\pi/2, \pi/2]$ the inequality $\sin x \geq x\sqrt{\cos x}$, which was established by J. Sándor [203].

3.7.4 Extensions of Jordan's Inequality

The following inequality is known in literature as Jordan's inequality (see the book of D.S. Mitrinović [156, p. 33]):

$$1 \geq \frac{\sin x}{x} \geq \frac{2}{\pi} \quad \text{for all} \quad x \in \left[-\frac{\pi}{2}, \frac{\pi}{2}\right]. \tag{3.125}$$

It has been studied by several mathematicians in order to sharpen this basic analytic inequality. In this section we present two recent results related to Jordan's inequality and we extend them to generalized Bessel functions, in order to obtain other lower and upper bounds for the function λ_p. Recall that Redheffer's inequality (3.106) and Jordan's inequality (3.125) do not imply each other. J. Sándor [202–204] proved that the function $\mathscr{J}_{1/2}(x) = (\sin x)/x$ is concave on $[0, \pi/2]$ and from this he deduced the following improvement of Jordan's inequality:

$$\frac{2}{\pi} + \frac{2}{\pi^2}(\pi - 2x) \geq \frac{\sin x}{x} \geq \frac{2}{\pi} + \frac{\pi-2}{\pi^2}(\pi - 2x) \quad \text{for all} \quad x \in \left[0, \frac{\pi}{2}\right]. \tag{3.126}$$

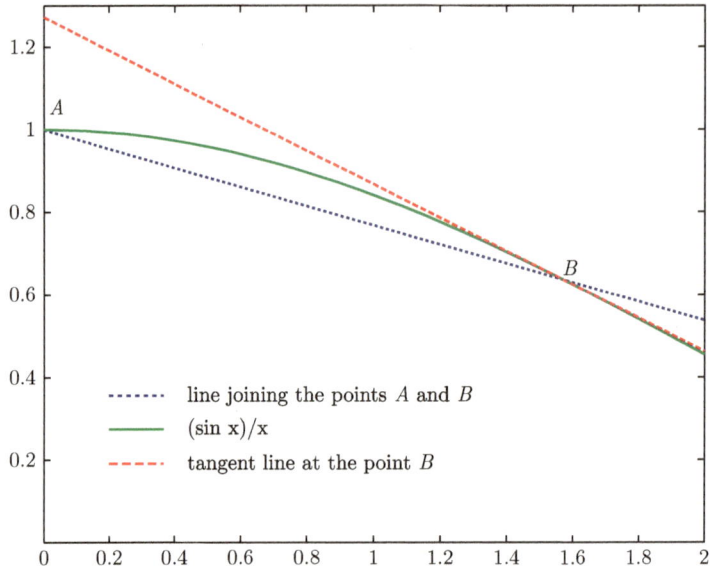

Fig. 3.1 The graph of the function $\mathscr{J}_{1/2}$ together with the line joining the points $A(0,1)$ and $B(\pi/2,2/\pi)$ and the tangent line to $\mathscr{J}_{1/2}$ at the point $B(\pi/2,2/\pi)$

Sándor's idea was that, since $\mathscr{J}_{1/2}$ is concave, its graph lies above the line segment joining the points $A(0,1)$ and $B(\pi/2,2/\pi)$ on the graph of $\mathscr{J}_{1/2}$ on $[0,\pi/2]$. From this follows the right-hand side of (3.126). Now from the left-hand side just consider the tangent line to $\mathscr{J}_{1/2}$ at the point $B(\pi/2,2/\pi)$, which line lies above the graph of $\mathscr{J}_{1/2}$ on $[0,\pi/2]$ (see Fig. 3.1).

The inequality (3.126) was also proved recently by X. Zhang et al. [236] by using Lemma 3.7, i.e. the monotone form of l'Hospital's rule established by G.D. Anderson et al. [18] (see also their book [19]). There is another improvement of Jordan's inequality proved by F. Qi and Q.D. Hao [191](see also the paper of F. Qi et al. [190, p. 522]) using the intricate technique of calculus, namely

$$\frac{2}{\pi}+\frac{\pi-2}{\pi^3}(\pi^2-4x^2) \geq \frac{\sin x}{x} \geq \frac{2}{\pi}+\frac{1}{\pi^3}(\pi^2-4x^2) \quad \text{for all} \quad x\in\left[-\frac{\pi}{2},\frac{\pi}{2}\right]. \quad (3.127)$$

This new refinement of Jordan's inequality was rediscovered too by X. Zhang et al. [236] using also the monotone form of l'Hospital's rule. The right-hand side of (3.127) was also proved by L. Debnath and C.J. Zhao [88] using a completely different method. Moreover, using also the monotone form of l'Hospital's rule L. Zhu [239] extended (3.127) in the following way for $-\pi/2 \leq x \leq r \leq \pi/2$:

$$\frac{\sin r}{r}+\frac{r-\sin r}{r^3}(r^2-x^2) \geq \frac{\sin x}{x} \geq \frac{\sin r}{r}+\frac{\sin r-r\cos r}{2r^3}(r^2-x^2). \quad (3.128)$$

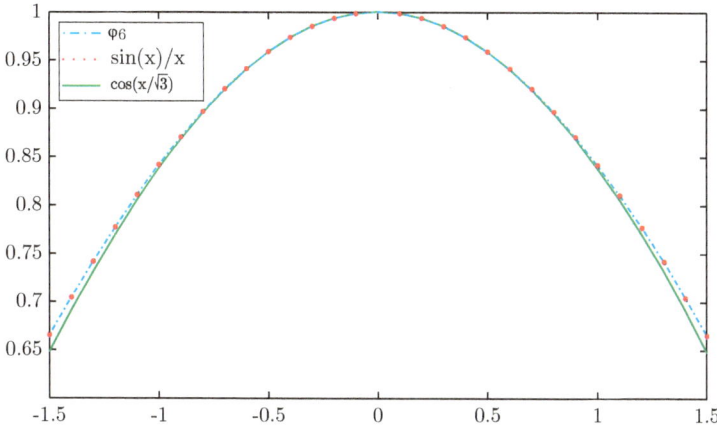

Fig. 3.2 The graph of the functions φ_6, $\mathscr{I}_{1/2}$ and $x \mapsto \cos(x/\sqrt{3})$

For these, and related details, see the survey papers about Jordan's inequality of J. Sándor [204], F. Qi [189], F. Qi and D.W. Niu [192] and F. Qi et al. [193]. It is known that the inequalities (3.126) and (3.127) cannot be compared on the whole interval $[0, \pi/2]$ (see the discussion below). Observe that (3.107) yields for $p = 1/2$ the following inequality (see Fig. 3.2):

$$\varphi_6(x) = \frac{2}{9}\left[2 + \frac{5}{2}\cos\left(\sqrt{\frac{3}{5}}x\right)\right] \geq \frac{\sin x}{x} \geq \cos\left(\frac{x}{\sqrt{3}}\right) \qquad (3.129)$$

for all $|x| < \pi/2$. This result of E. Neuman [161], which is actually a quite tight refinement of Jordan's inequality (3.125), does not appear in any paper related to sharpening of Jordan's inequality, even if direct computations and numerical experiments in Derive6 show the followings:

1. The second inequality in (3.129) is better than the second inequality in (3.127) for all $x \in [-x_4, x_4]$, where $x_4 \simeq 1.204850991\ldots$ is the root of the equation $\cos(x/\sqrt{3}) = 3/\pi - 4x^2/\pi^3$ on $[0, \pi/2]$. The situation is reversed when $x \in [-\pi/2, -x_4]$ or $x \in [x_4, \pi/2]$.
2. The second inequality in (3.129) is better than the second inequality in (3.126) for all $x \in [0, x_5]$, where $x_5 \simeq 1.475028163\ldots$ is the root of the equation $\cos(x/\sqrt{3}) = 1 - 2(\pi - 2)x/\pi^2$ on $[0, \pi/2]$. Further the situation is reversed when $x \in [x_5, \pi/2]$.
3. The second inequality in (3.127) is better than the second inequality in (3.126) for all $x \in [\pi(\pi - 3)/2, \pi/2]$. Moreover the situation is reversed for all $x \in [0, \pi(\pi - 3)/2]$. Note that $\pi(\pi - 3)/2 \simeq 0.2224132208\ldots$.
4. If $x \in [-x_6, x_6]$, then $\cos(x/\sqrt{3}) \geq 2/\pi$, where $x_6 \simeq 1.525398501\ldots$ is the root of the equation $\cos(x/\sqrt{3}) = 2/\pi$. When $x \in [-\pi/2, -x_6]$ or $x \in [x_6, \pi/2]$ the above inequality is reversed. Note that $\pi/2 \simeq 1.570796327\ldots$.

5. The first inequality in (3.129) is better than the first inequality in (3.127) for all $x \in [-x_7, x_7]$, where $x_7 \simeq 1.563220278\ldots$ is the root of the equation $\varphi_6(x) = 1 - 4(\pi - 2)x^2/\pi^3$ on $[0, \pi/2]$. The situation is reversed when $x \in [-\pi/2, -x_7]$ or $x \in [x_7, \pi/2]$.
6. The first inequality in (3.129) is better than the first inequality in (3.126) for all $x \in [0, x_8]$, where $x_8 \simeq 1.497945837\ldots$ is the root of the equation $\varphi_6(x) = 4(\pi - x)/\pi^2$ on $[0, \pi/2]$. The situation is reversed when $x \in [x_8, \pi/2]$.
7. The first inequality in (3.127) is better than the first inequality in (3.126) for all $x \in [0, (\pi/2)(4 - \pi)/(\pi - 2)]$. Note that $(\pi/2)(4 - \pi)/(\pi - 2) \simeq 1.181142066\ldots$. Further the situation is reversed when $x \in [(\pi/2)(4 - \pi)/(\pi - 2), \pi/2]$.
8. If $x \in [-\pi/2, \pi/2]$, then $\varphi_6(x) \leq 1$.

From the above discussion it follows that it is worth extending inequalities (3.126) and (3.127) for the generalized and normalized Bessel functions. Before we state our first main result in this section let us recall the monotone form of the l'Hospital rule.

Lemma 3.7. (G.D. Anderson, M.K. Vamanamurthy and M. Vuorinen [18]) *Given $\alpha, \beta \in \mathbb{R}$ such that $\alpha < \beta$, let $f, g : [\alpha, \beta] \to \mathbb{R}$ be continuous on $[\alpha, \beta]$, and differentiable on (α, β). Further let $g'(x) \neq 0$ for all $x \in (\alpha, \beta)$. If f'/g' is (strictly) increasing (decreasing) on (α, β), then so are*

$$x \in (\alpha, \beta) \mapsto \frac{f(x) - f(\alpha)}{g(x) - g(\alpha)} \text{ and } x \in (\alpha, \beta) \mapsto \frac{f(x) - f(\beta)}{g(x) - g(\beta)}.$$

Our first main result in this section reads as follows.

Theorem 3.33. (Á. Baricz [49]) *The following assertions are true:*

(a) *If $\kappa \geq 1/2$ and $c \in [0, 1]$, then for all $x \in [0, \pi/2]$ one has*

$$\left[1 - \lambda_p \left(\frac{\pi}{2} \right) \right] \frac{\pi - 2x}{\pi} \leq \lambda_p(x) - \lambda_p \left(\frac{\pi}{2} \right)$$

$$\leq \left[\left(\frac{c\pi}{2\kappa} \right) \lambda_{p+1} \left(\frac{\pi}{2} \right) \right] \frac{\pi - 2x}{\pi}. \tag{3.130}$$

(b) *If $\kappa > 0$ and $c \in [0, 1]$, then for all $x \in [-\pi/2, \pi/2]$ one has*

$$\left[\left(\frac{c}{4\kappa} \right) \lambda_{p+1} \left(\frac{\pi}{2} \right) \right] \frac{\pi^2 - 4x^2}{4} \leq \lambda_p(x) - \lambda_p \left(\frac{\pi}{2} \right)$$

$$\leq \left[1 - \lambda_p \left(\frac{\pi}{2} \right) \right] \frac{\pi^2 - 4x^2}{\pi^2}. \tag{3.131}$$

Proof. (a) Since for all $\kappa \geq 1/2$ and $c \in [0, 1]$ the function λ_p is concave on $[0, \pi/2]$ (see Theorem 3.29), it follows that its graph lies above the line segment joining the points $(0, 1)$ and $(\pi/2, 2/\pi)$ on the graph of λ_p on $[0, \pi/2]$. This property implies the left-hand side of (3.130). For the right-hand side it suffices to consider the tangent line to λ_p at the point $(\pi/2, 2/\pi)$, which lies above the graph of λ_p on $[0, \pi/2]$.

(b) Since the function λ_p is even, clearly it is enough to show (3.131) for $x \in [0, \pi/2]$. Let us consider the functions $\varphi_7, \varphi_8 : [0, \pi/2] \to [0, \infty)$, defined by $\varphi_7(x) = \lambda_p(x) - \lambda_p(\pi/2)$ and $\varphi_8(x) = \pi^2/4 - x^2$. From (3.105) we obtain $\varphi_7'(x)/\varphi_8'(x) = [c\lambda_{p+1}(x)]/(4\kappa)$. Because for all $c \in [0, 1]$ and $\kappa \geq 1/2$ the function λ_p is decreasing on $[0, \pi/2]$ (see Theorem 3.29), it follows that φ_7'/φ_8' is decreasing on $[0, \pi/2]$. Clearly $\varphi_7(\pi/2) = \varphi_8(\pi/2) = 0$, thus from Lemma 3.7 we get that φ_7/φ_8 is decreasing too on $[0, \pi/2]$. All that remains to show is that from l'Hospital's rule

$$\lim_{x \to \frac{\pi}{2}} \frac{\varphi_7(x)}{\varphi_8(x)} = \lim_{x \to \frac{\pi}{2}} \frac{\varphi_7'(x)}{\varphi_8'(x)} = \lim_{x \to \frac{\pi}{2}} \left(\frac{c}{4\kappa} \right) \lambda_{p+1}(x) = \left(\frac{c}{4\kappa} \right) \lambda_{p+1} \left(\frac{\pi}{2} \right).$$

Now the inequality (3.131) follows from the monotonicity and the limiting values of φ_7/φ_8.   □

Remark 3.17. Putting $c = 1$, $b = 1$ and $p = -1/2$ in (3.130) we get

$$1 - \frac{2}{\pi} x \leq \cos x \leq 2 \left(1 - \frac{2}{\pi} x \right) \text{ for all } x \in \left[0, \frac{\pi}{2} \right]. \tag{3.132}$$

The left-hand side of (3.132) is known as Kober's inequality [136]. Taking $c = 1$, $b = 1$ and $p = 1/2$ in (3.130) (in (3.131) respectively) we reobtain (3.126) ((3.127) respectively). Observe that changing x with $\pi/2 - x$ in (3.126) we obtain the following inequality for all $x \in [0, \pi/2]$

$$1 - \frac{2}{\pi} x + \frac{\pi - 2}{\pi^2} x(\pi - 2x) \leq \cos x \leq 1 - \frac{2}{\pi} x + \frac{2}{\pi^2} x(\pi - 2x), \tag{3.133}$$

which was proved by F. Qi and Q.D. Hao [191], X. Zhang et al. [236] using different methods. Now taking $c = 1$, $b = 1$ and $p = -1/2$ in (3.131) we obtain

$$\frac{\pi}{4} - \frac{1}{\pi} x^2 \leq \cos x \leq 1 - \frac{4}{\pi^2} x^2 \quad \text{for all} \quad x \in \left[-\frac{\pi}{2}, \frac{\pi}{2} \right]. \tag{3.134}$$

Straightforward simplifications and easy computations show that the right-hand side of (3.134) is exactly the right-hand side of (3.133), but the left-hand side of (3.133) (which is a refinement of Kober's inequality) is better than the left-hand side of (3.134) for all $x \in [0, \pi/2]$. Further the right-hand side of (3.134) is better than the right-hand side of (3.132) for all $x \in [0, \pi/2]$.

Finally, note that (3.130) can be proved also using Lemma 3.7. Just consider the functions φ_7 and $\varphi_9 : [0, \pi/2] \to [0, \infty)$, defined by $\varphi_9(x) = \pi/2 - x$. We know that for all $c \in [0, 1]$ and $\kappa \geq 1/2$, the function λ_p is concave on $[0, \pi/2]$, thus $x \mapsto \varphi_7'(x)/\varphi_9'(x) = -\lambda_p'(x)$ is increasing on $[0, \pi/2]$. Application of Lemma 3.7 gives that φ_7/φ_9 is increasing too on $[0, \pi/2]$, thus the required inequality (3.130) follows.

We end this section with the following extension of (3.128), which provides a generalization of (3.131). As we can see the result of L. Zhu [239] is in fact a typical result for Bessel functions. Since the proofs of the next inequalities go along the lines in the proof of (3.131), we state the following result without proof.

Theorem 3.34. (Á. Baricz [49]) *If $\kappa > 0$ and $c \in [0,1]$, then for all $-\pi/2 \le x \le r \le \pi/2$ we have*

$$\lambda_p(r) + \left[\left(\frac{c}{4\kappa}\right)\lambda_{p+1}(r)\right](r^2 - x^2) \le \lambda_p(x) \le \lambda_p(r) + \left[\frac{1 - \lambda_p(r)}{r^2}\right](r^2 - x^2).$$

$$\tag{3.135}$$

Moreover, if $\kappa > 0$ and $c \le 0$, then (3.135) holds for all $-\infty < x \le r < \infty$.

Remark 3.18. Choosing $c = 1$, $b = 1$ and $p = 1/2$ in (3.135) we obtain (3.128). Analogously, taking $c = -1$, $b = 1$ and $p = 1/2$ in (3.135) we obtain the hyperbolic counterpart of (3.128)

$$\frac{\sinh r}{r} + \frac{r - \sinh r}{r^3}(r^2 - x^2) \ge \frac{\sinh x}{x} \ge \frac{\sinh r}{r} + \frac{\sinh r - r\cosh r}{2r^3}(r^2 - x^2),$$

whenever $-\infty < x \le r < \infty$. Here we used the fact that

$$\mathscr{J}_{3/2}(x) = 3\left(\frac{\sin x}{x^3} - \frac{\cos x}{x^2}\right) \quad \text{and} \quad \mathscr{I}_{3/2}(x) = -3\left(\frac{\sinh x}{x^3} - \frac{\cosh x}{x^2}\right).$$

It is also worth mentioning that motivated by the paper of S. Wu and L. Debnath [230] the results of Theorems 3.33 and 3.34 were improved recently by the author in the forthcoming paper [53]. Moreover, the results of Theorems 3.30 and 3.31 were improved recently too by Á. Baricz and L. Zhu in another forthcoming paper [64], which is the direct continuation of L. Zhu's paper [238]. For other inequalities involving Bessel and modified Bessel functions which extend some other trigonometric inequalities we refer to the papers of Á. Baricz [51], Á. Baricz and J. Sándor [60].

3.7.5 Sharp Jordan Type Inequalities for Bessel Functions

During the past several years there has been a great interest in Jordan type inequalities. Among the results mentioned in the previous section many new results appeared in the last three years. Many refinements have been proved for Jordan's inequality (3.125) by D.W. Niu [164], D.W. Niu et al. [165], S. Wu and L. Debnath [231–233], S. Wu and H.M. Srivastava [234], S. Wu et al. [235], L. Zhu [240–242] and others. For a long list of recent papers on this topic we refer to the survey papers of F. Qi and D.W. Niu [192], F. Qi et al. [193]. In most of these papers the key tool in the proofs is the monotone form of l'Hospital's rule, i.e. Lemma 3.7. By using a similar method our aim in this section is to extend to Bessel functions all of the results of L. Zhu [240–242] concerning Jordan and Kober type inequalities and the new power series expansion for the sine function. We note that our approach is similar to those given in the above mentioned papers, however an important step in the proofs is simplified. Moreover, our approach in each cases gives us larger intervals of validity. At the end of this section we formulate an equivalent form of the main results, namely Theorems 3.35 and 3.36, in order to point out the connection between the results of D.W. Niu [164] and L. Zhu [240–242]. We note also that the approach in [164] is somewhat different to that given in this paper, and our range of validity is better than in [164]. Finally, it should be mentioned here that during the course of writing this monograph we have found the papers of L. Zhu [244, 245] in which the author by using a slightly different method extended also to generalized Bessel functions the results from [240–242] by rediscovering some of the main results of this section. However, in our main results the range of validity is better than in [244, 245] and in the proofs an important step is simplified. For more details see the paper of L. Zhu [244, p. 727].

Our first main result reads as follows.

Theorem 3.35. (Á. Baricz and S. Wu [62]) *Let* $p > -1$ *and let* $j_{p,1}$ *be the first positive zero of the Bessel function of the first kind* J_p. *Then for all* $0 < x \leq r \leq j_{p+1,1}$ *the following sharp Jordan-type inequalities hold*

$$S_{p,n}(x) + \alpha_p(r)(r^2 - x^2)^{n+1} \leq \mathcal{J}_p(x) \leq S_{p,n}(x) + \beta_p(r)(r^2 - x^2)^{n+1}, \quad (3.136)$$

where

$$S_{p,n}(x) = \sum_{k=0}^{n} a_{p,k}(r)(r^2 - x^2)^k,$$

n is a natural number and the coefficients $a_{p,k}(r)$ *are defined explicitly by*

$$a_{p,k}(r) = \frac{\mathcal{J}_{p+k}(r)}{4^k k!(p+1)_k} \quad (3.137)$$

for all $k \in \{0, 1, \ldots, n+1\}$ *or recursively by*

$$a_{p,0}(r) = \mathcal{J}_p(r), \quad a_{p,1}(r) = \frac{1}{4(p+1)}\mathcal{J}_{p+1}(r),$$

$$a_{p,k+1}(r) = \frac{p+k}{(k+1)r^2}a_{p,k}(r) - \frac{1}{4k(k+1)r^2}a_{p,k-1}(r) \quad (3.138)$$

for all $k \in \{1, 2, \ldots, n\}$. *Moreover, the constants*

$$\alpha_p(r) = a_{p,n+1}(r) \quad \text{and} \quad \beta_p(r) = \frac{1 - \sum_{k=0}^{n} a_{p,k}(r)r^{2k}}{r^{2(n+1)}}$$

are the best possible. In addition, there exist $\zeta \in (x, r)$ depending on n such that

$$\mathscr{I}_p(x) = \sum_{k=0}^{n} a_{p,k}(r)(r^2 - x^2)^k + \frac{\mathscr{I}_{p+n+1}(\zeta)}{4^{n+1}(n+1)!(p+1)_{n+1}}(r^2 - x^2)^{n+1}, \quad (3.139)$$

which leads to the power series expansion

$$\mathscr{I}_p(x) = \sum_{n \geq 0} a_{p,n}(r)(r^2 - x^2)^n. \quad (3.140)$$

Proof. Let $m \in \{1, 2, \ldots, n+1\}$ and consider the functions $f_m, g_m : [0, r] \to \mathbb{R}$, defined by

$$f_1(x) = \mathscr{I}_p(x) - \sum_{k=0}^{n} a_{p,k}(r)(r^2 - x^2)^k,$$

$$f_m(x) = \frac{\mathscr{I}_{p+m-1}(x)}{4^{m-1}(p+1)_{m-1}} - \sum_{k=m-1}^{n} k(k-1)\ldots(k-m+2)a_{p,k}(r)(r^2 - x^2)^{k-m+1}$$

for all $m \geq 2$ and

$$g_1(x) = (r^2 - x^2)^{n+1},$$

$$g_m(x) = (n+1)n(n-1)\ldots(n-m+3)(r^2 - x^2)^{n-m+2}, \quad m \geq 2.$$

Now consider the function $Q_1 : (0, r) \to \mathbb{R}$, defined by $Q_1(x) = f_1(x)/g_1(x)$. Observe that the inequality (3.136) is equivalent to $\alpha_p(r) \leq Q_1(x) \leq \beta_p(r)$. Thus, to prove (3.136), in what follows we show that Q_1 is decreasing. In view of

$$\mathscr{I}'_p(x) = -\frac{x}{2(p+1)}\mathscr{I}_{p+1}(x)$$

it is easy to verify that for all $s \in \{0, 1, 2, \ldots, n\}$ we have $f_{s+1}(r) = g_{s+1}(r) = 0$ and consequently for each $x \in (0, r)$ and $s \in \{1, 2, \ldots, n\}$ one has

$$\frac{f'_s(x)}{g'_s(x)} = \frac{f_{s+1}(x)}{g_{s+1}(x)} = \frac{f_{s+1}(x) - f_{s+1}(r)}{g_{s+1}(x) - g_{s+1}(r)}.$$

Taking into account the above relation and using the monotone form of l'Hospital's rule (see Lemma 3.7) $n+1$ times it is clear that to prove that Q_1 is decreasing it is enough to show that the function

$$x \mapsto \frac{f'_{n+1}(x)}{g'_{n+1}(x)} = a_{p,n+1}(r)\frac{\mathscr{I}_{p+n+1}(x)}{\mathscr{I}_{p+n+1}(r)}$$

is decreasing on $(0,r)$. But, it is known (see Á. Baricz [51, Theorem 3]) that for all $p > -1$ the function $x \mapsto \mathscr{J}_p(x)$ is decreasing on $(0, j_{p,1})$. Changing p with $p+n+1$ we obtain that the function $x \mapsto \mathscr{J}_{p+n+1}(x)$ is decreasing on $(0, j_{p+2,1}) \subset (0, j_{p+n+1,1})$ and with this the monotonicity of Q_1 is proved. Here we used the known fact that the function $p \mapsto j_{p,1}$ is increasing on $(-1, \infty)$. To complete the proof of (3.136) all that remains is to prove that the constants $\alpha_p(r)$ and $\beta_p(r)$ in (3.136) are the best possible. For this observe that

$$\lim_{x \searrow 0} Q_1(x) = \frac{1 - \sum\limits_{k=0}^{n} a_{p,k}(r) r^{2k}}{r^{2(n+1)}} = \beta_p(r)$$

and by using the l'Hospital's rule $n+1$ times

$$\lim_{x \nearrow r} Q_1(x) = \lim_{x \nearrow r} \frac{f_1'(x)}{g_1'(x)} = \lim_{x \nearrow r} \frac{f_2'(x)}{g_2'(x)} = \ldots = \lim_{x \nearrow r} \frac{f_{n+1}'(x)}{g_{n+1}'(x)} = a_{p,n+1}(r) = \alpha_p(r).$$

Finally, to prove the recursive relation for the coefficients $a_{p,k}(r)$, recall the recurrence relation (see G.N. Watson [227, p. 45])

$$J_{p-1}(x) + J_{p+1}(x) = (2p/x) J_p(x),$$

from which we easily obtain

$$4p(p+1) \mathscr{J}_{p-1}(x) + x^2 \mathscr{J}_{p+1}(x) = 4p(p+1) \mathscr{J}_p(x).$$

Changing in the above relation p with $p+k$, and using (3.137), we obtain (3.138).

Now, let us focus on (3.139). By the Cauchy mean value theorem (see the book of K.R. Stromberg [216, p. 178]) there exist a constant $\xi_1 \in (x,r)$ such that

$$Q_1(x) = \frac{f_1(x)}{g_1(x)} = \frac{f_1(x) - f_1(r)}{g_1(x) - g_1(r)} = \frac{f_1'(\xi_1)}{g_1'(\xi_1)},$$

but as we have mentioned above we have

$$\frac{f_1'(\xi_1)}{g_1'(\xi_1)} = \frac{f_2(\xi_1)}{g_2(\xi_1)} = \frac{f_2(\xi_1) - f_2(r)}{g_2(\xi_1) - g_2(r)}.$$

Thus, using again the Cauchy mean value theorem n times we obtain that there exist $\xi_{n+1} \in (\xi_n, r)$, where $\xi_i \in (\xi_{i-1}, r)$ for all $i \in \{2,3,\ldots,n\}$, such that

$$Q_1(x) = \frac{f_1'(\xi_1)}{g_1'(\xi_1)} = \ldots = \frac{f_n'(\xi_n)}{g_n'(\xi_n)} = \frac{f_{n+1}'(\xi_{n+1})}{g_{n+1}'(\xi_{n+1})} = a_{p,n+1}(r) \frac{\mathscr{J}_{p+n+1}(\xi_{n+1})}{\mathscr{J}_{p+n+1}(r)}.$$

Now denoting ξ_{n+1} with ζ and by using (3.137) we obtain (3.139). On the other hand when n tends to infinity ξ_{n+1} tends to r, and thus $\mathscr{J}_{p+n+1}(\xi_{n+1})/\mathscr{J}_{p+n+1}(r)$ tends to 1. Moreover, it is easy to show that $a_{p,n+1}(r)$ and $a_{p,n+1}(r)(r^2 - x^2)^{n+1}$ tends to zero as n tends to infinity. In other words, the reminder term in (3.139) tends to zero as n tends to infinity, which leads to the series expansion (3.140).   □

Remark 3.19. First let us focus on inequality (3.136) and suppose that $p = -1/2$. Then by using the relations (see (2.28) and (2.29) for $z = x$)

$$\mathscr{J}_{-1/2}(x) = \cos x, \quad \mathscr{J}_{1/2}(x) = \frac{\sin x}{x}$$

and the notation

$$S_{p,n}(x) = \sum_{k=0}^{n} a_{p,k}(r)(r^2 - x^2)^k,$$

from Theorem 3.35 we obtain for all $0 < x \leq r \leq \pi$ the following sharp Kober-type inequality:

$$S_{-1/2,n}(x) + \alpha_{-1/2}(r)(r^2 - x^2)^{n+1} \leq \cos x$$
$$\leq S_{-1/2,n}(x) + \beta_{-1/2}(r)(r^2 - x^2)^{n+1}, \quad (3.141)$$

where

$$S_{-1/2,n}(x) = \sum_{k=0}^{n} a_{-1/2,k}(r)(r^2 - x^2)^k,$$

n is a natural number and the coefficients $a_{-1/2,k}(r)$ are defined recursively by

$$a_{-1/2,0}(r) = \cos r, \quad a_{-1/2,1}(r) = \frac{\sin r}{2r},$$
$$a_{-1/2,k+1}(r) = \frac{2k-1}{2(k+1)r^2}a_{-1/2,k}(r) - \frac{1}{4k(k+1)r^2}a_{-1/2,k-1}(r)$$

for all $k \in \{1, 2, \ldots, n\}$. Moreover, the constants

$$\alpha_{-1/2}(r) = a_{-1/2,n+1}(r) \quad \text{and} \quad \beta_{-1/2}(r) = \frac{1 - \sum_{k=0}^{n} a_{-1/2,k}(r)r^{2k}}{r^{2(n+1)}}$$

are the best possible. This completes the result of L. Zhu [240, Theorem 13] and improves the other known results from the literature. For more details on improved Kober's inequalities we refer to the paper of F. Qi and D.W. Niu [192]. It is worth mentioning that although the technique is similar our approach is much simpler than in the paper of L. Zhu [240]. This is because in [240] there are used spherical Bessel functions instead of Bessel functions, for which the monotonicity of their higher order derivative requires some complicated preliminary results. Our proof is based just on the simple monotonicity property of the function $\mathscr{J}_p$ established by the author [51, Theorem 3].

Remark 3.20. Now, suppose that $p = 1/2$. Then, by using (2.29) and (2.30) for $z = x$, i.e.

$$\mathcal{J}_{1/2}(x) = \frac{\sin x}{x}, \quad \mathcal{J}_{3/2}(x) = 3\left(\frac{\sin x}{x^3} - \frac{\cos x}{x^2}\right),$$

from (3.136) we obtain for all $0 < x \le r \le j_{3/2,1}$ the following sharp Jordan-type inequality:

$$S_{1/2,n}(x) + \alpha_{1/2}(r)(r^2 - x^2)^{n+1} \le \frac{\sin x}{x}$$

$$\le S_{1/2,n}(x) + \beta_{1/2}(r)(r^2 - x^2)^{n+1}, \qquad (3.142)$$

where

$$S_{1/2,n}(x) = \sum_{k=0}^{n} a_{1/2,k}(r)(r^2 - x^2)^k,$$

n is a natural number and the coefficients $a_{1/2,k}(r)$ are defined recursively by

$$a_{1/2,0}(r) = \frac{\sin r}{r}, \quad a_{1/2,1}(r) = \frac{\sin r - r\cos r}{2r^3},$$

$$a_{1/2,k+1}(r) = \frac{2k+1}{2(k+1)r^2}a_{1/2,k}(r) - \frac{1}{4k(k+1)r^2}a_{1/2,k-1}(r)$$

for all $k \in \{1, 2, \ldots, n\}$. Here $j_{3/2,1} = 4.493409457$ in view of (2.30) is in fact the first positive zero of the equation $\tan x = x$. Moreover, the constants

$$\alpha_{1/2}(r) = a_{1/2,n+1}(r) \quad \text{and} \quad \beta_{1/2}(r) = \frac{1 - \displaystyle\sum_{k=0}^{n} a_{1/2,k}(r)r^{2k}}{r^{2(n+1)}}$$

are the best possible. We note that this result was obtained by L. Zhu in [240, Theorem 6] and [242, Theorem 1], however our approach is simpler than in [240, 242]. Moreover, it is worth mentioning that in [242] the inequality (3.142) is proved just for the case when $0 < x \le r \le j_{1/2,1} = \pi$, while in [240] just for the case when $0 < x \le r \le j_{-1/2,1} = \pi/2$. Thus our result (3.136) not only extend (3.142) to Bessel functions, but even improves the range of validity. Now, if we choose $r = \pi/2$ in (3.142), then we get the following sharp Jordan-type inequalities

$$\sum_{k=0}^{n} b_k(\pi^2 - 4x^2)^k + b_{n+1}(\pi^2 - 4x^2)^{n+1} \le \frac{\sin x}{x},$$

$$\frac{\sin x}{x} \le \sum_{k=0}^{n} b_k(\pi^2 - 4x^2)^k + \frac{1 - \displaystyle\sum_{k=0}^{n} b_k\pi^{2k}}{\pi^{2(n+1)}}(\pi^2 - 4x^2)^{n+1},$$

where n is a natural number, as above, and the coefficients b_k are defined as follows

$$b_0 = \frac{2}{\pi}, \; b_1 = \frac{1}{\pi^3}, \; b_{k+1} = \frac{2k+1}{2(k+1)\pi^2}b_k - \frac{1}{16k(k+1)\pi^2}b_{k-1}$$

for all $k \in \{1,2,\dots,n\}$. These results appear on three different papers of L. Zhu, namely in [240, Theorem 7], [242, Corollary 1] and [241, Theorem 5].

Remark 3.21. Let focus on the expansions (3.139) and (3.140) and suppose that $p = -1/2$. Then for all $0 < x \le r \le \pi$ we obtain

$$\cos x = \sum_{k=0}^{n} a_{-1/2,k}(r)(r^2 - x^2)^k + \frac{\mathscr{I}_{n+1/2}(\zeta)}{4^{n+1}(n+1)!(1/2)_{n+1}}(r^2 - x^2)^{n+1},$$

which leads to the new power series expansion of the cosine function

$$\cos x = \sum_{k \ge 0} a_{-1/2,k}(r)(r^2 - x^2)^k,$$

where the coefficients $a_{-1/2,k}(r)$ are defined recursively by

$$a_{-1/2,0}(r) = \cos r, \; a_{-1/2,1}(r) = \frac{\sin r}{2r},$$

$$a_{-1/2,k+1}(r) = \frac{2k-1}{2(k+1)r^2}a_{-1/2,k}(r) - \frac{1}{4k(k+1)r^2}a_{-1/2,k-1}(r)$$

for all $k \in \{1,2,\dots\}$. Moreover, in this case $\beta_{-1/2}(r)$ becomes zero, i.e. the coefficients $a_{-1/2,k}(r)$ have the following property

$$\sum_{k=0}^{n} a_{-1/2,k}(r)r^{2k} = 1.$$

Remark 3.22. Suppose that $p = 1/2$. Then for all $0 < x \le r \le j_{3/2,1}$ we obtain

$$\frac{\sin x}{x} = \sum_{k=0}^{n} a_{1/2,k}(r)(r^2 - x^2)^k + \frac{\mathscr{I}_{n+3/2}(\zeta)}{4^{n+1}(n+1)!(3/2)_{n+1}}(r^2 - x^2)^{n+1},$$

which leads to the new power series expansion of the sine function

$$\frac{\sin x}{x} = \sum_{k \ge 0} a_{1/2,k}(r)(r^2 - x^2)^k,$$

where the coefficients $a_{1/2,k}(r)$ are defined recursively by

$$a_{1/2,0}(r) = \frac{\sin r}{r}, \quad a_{1/2,1}(r) = \frac{\sin r - r\cos r}{2r^3},$$

$$a_{1/2,k+1}(r) = \frac{2k+1}{2(k+1)r^2}a_{1/2,k}(r) - \frac{1}{4k(k+1)r^2}a_{1/2,k-1}(r)$$

for all $k \in \{1,2,\dots\}$. Moreover, in this case $\beta_{1/2}(r)$ becomes zero, i.e. the coefficients $a_{1/2,k}(r)$ have the following property

$$\sum_{k=0}^{n} a_{1/2,k}(r)r^{2k} = 1.$$

We note here that the above series expansions are in fact equivalent to the result of L. Zhu (see [240, Theorem 8] and [240, Theorem 9]), however, in [240] the results in the question were proved just for $0 < x \leq r \leq \pi/2$. See also the paper of L. Zhu [241, Theorem 7], where [240, Theorem 8] is reproduced for $r = \pi/2$. Now, choose in the above relations $r = \pi/2$. Then we obtain (see L. Zhu [241, Theorem 8], [240, Theorem 10]):

$$\frac{\sin x}{x} = \sum_{k \geq 0} b_k(\pi^2 - 4x^2)^k,$$

where

$$b_0 = \frac{2}{\pi}, \quad b_1 = \frac{1}{\pi^3}, \quad b_{k+1} = \frac{2k+1}{2(k+1)\pi^2}b_k - \frac{1}{16k(k+1)\pi^2}b_{k-1}$$

for all $k \in \{1,2,\dots\}$.

Remark 3.23. Finally, choose $r = j_{p,1}$ in (3.139) and (3.140). Then, for all $0 < x \leq j_{p,1}$, we obtain the following new expansion of the Bessel functions of the first kind:

$$J_p(x) = \sum_{k=1}^{n} \left(\frac{x}{j_{p,1}}\right)^p \frac{J_{p+k}(j_{p,1})}{2^k k! j_{p,1}^k}(j_{p,1}^2 - x^2)^k$$

$$+ \left(\frac{x}{\zeta}\right)^p \frac{J_{p+n+1}(\zeta)}{2^{n+1}(n+1)!\zeta^{n+1}}(j_{p,1}^2 - x^2)^{n+1}, \tag{3.143}$$

where $\zeta \in (x, j_{p,1})$. This leads to the new series expansion

$$J_p(x) = \sum_{n \geq 1} \left(\frac{x}{j_{p,1}}\right)^p \frac{J_{p+n}(j_{p,1})}{2^n n! j_{p,1}^n}(j_{p,1}^2 - x^2)^n. \tag{3.144}$$

First observe that this new series expansion is consistent with the fact that $j_{p,1}$ is a simple zero of J_p. More precisely, from (3.144) it follows that $J_p(j_{p,1}) = 0$ and

$J'_p(j_{p,1}) = -J_{p+1}(j_{p,1}) \neq 0$, which shows that $j_{p,1}$ is indeed a simple zero of J_p. Here we used the known fact that the zeros of J_p and J_{p+1} are interlaced and then we have $J_{p+1}(j_{p,1}) \neq 0$. Secondly, it is worth mentioning here that the new formulas (3.139) and (3.140), and consequently (3.143) and (3.144), follows easily from the classical Taylor theorem with Lagrange's form of the remainder (see the book of T.M. Apostol [27]). For this, consider the function

$$x \mapsto \varphi_p(x) = \mathscr{I}_p(\sqrt{x}) = 2^p \Gamma(p+1) x^{-p/2} J_p(\sqrt{x}),$$

which is continuously differentiable $n+1$ times on the whole real line and satisfies the differentiation formula $4(p+1)\varphi'_p(x) = -\varphi_{p+1}(x)$ for all $x \in \mathbb{R}$ and $p > -1$. Then clearly we have

$$\varphi_p^{(n)}(x) = \frac{(-1)^n}{4^n(p+1)_n} \varphi_{p+n}(x)$$

for all $x \in \mathbb{R}$, $n \in \{0,1,2,\dots\}$ and $p > -1$. Consequently, from Taylor's theorem with Lagrange's form of the remainder we conclude that there exists $\xi \in (x,r)$ such that

$$\varphi_p(x) = \sum_{k=0}^{n} \frac{(-1)^k \varphi_{p+k}(r)}{4^k k!(p+1)_k}(x-r)^k + \frac{(-1)^{n+1} \varphi_p(\xi)}{4^{n+1}(n+1)!(p+1)_{n+1}}(x-r)^{n+1},$$

which leads to the power series expansion

$$\varphi_p(x) = \sum_{n \geq 0} \frac{(-1)^n \varphi_{p+n}(r)}{4^n n!(p+1)_n}(x-r)^n.$$

Now, changing x with x^2 and r with r^2 we reobtain the expansions (3.139) and (3.140). Since Taylor's theorem (with the integral formulation of the remainder term) is also valid if the corresponding function has complex values, from the above discussion we conclude that the new series expansions (3.140) and (3.144) are in fact valid for wider range of x and p, i.e. for $x, p \in \mathbb{C}$ such that $p \neq -1, -2, \dots$.

Corresponding to Theorem 3.35 we have the following results for the function $\mathscr{I}_p$. We note that the proof of this theorem is similar to that of Theorem 3.35, however we have included below its proof only for the sake of completeness.

Theorem 3.36. (Á. Baricz and S. Wu [62]) *If $p > -1$ and $0 < x \leq r$, then the following sharp Jordan-type inequalities hold*

$$P_{p,n}(x) + \varepsilon_p(r)(r^2 - x^2)^{n+1} \leq \mathscr{I}_p(x) \leq P_{p,n}(x) + \delta_p(r)(r^2 - x^2)^{n+1}, \quad (3.145)$$

where

$$P_{p,n}(x) = \sum_{k=0}^{n} c_{p,k}(r)(r^2 - x^2)^k,$$

n is an even natural number and the coefficients $c_{p,k}(r)$ are defined explicitly by

$$c_{p,k}(r) = \frac{(-1)^k \mathscr{I}_{p+k}(r)}{4^k k! (p+1)_k} \tag{3.146}$$

for all $k \in \{0,1,\dots,n+1\}$ or recursively by

$$c_{p,0}(r) = \mathscr{I}_p(r), \; c_{p,1}(r) = -\frac{1}{4(p+1)} \mathscr{I}_{p+1}(r),$$

$$c_{p,k+1}(r) = \frac{p+k}{(k+1)r^2} c_{p,k}(r) + \frac{1}{4k(k+1)r^2} c_{p,k-1}(r) \tag{3.147}$$

for all $k \in \{1,2,\dots,n\}$. Moreover, the Jordan-type inequality (3.145) is reversed when n is odd and in both of cases the constants

$$\varepsilon_p(r) = c_{p,n+1}(r) \; \text{ and } \; \delta_p(r) = \frac{1 - \sum\limits_{k=0}^{n} c_{p,k}(r) r^{2k}}{r^{2(n+1)}}$$

are the best possible. In addition, there exist $\zeta \in (x,r)$ depending on m such that

$$\mathscr{I}_p(x) = \sum_{k=0}^{m} c_{p,k}(r)(r^2 - x^2)^k + \frac{(-1)^{m+1} \mathscr{I}_{p+m+1}(\zeta)}{4^{m+1}(m+1)!(p+1)_{m+1}} (r^2 - x^2)^{m+1}, \tag{3.148}$$

which leads to the power series expansion

$$\mathscr{I}_p(x) = \sum_{m \geq 0} c_{p,m}(r)(r^2 - x^2)^m. \tag{3.149}$$

Proof. Our strategy is as in the proof of Theorem 3.35. However, we distinguish here two cases. First we suppose that n is even. Let $m \in \{1,2,\dots,n+1\}$ and consider the functions $h_m, g_m : [0,r] \to \mathbb{R}$, defined by

$$h_1(x) = \mathscr{I}_p(x) - \sum_{k=0}^{n} c_{p,k}(r)(r^2 - x^2)^k,$$

$$h_m(x) = \frac{(-1)^{m-1} \mathscr{I}_{p+m-1}(x)}{4^{m-1}(p+1)_{m-1}} - \sum_{k=m-1}^{n} k(k-1)\dots(k-m+2)c_{p,k}(r)(r^2-x^2)^{k-m+1}$$

for all $m \geq 2$ and

$$g_1(x) = (r^2 - x^2)^{n+1},$$

$$g_m(x) = (n+1)n(n-1)\dots(n-m+3)(r^2-x^2)^{n-m+2}, \; m \geq 2.$$

Now consider the function $Q_2 : (0, r) \to \mathbb{R}$, defined by $Q_2(x) = h_1(x)/g_1(x)$. First observe that the inequality (3.145) is equivalent to $\varepsilon_p(r) \leq Q_2(x) \leq \delta_p(r)$. Thus, to prove (3.145), in what follows we show that Q_2 is decreasing. In view of the differentiation formula

$$\mathscr{I}_p'(x) = \frac{x}{2(p+1)} \mathscr{I}_{p+1}(x),$$

it is easy to verify that for all $s \in \{0, 1, 2, \ldots, n\}$ we have $h_{s+1}(r) = g_{s+1}(r) = 0$ and consequently for each $x \in (0, r)$ and $s \in \{1, 2, \ldots, n\}$ one has

$$\frac{h_s'(x)}{g_s'(x)} = \frac{h_{s+1}(x)}{g_{s+1}(x)} = \frac{h_{s+1}(x) - h_{s+1}(r)}{g_{s+1}(x) - g_{s+1}(r)}.$$

Using again the monotone form of l'Hospital's rule (see Lemma 3.7) $n + 1$ times it is clear that to show that Q_2 is decreasing it is enough to prove that the function

$$x \mapsto \frac{h_{n+1}'(x)}{g_{n+1}'(x)} = -c_{p,n+1}(r) \frac{\mathscr{I}_{p+n+1}(x)}{\mathscr{I}_{p+n+1}(r)}$$

is decreasing on $(0, r)$. But, it is known that for all $p > -1$ the function $x \mapsto \mathscr{I}_p(x)$ is increasing on $(0, \infty)$. Now changing p with $p + n + 1$ we obtain that the function $x \mapsto \mathscr{I}_{p+n+1}(x)$ is increasing too on $(0, \infty)$ and with this the monotonicity of Q_2 is proved.

To complete the proof of (3.145) all that remains is to prove that the constants $\varepsilon_p(r)$ and $\delta_p(r)$ in (3.145) are the best possible. For this observe that

$$\lim_{x \searrow 0} Q_2(x) = \frac{1 - \sum\limits_{k=0}^{n} c_{p,k}(r) r^{2k}}{r^{2(n+1)}} = \delta_p(r)$$

and by using the l'Hospital's rule $n + 1$ times

$$\lim_{x \nearrow r} Q_2(x) = \lim_{x \nearrow r} \frac{h_1'(x)}{g_1'(x)} = \lim_{x \nearrow r} \frac{h_2'(x)}{g_2'(x)} = \ldots = \lim_{x \nearrow r} \frac{h_{n+1}'(x)}{g_{n+1}'(x)} = c_{p,n+1}(r) = \varepsilon_p(r).$$

Now, suppose that n is odd. A similar argument to that presented above yields that in this case that the function

$$x \mapsto \frac{h_{n+1}'(x)}{g_{n+1}'(x)} = c_{p,n+1}(r) \frac{\mathscr{I}_{p+n+1}(x)}{\mathscr{I}_{p+n+1}(r)}$$

is increasing on $(0, r)$. Consequently Q_2 is increasing, and thus we have $\varepsilon_p(r) \geq Q_2(x) \geq \delta_p(r)$ for all $p > -1$ and $0 < x \leq r$, i.e. the inequality (3.145) is reversed.

Finally, to prove the recursive relation for the coefficients $c_{p,k}(r)$ recall the formula (see G.N. Watson [227, p. 79])

$$I_{p-1}(x) - I_{p+1}(x) = (2p/x)I_p(x),$$

from which we easily obtain

$$4p(p+1)\mathscr{I}_{p-1}(x) - x^2\mathscr{I}_{p+1}(x) = 4p(p+1)\mathscr{I}_p(x).$$

Changing in the above relation p with $p+k$ and using (3.146) we easily obtain (3.147).

Now, let us focus on (3.148). Using the Cauchy mean value theorem (see the book of K.R. Stromberg [216, p. 178]) there exist a constant $\xi_1 \in (x,r)$ such that

$$Q_2(x) = \frac{h_1(x)}{g_1(x)} = \frac{h_1(x) - f_1(r)}{g_1(x) - g_1(r)} = \frac{h_1'(\xi_1)}{g_1'(\xi_1)},$$

but as we have mentioned above we have

$$\frac{h_1'(\xi_1)}{g_1'(\xi_1)} = \frac{h_2(\xi_1)}{g_2(\xi_1)} = \frac{h_2(\xi_1) - h_2(r)}{g_2(\xi_1) - g_2(r)}.$$

Thus using again the Cauchy mean value theorem m times we obtain that there exist $\xi_{m+1} \in (\xi_m, r)$, where $\xi_i \in (\xi_{i-1}, r)$ for all $i \in \{2,3,\dots,m\}$, such that

$$Q_2(x) = \frac{h_1'(\xi_1)}{g_1'(\xi_1)} = \frac{h_2'(\xi_2)}{g_2'(\xi_2)} = \dots = \frac{h_m'(\xi_m)}{g_m'(\xi_m)}$$

$$= \frac{h_{m+1}'(\xi_{m+1})}{g_{m+1}'(\xi_{m+1})} = (-1)^{m+1} c_{p,m+1}(r) \frac{\mathscr{I}_{p+m+1}(\xi_{m+1})}{\mathscr{I}_{p+m+1}(r)}.$$

Now denoting ξ_{m+1} with ζ and by using (3.146) we obtain (3.148). On the other hand when m tends to infinity ξ_{m+1} tends to r, and thus $\mathscr{I}_{p+m+1}(\xi_{m+1})/\mathscr{I}_{p+m+1}(r)$ tends to 1. Moreover, $(-1)^{m+1}c_{p,m+1}(r)$ as well as $(-1)^{m+1}c_{p,m+1}(r)(r^2 - x^2)^{m+1}$ tends to zero as m tends to infinity. In other words, the reminder term in (3.148) tends to zero as m tends to infinity, which leads to the series expansion (3.149). □

Remark 3.24. First let us focus on inequality (3.145) and suppose that $p = -1/2$. Then by using

$$\mathscr{I}_{-1/2}(x) = \cosh x, \quad \mathscr{I}_{1/2}(x) = \frac{\sinh x}{x}$$

and the notation

$$P_{p,n}(x) = \sum_{k=0}^{n} c_{p,k}(r)(r^2 - x^2)^k,$$

from Theorem 3.36 for all $0 < x \leq r$ we get the following Kober-type inequality:

$$P_{-1/2,n}(x) + \varepsilon_{-1/2}(r)(r^2 - x^2)^{n+1} \leq \cosh x$$
$$\leq P_{-1/2,n}(x) + \delta_{-1/2}(r)(r^2 - x^2)^{n+1}, \quad (3.150)$$

where

$$P_{-1/2,n}(x) = \sum_{k=0}^{n} c_{-1/2,k}(r)(r^2 - x^2)^k,$$

n is an even natural number and the coefficients $c_{-1/2,k}(r)$ are defined recursively by

$$c_{-1/2,0}(r) = \cosh r, \quad c_{-1/2,1}(r) = -\frac{\sinh r}{2r},$$

$$c_{-1/2,k+1}(r) = \frac{2k-1}{2(k+1)r^2} c_{-1/2,k}(r) + \frac{1}{4k(k+1)r^2} c_{-1/2,k-1}(r)$$

for all $k \in \{1, 2, \ldots, n\}$. Moreover, the constants

$$\varepsilon_{-1/2}(r) = c_{-1/2,n+1}(r) \quad \text{and} \quad \delta_{-1/2}(r) = \frac{1 - \sum_{k=0}^{n} c_{-1/2,k}(r)r^{2k}}{r^{2(n+1)}}$$

are the best possible. Note that, when n is odd, the inequality (3.150) is reversed.

Remark 3.25. Now, suppose that $p = 1/2$. Then, by using

$$\mathscr{I}_{1/2}(x) = \frac{\sinh x}{x}, \quad \mathscr{I}_{3/2}(x) = -3\left(\frac{\sinh x}{x^3} - \frac{\cosh x}{x^2}\right)$$

from (3.145) we obtain for all $0 < x \leq r$ the following sharp Jordan-type inequality:

$$P_{1/2,n}(x) + \varepsilon_{1/2}(r)(r^2 - x^2)^{n+1} \leq \frac{\sinh x}{x}$$
$$\leq P_{1/2,n}(x) + \delta_{1/2}(r)(r^2 - x^2)^{n+1}, \quad (3.151)$$

where

$$P_{1/2,n}(x) = \sum_{k=0}^{n} c_{1/2,k}(r)(r^2 - x^2)^k,$$

n is an even natural number and the coefficients $c_{1/2,k}(r)$ are defined recursively by

$$c_{1/2,0}(r) = \frac{\sinh r}{r}, \quad c_{1/2,1}(r) = \frac{\sinh r - r\cosh r}{2r^3},$$

$$c_{1/2,k+1}(r) = \frac{2k+1}{2(k+1)r^2} c_{1/2,k}(r) + \frac{1}{4k(k+1)r^2} c_{1/2,k-1}(r)$$

for all $k \in \{1, 2, \ldots, n\}$. Moreover, the constants

$$\varepsilon_{1/2}(r) = c_{1/2, n+1}(r) \quad \text{and} \quad \delta_{1/2}(r) = \frac{1 - \sum\limits_{k=0}^{n} c_{1/2, k}(r) r^{2k}}{r^{2(n+1)}}$$

are the best possible. We note that when n is odd then the inequality (3.151) is reversed.

Remark 3.26. Finally, choose $r = j_{p,1}$ in (3.148) and (3.149). Then, for all $0 < x \leq j_{p,1}$, we obtain the following new expansion of the modified Bessel functions of the first kind:

$$I_p(x) = \sum_{k=0}^{m} \left(\frac{x}{j_{p,1}} \right)^p \frac{(-1)^k I_{p+k}(j_{p,1})}{2^k k! j_{p,1}^k} (j_{p,1}^2 - x^2)^k$$

$$+ \left(\frac{x}{\zeta} \right)^p \frac{(-1)^{m+1} I_{p+m+1}(\zeta)}{2^{m+1}(m+1)! \zeta^{m+1}} (j_{p,1}^2 - x^2)^{m+1},$$

where $\zeta \in (x, j_{p,1})$. This leads to the new series expansion

$$I_p(x) = \sum_{m \geq 0} \left(\frac{x}{j_{p,1}} \right)^p \frac{(-1)^m I_{p+m}(j_{p,1})}{2^m m! j_{p,1}^m} (j_{p,1}^2 - x^2)^m. \tag{3.152}$$

Observe that, analogously to the results of the previous section, (3.148), (3.149) and the above formulas are valid for all $x, p \in \mathbb{C}$ such that $p \neq -1, -2, \ldots$. This is an immediate consequences of Taylor's theorem with Lagrange's form of the remainder (see the book of T.M. Apostol [27]) applied to the function

$$x \mapsto \gamma_p(x) = \mathscr{I}_p(\sqrt{x}) = 2^p \Gamma(p+1) x^{-p/2} I_p(\sqrt{x}).$$

We note that this simple idea to use instead of $\mathscr{I}_p$ the function γ_p is also useful in the problem of finding a finite sum formula for the probability density function of the non-central chi-squared distribution. See the paper of S. András and Á. Baricz [23] for further details.

In what follows, motivated by the work of D.W. Niu [164] we reformulate the main results of this section for generalized Bessel functions of the first kind. We show that the main results of this section, namely Theorems 3.35 and 3.36, can be unified as follows, which improves significantly the earlier results of Á. Baricz [49, Theorem 14] (see Theorem 3.34) and [53, Theorem 2.2]. We note that when $b = c = 1$ then Theorem 3.37 reduces to Theorem 1, while for $b = 1$ and $c = -1$ Theorem 3.37 becomes Theorem 3.36.

Theorem 3.37. (Á. Baricz and S. Wu [62]) *Let $\kappa > 0$. Then for all $c \in [0, 1]$ and $0 < x \leq r \leq j_{\kappa, 1}$ the following sharp Jordan-type inequalities hold*

$$G_{p,m}(x) + \varpi_p(r)(r^2 - x^2)^{m+1} \leq \lambda_p(x) \leq G_{p,m}(x) + \tau_p(r)(r^2 - x^2)^{m+1}, \tag{3.153}$$

where

$$G_{p,m}(x) = \sum_{i=0}^{m} d_{p,i}(r)(r^2 - x^2)^i,$$

m is a natural number and the coefficients $d_{p,i}(r)$ are defined explicitly by

$$d_{p,i}(r) = \left(\frac{c}{4}\right)^i \frac{\lambda_{p+i}(r)}{i!(\kappa)_i}$$

for all $i \in \{0, 1, \ldots, m+1\}$ or recursively by

$$d_{p,0}(r) = \lambda_p(r), \; d_{p,1}(r) = \frac{c}{4\kappa}\lambda_{p+1}(r),$$

$$d_{p,i+1}(r) = \frac{\kappa + i - 1}{(i+1)r^2}d_{p,i}(r) - \frac{c}{4i(i+1)r^2}d_{p,i-1}(r)$$

for all $i \in \{1, 2, \ldots, n\}$. Moreover, if $c \leq 0$, $0 < x \leq r$ and m is even, then the Jordan-type inequality (3.153) holds true, while if $c \leq 0$, $0 < x \leq r$ and m is odd, then the Jordan-type inequality (3.153) is reversed. In each of cases the constants

$$\varpi_p(r) = d_{p,m+1}(r) \; and \; \tau_p(r) = \frac{1 - \sum\limits_{k=0}^{m} d_{p,k}(r)r^{2k}}{r^{2(m+1)}}$$

are the best possible. In addition, for all $c \leq 1$ there exist $\zeta \in (x, r)$ depending on m such that

$$\lambda_p(x) = \sum_{i=0}^{m} d_{p,i}(r)(r^2 - x^2)^i + \frac{c^{m+1}\lambda_{p+m+1}(\zeta)}{4^{m+1}(m+1)!(\kappa)_{m+1}}(r^2 - x^2)^{m+1},$$

which leads to the power series expansion

$$\lambda_p(x) = \sum_{m \geq 0} d_{p,m}(r)(r^2 - x^2)^m.$$

Proof. First observe that if $c \geq 0$, then $\mathscr{I}_{\kappa-1}(t\sqrt{c}) = \lambda_p(t)$, while for $c \leq 0$ we have $\mathscr{I}_{\kappa-1}(t\sqrt{-c}) = \lambda_p(t)$. Thus, if we suppose that $c \in [0, 1]$ and we change in (3.136) p with $\kappa - 1$, x with $x\sqrt{c}$ and r with $r\sqrt{c}$, then we obtain (3.153). Similarly, if we suppose that $c \leq 0$ and we change in (3.145) p with $\kappa - 1$, x with $x\sqrt{-c}$ and r with $r\sqrt{-c}$, then we obtain (3.153). $\quad\square$

Remark 3.27. During the course of writing the thesis [54] we have found the work of D.W. Niu [164], where among other things, motivated by [49, Theorem 14] the inequality (3.153) is proved for $m = n - 1$. See also the survey papers of F. Qi and D.W. Niu [192], F. Qi et al. [193] for more details. However, in the case when

$c \in [0,1]$, it is assumed that $\kappa \geq 1/2$ and $r \leq \pi/2$. As we have seen above these conditions can be relaxed to $\kappa > 0$ and $r \leq j_{\kappa,1}$. Moreover, it is important to note here that since

$$1.570796327 = \pi/2 = j_{-1/2,1} < 2.4048255577 = j_{0,1} < j_{\kappa,1},$$

the interval $(0, j_{\kappa,1}]$ is larger than the interval $(0, \pi/2]$. We note that, since in [164] the approach is somewhat different to that given in the papers of L. Zhu [240–242] apparently it is not clear what is the connection between the inequality (3.142) and inequality (3.153) for $m = n - 1$. This is because in [240–242] the corresponding coefficients are defined recursively, while in [164] explicitly. However, from our discussion it is clear that in fact the inequality (3.142) is a particular case of (3.153), just taking $p = -1/2$ and $b = c = 1$ in Theorem 3.37. Finally, recall that when $b = 2$ and $c = 1$ then the generalized Bessel function of the first kind w_p becomes $x \mapsto (2/\sqrt{\pi})j_p(x)$, where

$$x \mapsto j_p(x) = \sqrt{\pi/(2x)}J_{p+1/2}(x)$$

is the spherical Bessel function of the first kind [1, p. 437], and in this case λ_p reduces to the function

$$x \mapsto \mathscr{J}_{p+1/2}(x) = 2^{p+1/2}\Gamma(p+3/2)x^{-(p+1/2)}J_{p+1/2}(x).$$

Similarly, when $b = 2$ and $c = -1$ then w_p reduces to $x \mapsto (2/\sqrt{\pi})i_p(x)$, where

$$x \mapsto i_p(x) = \sqrt{\pi/(2x)}I_{p+1/2}(x)$$

is the modified spherical Bessel function of the first kind [1, p. 443] and λ_p in this case becomes

$$x \mapsto \mathscr{I}_{p+1/2}(x) = 2^{p+1/2}\Gamma(p+3/2)x^{-(p+1/2)}I_{p+1/2}(x).$$

If we choose $b = 2$ and $c = \pm 1$ in Theorem 3.37 then we can obtain the corresponding results to Theorems 3.35 and 3.36 for spherical and modified spherical Bessel functions of the first kind.

3.7.6 The Sine and Hyperbolic Sine Integral

Let

$$\mathrm{Si}(x) = \int_0^x \frac{\sin t}{t}\, dt \quad \text{and} \quad \mathrm{Shi}(x) = \int_0^x \frac{\sinh t}{t}\, dt$$

be the sine and hyperbolic sine integral, which play an important role in various topics of Fourier analysis (see for example the book of A. Zygmund [246]).

S. Koumandos [137] recently proved that the sine integral is sub-additive on $[0, \infty)$. In this section we prove that the hyperbolic sine integral is super-additive on $[0, \infty)$. Moreover, we show that these properties also hold for the following integral, which generalizes the sine and hyperbolic sine integrals. For $c \in \mathbb{R}$ and $\kappa > 0$ let us consider

$$\varsigma_p(x) = \int_0^x \lambda_p(t)\,dt = \sum_{n \geq 0} \frac{(-c/4)^n}{(\kappa)_n n!} \cdot \frac{x^{2n+1}}{2n+1}. \tag{3.154}$$

Clearly, when $b = 1$ and $c = 1$, then $\varsigma_{-1/2}(x) = \sin x$, $\varsigma_{1/2}(x) = \mathrm{Si}(x)$, while when $b = 1$ and $c = -1$, then $\varsigma_{-1/2}(x) = \sinh x$, $\varsigma_{1/2}(x) = \mathrm{Shi}(x)$. Our main result in this section reads as follows.

Theorem 3.38. (Á. Baricz [49]) *If $c \in [0, 1]$ and $\kappa \geq 1/2$, then ς_p is sub-additive on $[0, \pi]$. Moreover, if $c \leq 0$ and $\kappa > 0$, then ς_p is super-additive on $[0, \infty)$.*

Proof. It is known that if a function $f : [0, \infty) \to \mathbb{R}$, where $f(0) = 0$, has a continuous derivative on $[0, \infty)$ and the continuous function $x \in [0, \infty) \mapsto f(x)/x \in \mathbb{R}$ is decreasing (increasing), then f is sub-additive (super-additive). An easy calculation from (3.154) yields

$$\frac{d}{dx}\left(\frac{\varsigma_p(x)}{x}\right) = \frac{x\lambda_p(x) - \varsigma_p(x)}{x^2}. \tag{3.155}$$

Let us consider the function $\varphi_{10}(x) = x\lambda_p(x) - \varsigma_p(x)$. It follows that $\varphi'_{10}(x) = x\lambda'_p(x)$.

If $c \in [0, 1]$ and $\kappa \geq 1/2$, then it is known (see Theorem 3.29) that λ_p is decreasing on $[0, \pi]$, thus φ_{10} is decreasing too on $[0, \pi]$. Hence $\varphi_{10}(x) \leq \varphi_{10}(0) = 0$ and therefore from (3.155) we obtain that $x \mapsto \varsigma_p(x)/x$ is decreasing on $[0, \pi]$. From this follows the sub-additivity of ς_p.

If $c \leq 0$ and $\kappa > 0$, then from the series representation of λ_p, clearly λ_p has positive coefficients, and consequently it is increasing on $[0, \infty)$. Thus the functions φ_{10} and $x \mapsto \varsigma_p(x)/x$ are also increasing on $[0, \infty)$, which completes the proof. $\square$

Remark 3.28. We note that using the Chebyshev integral inequality (1.40) we can deduce other inequalities involving ς_p and λ_p. For example, choosing $p(t) = 1$, $f(t) = \lambda_{p+1}(t)$ ($f(t) = t\lambda_{p+1}(t)$ respectively) and $g(t) = t$, $t \in [a, b] = [0, x]$, $x \in [0, \infty)$ in (1.40), we have that if $\kappa > 0$ and $c < 0$, then

$$(-c)x\varsigma_{p+1}(x) \leq (4\kappa)\lambda_p(x) \quad \text{and} \quad 2\varsigma_p(x) \leq x\lambda_p(x) \quad \text{for all } x \geq 0.$$

Here we used the differentiation formula (3.105) and the fact that when $\kappa > 0$ and $c < 0$, then the functions $x \mapsto \lambda_p(x)$ and $x \mapsto x\lambda_p(x)$ are increasing on $[0, \infty)$.

Remark 3.29. In a recent paper H. Alzer and S. Koumandos [8] established sub- and super-additive properties of Fejér's sine polynomial $\sum_{m=1}^n (\sin mx)/m$. We note that if we consider the sum $\sum_{m=1}^{n} x\lambda_p(mx)$, then it is easy to verify that it is sub-additive (super-additive respectively) on $[0,\pi/m]$ (on $[0,\infty)$ respectively) if $c \in [0,1]$ and $\kappa \geq 1/2$ (if $c \leq 0$ and $\kappa > 0$ respectively).

3.8 Redheffer Type Inequalities for Bessel Functions

In Theorem 3.28 we extended Redheffer's inequality (3.106) for generalized Bessel functions. Our aim in this section is to continue this investigation. We note that motivated by this inequality C.P. Chen et al. [81] using mathematical induction and infinite product representation of $\cosh x$ established the following Redheffer-type inequality, which is the hyperbolic analogue of (3.111):

$$\cosh x \leq \frac{\pi^2 + 4x^2}{\pi^2 - 4x^2} \quad \text{for all} \quad x \in \left[-\frac{\pi}{2}, \frac{\pi}{2}\right]. \tag{3.156}$$

Moreover, they found the hyperbolic analogue of inequality (3.106), by showing that

$$\frac{\sinh x}{x} \leq \frac{\pi^2 + x^2}{\pi^2 - x^2} \quad \text{for all} \quad x \in [-\pi, \pi]. \tag{3.157}$$

As we mentioned above the proofs of the inequalities (3.111), (3.156) and (3.157) by C.P. Chen et al. [81] are based on the following representations (see the book of M. Abramowitz and I.A. Stegun [1, pp. 75 and 85]) of $\cos x$, $\sinh x$ and $\cosh x$:

$$\cos x = \prod_{n \geq 1} \left[1 - \frac{4x^2}{(2n-1)^2\pi^2}\right], \quad \cosh x = \prod_{n \geq 1} \left[1 + \frac{4x^2}{(2n-1)^2\pi^2}\right] \tag{3.158}$$

and

$$\frac{\sinh x}{x} = \prod_{n \geq 1} \left(1 + \frac{x^2}{n^2\pi^2}\right) \tag{3.159}$$

respectively. We note that using their method mutatis mutandis, in view of the following infinite product representation [1, p. 75] of $\sin x$, namely

$$\frac{\sin x}{x} = \prod_{n \geq 1} \left(1 - \frac{x^2}{n^2\pi^2}\right), \tag{3.160}$$

it can be easily shown that (3.106) holds for all $x \in [-\pi, \pi]$ (see the survey article of F. Qi [189]). Moreover, as we can see in the following sections the idea of using mathematical induction and infinite product representation is also fruitful for Bessel functions as well as for the Γ function. All results of this section may be found in the recent papers of Á. Baricz [47], Á. Baricz and S. Wu [63].

3.8.1 An Extension of Redheffer's Inequality and Its Hyperbolic Analogue

Our first main result of this section reads as follows.

Theorem 3.39. (Á. Baricz [47]) *Let us consider the functions* $\mathscr{J}_p : \mathbb{R} \to (-\infty, 1]$ *and* $\mathscr{I}_p : \mathbb{R} \to [1, \infty)$, *defined by (1.23) and (1.24), i.e.*

$$\mathscr{J}_p(x) = 2^p \Gamma(p+1) x^{-p} J_p(x) \text{ and } \mathscr{I}_p(x) = 2^p \Gamma(p+1) x^{-p} I_p(x),$$

where J_p *and* I_p *is the Bessel function, modified Bessel function respectively. Furthermore, suppose that* $p > -1$ *and let* $j_{p,n}$ *be the nth positive zero of the Bessel function* J_p. *If* $\Delta_p(n) = j_{p,n+1}^2 - j_{p,1} j_{p,n} - j_{p,n} j_{p,n+1} \geq 0$, *where* $n \in \{1, 2, \ldots\}$, *then the following inequalities hold:*

$$\mathscr{J}_p(x) \geq \frac{j_{p,1}^2 - x^2}{j_{p,1}^2 + x^2} \text{ for all } x \in [-\xi_p, \xi_p], \tag{3.161}$$

$$\mathscr{I}_p(x) \leq \frac{j_{p,1}^2 + x^2}{j_{p,1}^2 - x^2} \text{ for all } x \in (-j_{p,1}, j_{p,1}), \tag{3.162}$$

where

$$\xi_p = \min_{n \geq 1, p > -1} \left\{ j_{p,1}, \sqrt{\Delta_p(n)} \right\}.$$

Proof. In order to prove (3.161) it is enough to show that

$$\mathscr{J}_p(x j_{p,1}) \geq \frac{1 - x^2}{1 + x^2} \tag{3.163}$$

for all $x \in [-\xi_p / j_{p,1}, \xi_p / j_{p,1}]$. It is known that for J_p the following infinite product formula (see G.N. Watson [227, p. 498]) is valid for arbitrary x and $p \neq -1, -2, \ldots$:

$$\mathscr{J}_p(x) = 2^p \Gamma(p+1) x^{-p} J_p(x) = \prod_{n \geq 1} \left(1 - \frac{x^2}{j_{p,n}^2} \right).$$

From this we deduce that

$$\mathscr{J}_p(x j_{p,1}) = \frac{1 - x^2}{1 + x^2} \left[(1 + x^2) \lim_{n \to \infty} Q_{p,n} \right], \tag{3.164}$$

where

$$Q_{p,n} = \prod_{k=2}^{n} \left(1 - \frac{x^2 j_{p,1}^2}{j_{p,k}^2} \right).$$

In what follows we prove by mathematical induction that

$$(1+x^2)Q_{p,n} \geq 1 + \frac{x^2 j_{p,1}}{j_{p,n}} \tag{3.165}$$

for all $p > -1$, $n \geq 2$ and $x \in [-\xi_p/j_{p,1}, \xi_p/j_{p,1}]$. For $n = 2$ clearly we have

$$(1+x^2)Q_{p,2} - \left(1 + \frac{x^2 j_{p,1}}{j_{p,2}}\right) = \frac{x^2}{j_{p,2}^2}\left[\Delta_p(1) - j_{p,1}^2 x^2\right] \geq 0.$$

Now suppose that (3.165) holds for some $m \geq 2$. From the definition of $Q_{p,m}$, we easily get

$$Q_{p,m+1} = Q_{p,m} \cdot \left(1 - \frac{x^2 j_{p,1}^2}{j_{p,m+1}^2}\right) \quad \text{for all } m \in \{2,3,4,\dots\},$$

thus

$$(1+x^2)Q_{p,m+1} - \left(1 + \frac{x^2 j_{p,1}}{j_{p,m+1}}\right)$$

$$= (1+x^2)Q_{p,m}\left(1 - \frac{x^2 j_{p,1}^2}{j_{p,m+1}^2}\right) - \left(1 + \frac{x^2 j_{p,1}}{j_{p,m+1}}\right)$$

$$\geq \left(1 + \frac{x^2 j_{p,1}}{j_{p,m}}\right)\left(1 - \frac{x^2 j_{p,1}^2}{j_{p,m+1}^2}\right) - \left(1 + \frac{x^2 j_{p,1}}{j_{p,m+1}}\right)$$

$$= \frac{x^2 j_{p,1}}{j_{p,m} j_{p,m+1}^2}\left[\Delta_p(m) - j_{p,1}^2 x^2\right] \geq 0,$$

and hence by induction (3.165) follows. Here we used that the hypothesis implies that $x \in [-\sqrt{\Delta_p(m)}/j_{p,1}, \sqrt{\Delta_p(m)}/j_{p,1}] \subseteq [-j_{p,m+1}/j_{p,1}, j_{p,m+1}/j_{p,1}]$. On the other hand, from the MacMahon expansion (see G.N. Watson [227, p. 506]),

$$j_{p,n} = (n + p/2 - 1/4)\pi + \mathcal{O}(n^{-1}), \quad n \to \infty,$$

we have $j_{p,n} \to \infty$, as n tends to infinity. Finally, we obtain from (3.165) that

$$\lim_{n \to \infty}(1+x^2)Q_{p,n} \geq \lim_{n \to \infty}\left(1 + \frac{x^2 j_{p,1}}{j_{p,n}}\right) = 1,$$

which in view of (3.164) implies (3.163). This completes the proof of (3.161).

Proceeding similarly as in the proof of (3.161) we can prove (3.162). It suffices to show that

$$\mathscr{I}_p(x j_{p,1}) \leq \frac{1+x^2}{1-x^2} \tag{3.166}$$

for all $x \in (-1,1)$. It is known that for I_p the following infinite product formula is also valid for arbitrary x and $p \neq -1, -2, \ldots$:

$$\mathscr{I}_p(x) = 2^p \Gamma(p+1) x^{-p} I_p(x) = \prod_{n \geq 1} \left(1 + \frac{x^2}{j_{p,n}^2} \right).$$

From this we get that

$$\mathscr{I}_p(x j_{p,1}) = \frac{1+x^2}{1-x^2} \left[(1-x^2) \lim_{n \to \infty} R_{p,n} \right], \tag{3.167}$$

where

$$R_{p,n} = \prod_{k=2}^{n} \left(1 + \frac{x^2 j_{p,1}^2}{j_{p,k}^2} \right).$$

In what follows we show by using mathematical induction that

$$(1-x^2) R_{p,n} \leq 1 - \frac{x^2 j_{p,1}}{j_{p,n}} \tag{3.168}$$

for all $p > -1$, $n \geq 2$ and $x \in (-1,1)$. For $n = 2$ clearly we have

$$(1-x^2) R_{p,2} - \left(1 - \frac{x^2 j_{p,1}}{j_{p,2}} \right) = \frac{x^2}{j_{p,2}^2} \left[-\Delta_p(1) - j_{p,1}^2 x^2 \right] \leq 0.$$

Now suppose that (3.168) holds for some $m \geq 2$. From the definition of $R_{p,m}$, we easily get

$$R_{p,m+1} = R_{p,m} \cdot \left(1 + \frac{x^2 j_{p,1}^2}{j_{p,m+1}^2} \right) \quad \text{for all } m \in \{2,3,4,\ldots\},$$

thus

$$(1-x^2) R_{p,m+1} - \left(1 - \frac{x^2 j_{p,1}}{j_{p,m+1}} \right)$$

$$= (1-x^2) R_{p,m} \left(1 + \frac{x^2 j_{p,1}^2}{j_{p,m+1}^2} \right) - \left(1 - \frac{x^2 j_{p,1}}{j_{p,m+1}} \right)$$

$$\leq \left(1 - \frac{x^2 j_{p,1}}{j_{p,m}} \right) \left(1 + \frac{x^2 j_{p,1}^2}{j_{p,m+1}^2} \right) - \left(1 - \frac{x^2 j_{p,1}}{j_{p,m+1}} \right)$$

$$= \frac{x^2 j_{p,1}}{j_{p,m} j_{p,m+1}^2} \left[-\Delta_p(m) - j_{p,1}^2 x^2 \right] \leq 0,$$

and hence by induction (3.168) follows. Finally using again the fact that $j_{p,n} \to \infty$, as n tends to infinity, from (3.168) we obtain that

$$\lim_{n\to\infty}(1-x^2)R_{p,n} \leq \lim_{n\to\infty}\left(1-\frac{x^2 j_{p,1}}{j_{p,n}}\right) = 1,$$

which in view of (3.167) implies (3.166). Thus the proof is complete.   □

Remark 3.30. For later use let us recall that for $p = -1/2$ and $p = 1/2$, respectively the functions $\mathscr{J}_p$ and $\mathscr{I}_p$ reduce to some elementary functions, like cosine, hyperbolic cosine, sine and hyperbolic sine. In view of the infinite product representations (3.158) and (3.160) of the functions $\mathscr{J}_{-1/2}$ and $\mathscr{J}_{1/2}$ it is clear that $j_{-1/2,n} = (2n-1)\pi/2$ and $j_{1/2,n} = n\pi$ for all $n \in \{1,2,\ldots\}$. Consequently, for all $n = 1,2,\ldots$ we have

$$\sqrt{\Delta_{1/2}(n)} = j_{1/2,1} = \pi$$

and

$$\sqrt{\Delta_{-1/2}(n)} = \frac{\pi}{2}\sqrt{2n+3} \geq \frac{\pi}{2}\sqrt{5} > \frac{\pi}{2} = j_{-1/2,1},$$

which imply that $\xi_{-1/2} = \pi/2$ and $\xi_{1/2} = \pi$. So, if we take in (3.161) $p = -1/2$ and $p = 1/2$, respectively, then we reobtain the first inequality from (3.156) and Redheffer's inequality (3.106) respectively, with the intervals of validity $[-\pi/2,\pi/2]$ and $[-\pi,\pi]$, respectively. The situation is similar with inequality (3.162), namely, if we choose in (3.162) $p = -1/2$ and $p = 1/2$, respectively, then we get the second inequality from (3.156) and inequality (3.157), with the same intervals of validity, i.e. $[-\pi/2,\pi/2]$ and $[-\pi,\pi]$, respectively.

3.8.2 Sharp Exponential Redheffer-Type Inequalities for Bessel Functions

Recall that recently motivated by Redheffer's inequality (3.106), C.P. Chen et al. [81], by using mathematical induction and infinite product representation of $\cos x$, $\sinh x$ and $\cosh x$, established the Redheffer-type inequalities (3.111), (3.156) and (3.157). In order to sharpen and extend the above results, recently L. Zhu and J. Sun [243] presented six new Redheffer-type inequalities involving circular functions and hyperbolic functions. In this section our aim is to continue these investigations by extending all of the results of L. Zhu and J. Sun [243] and by improving the results of the previous section. First we present some sharp exponential Redheffer-type inequalities for Bessel and modified Bessel functions of the first kind, and after this we offer some immediate applications of these results. Moreover, in this section we present alternative proofs for some inequalities for Bessel functions of the first kind established by E.K. Ifantis and P.D. Siafarikas [118]. The key tools in our proofs are some known results on the zeros of Bessel functions of the first kind, like the Rayleigh bounds on the square of the first positive zero of Bessel functions, and some recent results of the author on Bessel functions. We note that all results of this section may be found in the recent paper of Á. Baricz and S. Wu [63].

The following preliminary result will be useful in the sequel.

Lemma 3.8. (Á. Baricz and S. Wu [63]) *Let $p > -1$ and let $j_{p,1}$ be the first positive zero of the Bessel function J_p. Then the equation $j_{p,1}^2 = 8(p+1)$ has exactly one positive root $p_0 \in (1,2)$. Moreover, if $p \in (-1,p_0]$, then $j_{p,1}^2 \leq 8(p+1)$, and if $p \geq p_0$, then the above inequality is reversed.*

Proof. Due to M.E.H. Ismail and M.E. Muldoon [120, Theorem 2] it is known that the function $p \mapsto j_{p,1}^2/(p+1)$ is increasing on $(-1,\infty)$ (see also the papers of M.E.H. Ismail and M.E. Muldoon [121, p. 9], Á. Elbert and P.D. Siafarikas [95, p. 57]). One the other hand, concerning the local behavior of $j_{p,1}$, in the neighborhood of $p = -1$ the following representation is valid (see the paper of R. Piessens [171])

$$j_{p,1} = 2(p+1)^{1/2}\left[1 + \frac{1}{4}(p+1) - \frac{7}{96}(p+1)^2 + \dots\right],$$

which implies that $j_{p,1}^2/(p+1) \to 4$ as $p \to -1$. It is also known (see [120, Eq. 6.8]) that for each $p > -1$ we have the following lower and upper bounds for the square of the first positive zero

$$4(p+1)(p+2)^{1/2} < j_{p,1}^2 < 2(p+1)(p+3).$$

Hence $j_{p,1}^2/(p+1) \to \infty$ as $p \to \infty$, and consequently indeed the equation $j_{p,1}^2 = 8(p+1)$ has an unique solution $p_0 > -1$. Now we prove that $p_0 \in (1,2)$. Using the above inequalities we get that $j_{p,1}^2 < 8(p+1)$ for each $p \in (-1,1)$ and $j_{p,1}^2 > 8(p+1)$ for each $p > 2$. Since $j_{1,1} = 3.83171\dots$ and $j_{2,1} = 5.13562\dots$ (see [1, p. 409]), it is easy to see that for $p \in \{1,2\}$ the equation $j_{p,1}^2 = 8(p+1)$ is not satisfied and thus p_0 is in the interval $(1,2)$. With this the proof is complete. $\square$

The following result improves and complements (3.161).

Theorem 3.40. (Á. Baricz and S. Wu [63]) *Let $p > -1$ and let*

$$\tau_p = [8(p+1) - j_{p,1}^2]^{1/2},$$

$$\omega_p = \begin{cases} j_{p,1}, & \text{if } p \in (-1,-1/2] \\ \tau_p, & \text{if } p \in (-1/2,0) \\ j_{p,1}, & \text{if } p \in [0,1/2] \\ \tau_p, & \text{if } p \in (1/2,p_0) \end{cases},$$

where $j_{p,1}$ is the first positive zero of the Bessel function of the first kind J_p and p_0 is the unique solution of the equation $j_{p,1}^2 = 8(p+1)$. Then the following sharp exponential Redheffer-type inequalities hold

$$\mathcal{I}_p(x) \geq \left(\frac{j_{p,1}^2 - x^2}{j_{p,1}^2 + x^2} \right)^{\alpha_p} \quad \text{for all } |x| < \omega_p \text{ and } p \in (-1, p_0), \quad (3.169)$$

$$\mathcal{I}_p(x) \leq \left(\frac{j_{p,1}^2 - x^2}{j_{p,1}^2 + x^2} \right)^{\beta_p} \quad \text{for all } |x| < j_{p,1} \text{ and } p \geq -7/8, \quad (3.170)$$

$$\frac{\mathcal{I}_{p+1}(x)}{\mathcal{I}_p(x)} \geq \left(\frac{j_{p,1}^2 + x^2}{j_{p,1}^2 - x^2} \right)^{\delta_p} \quad \text{for all } |x| < j_{p,1} \text{ and } p \geq -7/8, \quad (3.171)$$

$$\frac{\mathcal{I}_{p+1}(x)}{\mathcal{I}_p(x)} \leq \left(\frac{j_{p,1}^2 + x^2}{j_{p,1}^2 - x^2} \right)^{\alpha_p} \quad \text{for all } |x| < j_{p,1} \text{ and } p > -1, \quad (3.172)$$

with the best possible constants

$$\alpha_p = 1, \ \beta_p = \frac{j_{p,1}^2}{8(p+1)} \quad \text{and} \quad \delta_p = \frac{j_{p,1}^2}{8(p+1)(p+2)}.$$

Proof. Observe that due to Lemma 3.8 τ_p is well defined. On the other hand, since the function $p \mapsto j_{p,1}^2/(p+1)$ is increasing, clearly for all $p > -1$ one has $j_{p,1}^2 > 4(p+1)$, i.e. for all $p \in (-1, p_0)$ we have $\tau_p < j_{p,1}$. Since all functions which appear in inequalities (3.169), (3.170), (3.171) and (3.172) are even, without loss of generality, in what follows we assume that $x \in (0, \tau_p)$ or $x \in (0, j_{p,1})$, depending on the inequality in the question. First we show that, assuming that the inequalities (3.169), (3.170), (3.171) and (3.172) hold, then the constants α_p, β_p and δ_p are the best possible. For this consider the functions $f_p, g_p : (0, j_{p,1}) \to \mathbb{R}$, defined by

$$f_p(x) = \frac{\log \mathcal{I}_p(x)}{\log \left(\frac{j_{p,1}^2 - x^2}{j_{p,1}^2 + x^2} \right)} \quad \text{and} \quad g_p(x) = \frac{\log \frac{\mathcal{I}_{p+1}(x)}{\mathcal{I}_p(x)}}{\log \left(\frac{j_{p,1}^2 + x^2}{j_{p,1}^2 - x^2} \right)}.$$

Since for all $|x| < j_{p,1}$ and $p > -1$ we have $\mathcal{I}_p(x) > 0$ (see Á. Baricz [51, Theorem 3]), clearly the function f_p is well defined. Similarly, because for all $|x| < j_{p+1,1}$ and $p > -1$ we have $\mathcal{I}_{p+1}(x) > 0$ and $(-j_{p,1}, j_{p,1}) \subset (-j_{p+1,1}, j_{p+1,1})$, it follows that the function g_p is well defined too. Using the l'Hospital rule it is easy to verify that we have

$$\lim_{x \to 0} f_p(x) = \lim_{x \to 0} \frac{\mathcal{I}_p'(x)}{\mathcal{I}_p(x)} \cdot \frac{x^4 - j_{p,1}^4}{4x j_{p,1}^2} = \lim_{x \to 0} \frac{\mathcal{I}_{p+1}(x)}{\mathcal{I}_p(x)} \cdot \frac{j_{p,1}^4 - x^4}{8(p+1) j_{p,1}^2} = \beta_p,$$

$$\lim_{x \to j_{p,1}} f_p(x) = \lim_{x \to j_{p,1}} \frac{\mathscr{J}_p'(x)}{\mathscr{J}_p(x)} \cdot \frac{x^4 - j_{p,1}^4}{4xj_{p,1}^2} = \lim_{x \to j_{p,1}} \frac{\mathscr{J}_{p+1}(j_{p,1})}{8(p+1)j_{p,1}^2} \cdot \frac{j_{p,1}^4 - x^4}{\mathscr{J}_p(x)} = \alpha_p,$$

$$\lim_{x \to 0} g_p(x) = \lim_{x \to 0} \frac{j_{p,1}^4 - x^4}{8(p+1)(p+2)j_{p,1}^2} = \delta_p$$

and then

$$\lim_{x \to j_{p,1}} g_p(x) = -\lim_{x \to j_{p,1}} \frac{\log \mathscr{J}_p(x)}{\log \left(\frac{j_{p,1}^2 + x^2}{j_{p,1}^2 - x^2} \right)} = \lim_{x \to j_{p,1}} \frac{\log \mathscr{J}_p(x)}{\log \left(\frac{j_{p,1}^2 - x^2}{j_{p,1}^2 + x^2} \right)} = \lim_{x \to j_{p,1}} f_p(x) = \alpha_p,$$

where we have used the differentiation formula

$$\mathscr{J}_p'(x) = -\frac{x}{2(p+1)} \mathscr{J}_{p+1}(x), \tag{3.173}$$

which can be verified easily by using the series representation of the function $\mathscr{J}_p$. In other words, we have $f_p(0^+) = \beta_p$, $g_p(0^+) = \delta_p$ and $f_p(j_{p,1}^-) = g_p(j_{p,1}^-) = \alpha_p$, which show that the constants α_p, β_p and δ_p are the best possible.

Now let us focus on the inequality (3.169). Recall that this inequality was proved in Theorem 3.39 (see the inequality (3.161)), but under the assumption that $\Delta_p(n) = j_{p,n+1}^2 - j_{p,1}j_{p,n} - j_{p,n}j_{p,n+1} \geq 0$ for all $n \in \{1,2,3,\dots\}$ and $|x| < \xi_p$. Here we show that the above condition can be relaxed for $p \in [0,1/2]$ and $p \in (-1,-1/2]$. For this consider the (unique) solution $j = j_{p,k}$ of the differential equation

$$\frac{dj}{dp} = 2j \int_0^\infty K_0(2j\sinh t)e^{-2pt}\,dt,$$

which satisfies the condition $j \to 0$ as $p \to -k^+$. Here K_0 stands for the modified Bessel function of the second kind of zero order (see G.N. Watson [227]), defined by

$$K_0(x) = \int_0^\infty e^{-x\cosh t}\,dt,$$

and for $k \in \{1,2,3,\dots\}$ the $j_{p,k}$ becomes exactly the kth positive zero of the Bessel function J_p of the first kind. Due to Á. Elbert and A. Laforgia [94] it is known that if $p \in [0,1/2]$, then $j_{p,k}$ is convex with respect to k. In other words, in particular we have that the sequence $\{j_{p,n}\}_{n\geq 0}$, where $j_{p,0} = 0$, is convex when $p \in [0,1/2]$. This implies that for each $p \in [0,1/2]$ and $n \in \{1,2,3,\dots\}$ we have

$$\Delta_p(n) - j_{p,1}^2 = (j_{p,n+1} + j_{p,1})(j_{p,n+1} - j_{p,n} - j_{p,1}) \geq 0,$$

since

$$j_{p,n+1} - j_{p,n} \geq j_{p,n} - j_{p,n-1} \geq \dots \geq j_{p,2} - j_{p,1} \geq j_{p,1}.$$

Consequently, if $p \in [0,1/2]$, then $\xi_p = j_{p,1}$.

On the other hand it is known (see A. Deaño et al. [87, Theorem 21]) that if $|p| \geq 1/2$, then for all $n \in \{1,2,3,\dots\}$ we have $j_{p,n+1} - j_{p,n} \geq \pi$, which implies that for all $p \in (-1,-1/2]$ and all $n \in \{1,2,3,\dots\}$ one has

$$j_{p,n+1} - j_{p,n} \geq \pi > \pi/2 = j_{-1/2,1} \geq j_{p,1},$$

i.e. $\xi_p = j_{p,1}$. Here we used that every positive zero $j_{p,n}$ of J_p satisfies the inequality $dj_{p,n}/dp > 1$ for all $p > -1$ (see E.K. Ifantis and P.D. Siafarikas [116, Corollary 3.1]) and in particular the function $p \mapsto j_{p,1}$ is increasing on $(-1,\infty)$.

For the remained part, i.e. when $p \in (-1/2,0)$ or $p \in (1/2,p_0)$, consider the function $h_p : [0, j_{p,1}) \to \mathbb{R}$, defined by

$$h_p(x) = \log \mathscr{I}_p(x) - \log \left(\frac{j_{p,1}^2 - x^2}{j_{p,1}^2 + x^2} \right).$$

Taking into account the inequality (see E.K. Ifantis and P.D. Siafarikas [118])

$$\frac{\mathscr{I}_{p+1}(x)}{\mathscr{I}_p(x)} = \frac{2(p+1)}{x} \frac{J_{p+1}(x)}{J_p(x)} < \frac{j_{p,1}^2}{j_{p,1}^2 - x^2}, \tag{3.174}$$

which holds for all $p > -1$ and $x \in (0, j_{p,1})$, and using (3.173), clearly we have

$$\begin{aligned}
h_p'(x) &= \frac{\mathscr{I}_p'(x)}{\mathscr{I}_p(x)} + \frac{4xj_{p,1}^2}{j_{p,1}^4 - x^4} \\
&= \frac{4xj_{p,1}^2}{j_{p,1}^4 - x^4} - \frac{x}{2(p+1)} \frac{\mathscr{I}_{p+1}(x)}{\mathscr{I}_p(x)} \\
&\geq \frac{4xj_{p,1}^2}{j_{p,1}^2 - x^2} \left[\frac{1}{j_{p,1}^2 + x^2} - \frac{1}{8(p+1)} \right] \geq 0,
\end{aligned}$$

where $x \in [0, \tau_p)$ and $p \in (-1, p_0)$, i.e. the function h_p is increasing on $[0, \tau_p)$. This in turn implies that $h_p(x) \geq h_p(0) = 0$ for all $x \in [0, \tau_p)$ and $p \in (-1, p_0)$. With this the proof of (3.169) is complete.

Now we are going to prove (3.170). Let us consider the function $\varphi_p : [0, j_{p,1}) \to \mathbb{R}$, defined by

$$\varphi_p(x) = \frac{j_{p,1}^2}{8(p+1)} \log \left(\frac{j_{p,1}^2 - x^2}{j_{p,1}^2 + x^2} \right) - \log \mathscr{I}_p(x).$$

In what follows we show that for each $p \geq -7/8$ the function φ_p is increasing, and consequently $\varphi_p(x) \geq \varphi_p(0) = 0$, i.e. (3.170) holds. For this recall the Rayleigh inequalities (see G.N. Watson [227, p. 502])

$$\left[\sigma_p^{(2m)} \right]^{-1/m} < j_{p,1}^2 < \sigma_p^{(2m)} / \sigma_p^{(2m+2)}, \tag{3.175}$$

which hold for all $m \in \{1, 2, 3, \ldots\}$ and $p > -1$, where

$$\sigma_p^{(2m)} = \sum_{n \geq 1} j_{p,n}^{-2m}$$

is the Rayleigh function of order $2m$, and Kishore's formula (see N. Kishore [135])

$$\frac{x}{2} \frac{J_{p+1}(x)}{J_p(x)} = \sum_{m \geq 1} \sigma_p^{(2m)} x^{2m}, \tag{3.176}$$

where $|x| < j_{p,1}$ and $p > -1$. Using (3.173), (3.175) and (3.176) we obtain

$\varphi_p'(x)$

$$= -\frac{x}{2(p+1)} \frac{j_{p,1}^4}{j_{p,1}^4 - x^4} - \frac{\mathscr{I}_p'(x)}{\mathscr{I}_p(x)}$$

$$= \frac{x}{2(p+1)} \frac{\mathscr{I}_{p+1}(x)}{\mathscr{I}_p(x)} - \frac{x}{2(p+1)} \frac{j_{p,1}^4}{j_{p,1}^4 - x^4}$$

$$= \Theta_p(x) \left[4(p+1)(j_{p,1}^2 + x^2) \frac{x}{2} \frac{J_{p+1}(x)}{J_p(x)} - \frac{j_{p,1}^4 x^2}{j_{p,1}^2 - x^2} \right]$$

$$= \Theta_p(x) \left[4(p+1)(j_{p,1}^2 + x^2) \sum_{m \geq 1} \sigma_p^{(2m)} x^{2m} - j_{p,1}^2 x^2 \sum_{m \geq 0} \left(\frac{x}{j_{p,1}} \right)^{2m} \right]$$

$$= \Theta_p(x) \left[4(p+1) j_{p,1}^2 \sum_{m \geq 2} \sigma_p^{(2m)} x^{2m} + 4(p+1) \sum_{m \geq 2} \sigma_p^{(2m-2)} x^{2m} - \sum_{m \geq 2} \frac{1}{j_{p,1}^{2m-2}} x^{2m} \right]$$

$$= \Theta_p(x) \sum_{m \geq 2} \left[4(p+1) j_{p,1}^2 \sigma_p^{(2m)} + 4(p+1) \sigma_p^{(2m-2)} - \frac{1}{j_{p,1}^{2m-2}} \right] x^{2m}$$

$$\geq \Theta_p(x) \sum_{m \geq 2} \left[8(p+1) j_{p,1}^2 \sigma_p^{(2m)} - \frac{1}{j_{p,1}^{2m-2}} \right] x^{2m}$$

$$\geq \Theta_p(x) \sum_{m \geq 2} \frac{8p+7}{j_{p,1}^{2m-2}} x^{2m} \geq 0,$$

where

$$\Theta_p(x) = \frac{1}{2(p+1)x} \frac{1}{j_{p,1}^2 + x^2}$$

and we have used that due to the Rayleigh formula (see G.N. Watson [227, p. 502])

$$\sigma_p^{(2)} = \sum_{n \geq 1} \frac{1}{j_{p,n}^2} = \frac{1}{4(p+1)}$$

one has

$$4(p+1)j_{p,1}^2 \sigma_p^{(2)} x^2 = j_{p,1}^2 x^2.$$

Finally, we prove the inequalities (3.171) and (3.172). For this first we show that the inequality (3.171) is in fact an immediate consequence of the inequality (3.170). Recall that the function $p \mapsto [\mathscr{I}_p(x)]^{p+1}$ is increasing on $(-1, \infty)$ for each fixed $x \in (0, j_{p,1})$ (see Á. Baricz [51, Theorem 3]), and thus we have

$$\mathscr{I}_{p+1}(x) \geq [\mathscr{I}_p(x)]^{(p+1)/(p+2)}$$

for all $p > -1$ and $x \in (0, j_{p,1})$. This in turn together with (3.170) implies that

$$\frac{\mathscr{I}_{p+1}(x)}{\mathscr{I}_p(x)} \geq [\mathscr{I}_p(x)]^{(p+1)/(p+2)-1} = \frac{1}{[\mathscr{I}_p(x)]^{1/(p+2)}}$$

$$\geq \left(\frac{j_{p,1}^2 + x^2}{j_{p,1}^2 - x^2} \right)^{\beta_p/(p+2)} = \left(\frac{j_{p,1}^2 + x^2}{j_{p,1}^2 - x^2} \right)^{\delta_p},$$

as we required. Finally, observe that inequality (3.172) follows easily from (3.174).
□

Remark 3.31. First we note that using again the inequality (see M.E.H. Ismail and M.E. Muldoon [120, Eq. 6.8]) $j_{p,1}^2 < 2(p+1)(p+3)$ we obtain that $\delta_p < 1$ for all $p > -1$. Moreover, using the inequality (3.175) for $m = 1$, i.e.

$$4(p+1) < j_{p,1}^2 < 4(p+1)(p+2)$$

we obtain that $\delta_p < 1/2 < \beta_p$ for all $p > -1$. Here we used that [227, p. 502]

$$\sigma_p^{(2)} = \frac{1}{4(p+1)}$$

and

$$\sigma_p^{(4)} = \frac{1}{16(p+1)^2(p+2)}.$$

Remark 3.32. Observe that in (3.169) the condition $p < p_0$ is not only sufficient, but also necessary. More precisely, since for all $p > -1$ and $x \in (-j_{p,1}, j_{p,1})$ we have

$$\frac{j_{p,1}^2 - x^2}{j_{p,1}^2 + x^2} \leq 1,$$

from (3.169) and (3.170) it follows that β_p is less than $\alpha_p = 1$, i.e. $p \in (-1, p_0)$. Moreover, when $p \geq p_0$, then the inequality (3.169) is reversed and for $p > p_0$ is weaker than (3.170). This observation is in the agreement with the fact that for example

$$\Delta_2(2) = j_{2,3}^2 - j_{2,1}j_{2,2} - j_{2,2}j_{2,3} = -6.01404 < 0,$$

where we have used that $j_{2,1} = 5.13562$, $j_{2,2} = 8.41724$ and $j_{2,3} = 11.61984$ (see for example [1, p. 409]). With other words, for $p > p_0$ the expression $\Delta_p(n)$ is not necessarily positive for all $n \in \{1, 2, 3, \dots\}$.

Remark 3.33. Recall that in particular the function $\mathscr{J}_p$ reduces to some elementary functions, like sine and cosine. More precisely, in particular we have (see (2.28), (2.29) and (2.30) for $z = x$)

$$\mathscr{J}_{-1/2}(x) = \sqrt{\pi/2} \cdot x^{1/2} J_{-1/2}(x) = \cos x, \qquad (3.177)$$

$$\mathscr{J}_{1/2}(x) = \sqrt{\pi/2} \cdot x^{-1/2} J_{1/2}(x) = \frac{\sin x}{x}, \qquad (3.178)$$

$$\mathscr{J}_{3/2}(x) = 3\sqrt{\pi/2} \cdot x^{-3/2} J_{3/2}(x) = 3 \left(\frac{\sin x}{x^3} - \frac{\cos x}{x^2} \right), \qquad (3.179)$$

respectively, which can verified easily by using the series representation of the function $\mathscr{J}_p$ and of the cosine and sine functions, respectively. We note that by choosing in (3.169) and (3.170) $p = -1/2$, in view of (3.177) we obtain the following sharp Redheffer-type inequalities (see L. Zhu and J. Sun [243, Theorem 2])

$$\left(\frac{\pi^2 - 4x^2}{\pi^2 + 4x^2} \right)^{\alpha_{-1/2}} \leq \cos x \leq \left(\frac{\pi^2 - 4x^2}{\pi^2 + 4x^2} \right)^{\beta_{-1/2}} \quad \text{for all } x \in \left[-\frac{\pi}{2}, \frac{\pi}{2} \right],$$

with the best possible constants $\alpha_{-1/2} = 1$ and $\beta_{-1/2} = \pi^2/16$ (see Fig. 3.3).

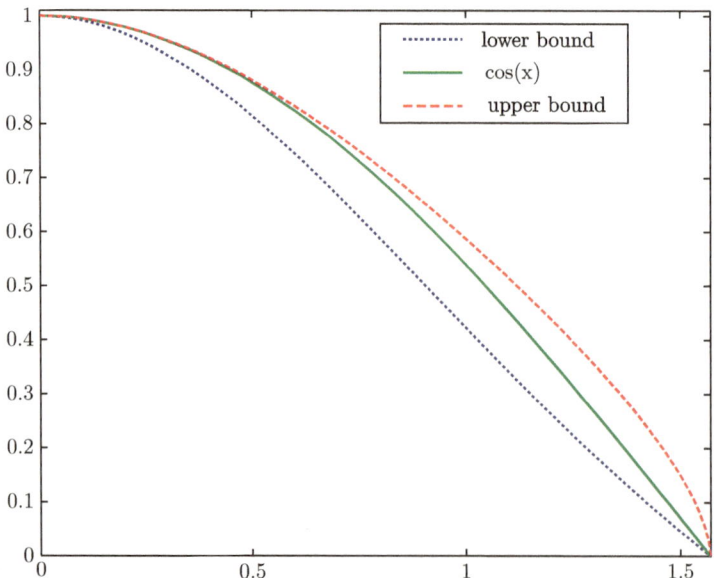

Fig. 3.3 The graph of the functions $\dfrac{\pi^2 - 4x^2}{\pi^2 + 4x^2}$, $\cos x$ and $\left(\dfrac{\pi^2 - 4x^2}{\pi^2 + 4x^2} \right)^{\pi^2/16}$ on $(0, \pi/2)$

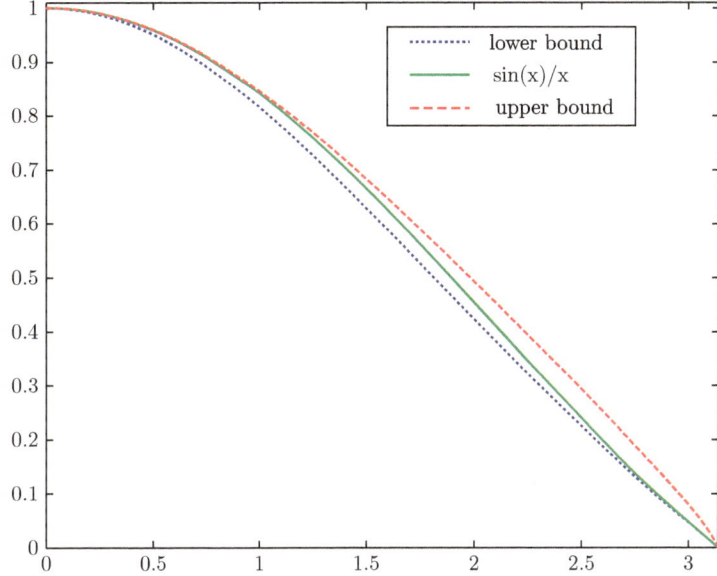

Fig. 3.4 The graph of the functions $\dfrac{\pi^2-x^2}{\pi^2+x^2}, \dfrac{\sin x}{x}$ and $\left(\dfrac{\pi^2-x^2}{\pi^2+x^2}\right)^{\pi^2/12}$ on $(0,\pi)$

Similarly, taking $p=1/2$ in (3.169) and (3.170), in view of (3.178), we reobtain the following sharp inequalities (see L. Zhu and J. Sun [243, Theorem 1])

$$\left(\frac{\pi^2-x^2}{\pi^2+x^2}\right)^{\alpha_{1/2}} \le \frac{\sin x}{x} \le \left(\frac{\pi^2-x^2}{\pi^2+x^2}\right)^{\beta_{1/2}} \quad \text{for all } x \in [-\pi,\pi],$$

with the best possible constants $\alpha_{1/2}=1$ and $\beta_{1/2}=\pi^2/12$ (see Fig. 3.4). Analogously, if we take $p=-1/2$ in (3.171) and (3.172), then in view of (3.177) and (3.178) we get the following sharp Redheffer-type inequalities (see L. Zhu and J.Sun [243, Theorem 3])

$$\left(\frac{\pi^2+4x^2}{\pi^2-4x^2}\right)^{\delta_{-1/2}} \le \frac{\tan x}{x} \le \left(\frac{\pi^2+4x^2}{\pi^2-4x^2}\right)^{\alpha_{-1/2}} \quad \text{for all } x \in \left[-\frac{\pi}{2},\frac{\pi}{2}\right],$$

with the best possible constants $\alpha_{-1/2}=1$ and $\delta_{-1/2}=\pi^2/24$ (see Fig. 3.5).

Here we used that $j_{-1/2,1}=\pi/2$ and $j_{1/2,1}=\pi$, which can be verified easily by using the infinite product representation of the cosine and sine functions, and of the function J_p, respectively.

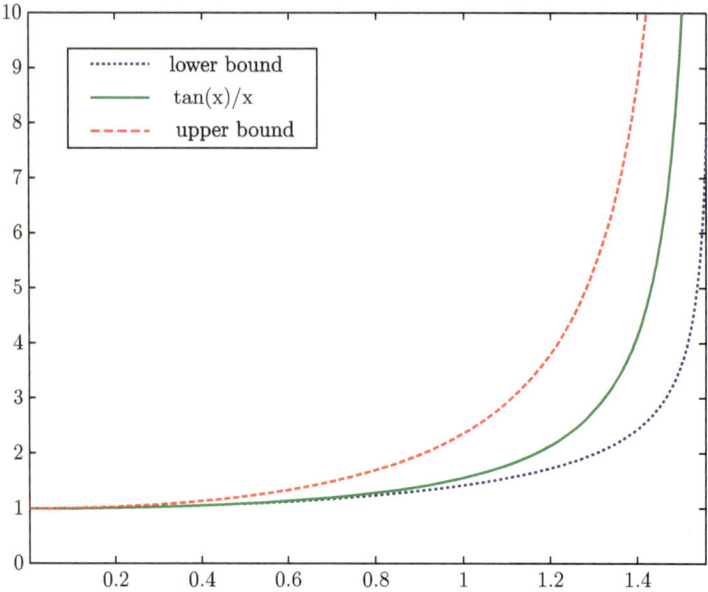

Fig. 3.5 The graph of the functions $\left(\dfrac{\pi^2+4x^2}{\pi^2-4x^2}\right)^{\pi^2/24}$, $\dfrac{\tan x}{x}$ and $\dfrac{\pi^2+4x^2}{\pi^2-4x^2}$ on $(0,\pi/2)$

Remark 3.34. We note that the proof of inequality (3.170) is similar to that given by L. Zhu and J. Sun [243] for Theorems 1 and 2. However, our approach in the proof of inequalities (3.171) and (3.172) is simpler than the method given in [243]. It is worth mentioning that N. Kishore [135] pointed out that, if B_m denotes the mth Bernoulli number in the even suffix notation (see [1]) and $G_m = 2(1-2^m)B_m$, then

$$\sigma_{-1/2}^{(2m)} = (-1)^m\frac{2^{2m-2}}{(2m)!}G_{2m} \text{ and } \sigma_{1/2}^{(2m)} = (-1)^{m-1}\frac{2^{2m-1}}{(2m)!}B_{2m}.$$

Using (3.177), (3.178), (3.179) and Kishore's expansion (3.176) for $p = -1/2$ and $p = 1/2$ these formulas leads to the well known power series expansions (see the book of M. Abramowitz and I.A. Stegun [1, p. 75])

$$\tan x = \sum_{m\geq 1}(-1)^m\frac{2^{2m}(1-2^{2m})}{(2m)!}B_{2m}x^{2m-1} = \sum_{m\geq 1}\frac{2^{2m}(2^{2m}-1)}{(2m)!}|B_{2m}|x^{2m-1}, \ |x| < \frac{\pi}{2},$$

$$x\cot x = 1 - \sum_{m\geq 1}(-1)^{m-1}\frac{2^{2m}}{(2m)!}B_{2m}x^{2m} = 1 - \sum_{m\geq 1}\frac{2^{2m}}{(2m)!}|B_{2m}|x^{2m}, \ |x| < \pi,$$

which were the chief tools in the proof of the main results in [243].

Remark 3.35. E.K. Ifantis and P.D. Siafarikas [118] in order to prove (3.174) used the power series representation (see E.K. Ifantis and P.D. Siafarikas [117, p. 95])

$$\frac{\mathscr{J}_{p+1}(x)}{\mathscr{J}_p(x)} = \frac{2(p+1)}{x}\frac{J_{p+1}(x)}{J_p(x)} = 1 + \sum_{m \geq 1} \frac{||S_p^m e_1||^2}{2^{2m}} x^{2m},$$

where $p > -1$ and $x \in (0, j_{p,1})$. Here $S_p = L_p^{1/2}(V + V^*)L_p^{1/2}$ is a compact and self-adjoint operator, V is an unilateral shift operator with respect to the orthonormal basis e_m, $m \in \{1, 2, 3, \dots\}$ in an abstract Hilbert space, L_p is the diagonal operator such that $L_p e_m = (1/(p+m))e_m$ and V^* is the adjoint of V. We show here that inequality (3.174) in fact can be deduced easily without of the above power series representation of the quotient $\mathscr{J}_{p+1}(x)/\mathscr{J}_p(x)$. For this we prove first that for all $p > -1$ and $m \in \{0, 1, 2, \dots\}$ the following inequality holds

$$4(p+1)\sigma_p^{(2m+2)} j_{p,1}^{2m} \leq 1, \tag{3.180}$$

and for $m \geq 1$ is strict. For $m = 0$ clearly we have equality, while for $m = 1$ the above inequality becomes exactly the right-hand side of (3.175) for the case $m = 1$, i.e. $j_{p,1}^2 < 4(p+1)(p+2)$. Now suppose that (3.180) holds for some $m = k - 1 \geq 2$, i.e. we have

$$4(p+1)\sigma_p^{(2k)} j_{p,1}^{2k-2} < 1.$$

Then by using the right-hand side of (3.175) we have

$$4(p+1)\sigma_p^{(2k+2)} j_{p,1}^{2k} < 4(p+1)\frac{\sigma_p^{(2k)}}{j_{p,1}^2} j_{p,1}^{2k} = 4(p+1)\sigma_p^{(2k)} j_{p,1}^{2k-2} < 1,$$

and thus by mathematical induction we have that indeed the inequality (3.180) is true. Consequently, by using the Kishore expansion (3.176) and inequality (3.180) it follows that

$$\frac{2(p+1)}{x}\frac{J_{p+1}(x)}{J_p(x)} - \frac{j_{p,1}^2}{j_{p,1}^2 - x^2}$$

$$= \frac{4(p+1)}{x^2}\frac{x}{2}\frac{J_{p+1}(x)}{J_p(x)} - \frac{j_{p,1}^2}{j_{p,1}^2 - x^2}$$

$$= \frac{4(p+1)}{x^2}\sum_{m \geq 1} \sigma_p^{(2m)} x^{2m} - \sum_{m \geq 0}\frac{1}{j_{p,1}^{2m}} x^{2m}$$

$$= \sum_{m \geq 1}\left[4(p+1)\sigma_p^{(2m+2)} - \frac{1}{j_{p,1}^{2m}}\right] x^{2m} < 0$$

for all $x \in (0, j_{p,1})$ and $p > -1$, i.e. the proof of (3.174) is complete. Moreover, using the above argument it can be proved that

$$\frac{2(p+1)}{x} \frac{J_{p+1}(x)}{J_p(x)} - \left[1 + \frac{x^2 j_{p,1}^2}{4(p+1)(p+2)(j_{p,1}^2 - x^2)}\right]$$

$$= \frac{4(p+1)}{x^2} \sum_{m \geq 1} \sigma_p^{(2m)} x^{2m} - \left[1 + \frac{x^2}{4(p+1)(p+2)} \sum_{m \geq 0} \frac{1}{j_{p,1}^{2m}} x^{2m}\right]$$

$$= \sum_{m \geq 2} \left[4(p+1)\sigma_p^{(2m+2)} - \frac{1}{4(p+1)(p+2)j_{p,1}^{2m-2}}\right] x^{2m} < 0$$

for all $x \in (0, j_{p,1})$ and $p > -1$. Here we used that from the right-hand side of (3.175) by using again mathematical induction it follows that

$$\sigma_p^{(2m+2)} \leq \sigma_p^{(4)} / j_{p,1}^{2m-2}$$

for all $m \in \{1,2,3,\dots\}$, and when $m \geq 2$ the above inequality is strict. This leads to the known inequality (see E.K. Ifantis and P.D. Siafarikas [118, Eq. 2.18])

$$\frac{2(p+1)}{x} \frac{J_{p+1}(x)}{J_p(x)} < 1 + \frac{x^2 j_{p,1}^2}{4(p+1)(p+2)(j_{p,1}^2 - x^2)},$$

which is better than (3.174).

It is also worth mentioning here that [116] the eigenvalues of S_p are precisely the values $\pm 2/j_{p,m}$ and $\|S_p\| = 2/j_{p,1}$. However, comparing (3.176) with the above expansion of E.K. Ifantis and P.D. Siafarikas, it follows that for all $m \in \{0,1,2,\dots\}$ we have

$$\|S_p^m e_1\|^2 = (p+1)2^{2m+2} \sigma_p^{(2m+2)}.$$

This complements the results of E.K. Ifantis and P.D. Siafarikas [116–118].

Corresponding to Theorem 3.40 we have the following results for the function $\mathscr{I}_p$. We note that this theorem improves and complements (3.162).

Theorem 3.41. (Á. Baricz and S. Wu [63]) *Let $p > -1$ and let $|x| < j_{p,1}$, where $j_{p,1}$ is the first positive zero of the Bessel function of the first kind J_p. Then the following sharp exponential Redheffer-type inequalities hold*

$$\left(\frac{j_{p,1}^2 + x^2}{j_{p,1}^2 - x^2}\right)^{\alpha_p} \leq \mathscr{I}_p(x) \leq \left(\frac{j_{p,1}^2 + x^2}{j_{p,1}^2 - x^2}\right)^{\beta_p}, \tag{3.181}$$

$$\left(\frac{j_{p,1}^2 - x^2}{j_{p,1}^2 + x^2}\right)^{\delta_p} \leq \frac{\mathscr{I}_{p+1}(x)}{\mathscr{I}_p(x)} \leq \left(\frac{j_{p,1}^2 - x^2}{j_{p,1}^2 + x^2}\right)^{\alpha_p}, \tag{3.182}$$

with the best possible constants

$$\alpha_p = 0, \ \beta_p = \frac{j_{p,1}^2}{8(p+1)} \ \text{and} \ \delta_p = \frac{j_{p,1}^2}{8(p+1)(p+2)}.$$

Proof. As in the proof of Theorem 3.40, since all functions which appear in inequalities (3.181) and (3.182) are even, without loss of generality, in what follows we assume that $x \in (0, j_{p,1})$. First we show that the constants α_p, β_p and δ_p are the best possible. For this consider the functions $f_p, g_p : (0, j_{p,1}) \to \mathbb{R}$, defined by

$$f_p(x) = \frac{\log \mathscr{I}_p(x)}{\log \left(\frac{j_{p,1}^2 + x^2}{j_{p,1}^2 - x^2} \right)} \ \text{and} \ g_p(x) = \frac{\log \frac{\mathscr{I}_{p+1}(x)}{\mathscr{I}_p(x)}}{\log \left(\frac{j_{p,1}^2 - x^2}{j_{p,1}^2 + x^2} \right)}.$$

Using the l'Hospital rule it is easy to verify that we have

$$\lim_{x \to 0} f_p(x) = \lim_{x \to 0} \frac{\mathscr{I}_p'(x)}{\mathscr{I}_p(x)} \cdot \frac{j_{p,1}^4 - x^4}{4x j_{p,1}^2} = \lim_{x \to 0} \frac{\mathscr{I}_{p+1}(x)}{\mathscr{I}_p(x)} \cdot \frac{j_{p,1}^4 - x^4}{8(p+1) j_{p,1}^2} = \beta_p$$

and

$$\lim_{x \to 0} g_p(x) = \lim_{x \to 0} \frac{j_{p,1}^4 - x^4}{8(p+1)(p+2) j_{p,1}^2} = \delta_p,$$

where we have used the differentiation formula

$$\mathscr{I}_p'(x) = \frac{x}{2(p+1)} \mathscr{I}_{p+1}(x), \tag{3.183}$$

which can be verified easily by using the series representation of the function $\mathscr{I}_p$. On the other hand we have

$$\lim_{x \to j_{p,1}} f_p(x) = \lim_{x \to j_{p,1}} g_p(x) = \alpha_p.$$

In other words, we have $f_p(0^+) = \beta_p$, $g_p(0^+) = \delta_p$ and $f_p(j_{p,1}^-) = g_p(j_{p,1}^-) = \alpha_p$, which show that the constants α_p, β_p and δ_p are the best possible.

Now let us focus on inequalities (3.181) and (3.182). Since the function $x \mapsto \mathscr{I}_p(x)$ is increasing on $(0, \infty)$ for each $p > -1$, we have $\mathscr{I}_p(x) \geq 1$, and thus the left-hand side of (3.181) is obvious. Similarly, since the function $p \mapsto \mathscr{I}_p(x)$ is decreasing on $(-1, \infty)$ for each fixed $x \in \mathbb{R}$ (see Á. Baricz [51, Theorem 1]), one has $\mathscr{I}_{p+1}(x) \leq \mathscr{I}_p(x)$, and thus the right-hand side of (3.182) is true. For the right-hand side of (3.181) consider the function $h_p : [0, j_{p,1}) \to \mathbb{R}$, defined by

$$h_p(x) = \frac{j_{p,1}^2}{8(p+1)} \log \left(\frac{j_{p,1}^2 + x^2}{j_{p,1}^2 - x^2} \right) - \log \mathscr{I}_p(x).$$

Then by using again the Rayleigh formula (see G.N. Watson [227, p. 502])

$$\sigma_p^{(2)} = \sum_{n\geq 1} \frac{1}{j_{p,n}^2} = \frac{1}{4(p+1)}$$

and the factorization

$$\mathscr{I}_p(x) = \prod_{n\geq 1} \left(1 + \frac{x^2}{j_{p,n}^2}\right),$$ (3.184)

which can be easily derived from [227, p. 498]

$$\mathscr{J}_p(x) = \prod_{n\geq 1} \left(1 - \frac{x^2}{j_{p,n}^2}\right),$$ (3.185)

we have

$$
\begin{aligned}
h_p'(x) &= \frac{1}{4(p+1)} \frac{2x j_{p,1}^4}{(j_{p,1}^4 - x^4)} - \sum_{n\geq 1} \frac{2x}{j_{p,n}^2 + x^2} \\
&= \frac{2x j_{p,1}^4}{(j_{p,1}^4 - x^4)} \sum_{n\geq 1} \frac{1}{j_{p,n}^2} - \sum_{n\geq 1} \frac{2x}{j_{p,n}^2 + x^2} \\
&= 2x \sum_{n\geq 1} \left[\frac{j_{p,1}^4}{(j_{p,1}^4 - x^4) j_{p,n}^2} - \frac{1}{j_{p,n}^2 + x^2} \right] \geq 0,
\end{aligned}
$$

for all $p > -1$ and $x \in [0, j_{p,1})$, i.e. the function h_p is increasing. Hence $h_p(x) \geq h_p(0) = 0$, and thus the proof of the right-hand side of (3.181) is done. However, there is another way to deduce (3.181). Namely, by the monotone form of l'Hospital's rule (see Lemma 3.7) to prove that the function f_p, defined above, is decreasing, it is enough to show that the function

$$x \mapsto \frac{\frac{d}{dx} \log \mathscr{I}_p(x)}{\frac{d}{dx} \log \left(\frac{j_{p,1}^2 + x^2}{j_{p,1}^2 - x^2}\right)} = \frac{1}{2 j_{p,1}^2} \sum_{n\geq 1} \frac{j_{p,1}^4 - x^4}{j_{p,n}^2 + x^2}$$

is decreasing too on $(0, j_{p,1})$, which is clearly true, since each terms in the above series are decreasing as functions of x. Consequently for each $x \in (0, j_{p,1})$ the inequalities

$$\alpha_p = f_p(j_{p,1}^-) < f_p(x) < f_p(0^+) = \beta_p$$

hold, as we required. Here we used on the one hand that the numerator and denominator of $f_p(x)$ vanishes at zero and on the other hand that from the infinite product formula (3.184) one has

$$\frac{\mathscr{I}_p'(x)}{\mathscr{I}_p(x)} = \sum_{n\geq 1} \frac{2x}{j_{p,n}^2 + x^2}.$$

Finally, we prove the left-hand side of (3.182). It is known that the function $p \mapsto [\mathscr{I}_p(x)]^{p+1}$ is increasing on $(-1, \infty)$ for each $x \in \mathbb{R}$ (see Á. Baricz [51, Theorem 1]), and thus we have

$$\mathscr{I}_{p+1}(x) \geq [\mathscr{I}_p(x)]^{(p+1)/(p+2)}$$

for all $p > -1$ and $x \in \mathbb{R}$. This in turn together with the right-hand side of (3.181) implies that

$$\frac{\mathscr{I}_{p+1}(x)}{\mathscr{I}_p(x)} \geq [\mathscr{I}_p(x)]^{(p+1)/(p+2)-1} = \frac{1}{[\mathscr{I}_p(x)]^{1/(p+2)}}$$

$$\geq \left(\frac{j_{p,1}^2 - x^2}{j_{p,1}^2 + x^2} \right)^{\beta_p/(p+2)} = \left(\frac{j_{p,1}^2 - x^2}{j_{p,1}^2 + x^2} \right)^{\delta_p},$$

and with this the proof is complete.   □

Remark 3.36. First note that in Theorem 3.39, by using mathematical induction and the infinite product representation (3.184), we proved that if $\Delta_p(n) \geq 0$ for each $n \in \{1, 2, 3, \dots\}$ and $p > -1$, then for all $|x| < j_{p,1}$ the Redheffer-type inequality (3.162) holds. Due to Lemma 3.8 if $p \in (-1, p_0)$, then $\beta_p < 1$, and thus the right-hand side of (3.181) is better than (3.162). When $p = p_0$, then $\beta_p = 1$, and thus the right-hand side of (3.181) and (3.162) are the same. However, when $p > p_0$ the inequality (3.162) does not hold necessarily, since for all $p > p_0$ and $n \in \{1, 2, 3, \dots\}$ the inequality $\Delta_p(n) \geq 0$ is not true. For example, as we have pointed out above, $\Delta_2(2) = -6.01404 < 0$.

Remark 3.37. It is worth mentioning that in particular the function $\mathscr{I}_p$ reduces to some elementary functions, like hyperbolic sine and hyperbolic cosine. More precisely, in particular we have (see also Remark 2.17 by choosing $z = x$)

$$\mathscr{I}_{-1/2}(x) = \sqrt{\pi/2} \cdot x^{1/2} I_{-1/2}(x) = \cosh x, \tag{3.186}$$

$$\mathscr{I}_{1/2}(x) = \sqrt{\pi/2} \cdot x^{-1/2} I_{1/2}(x) = \frac{\sinh x}{x}, \tag{3.187}$$

$$\mathscr{I}_{3/2}(x) = 3\sqrt{\pi/2} \cdot x^{-3/2} I_{3/2}(x) = -3 \left(\frac{\sinh x}{x^3} - \frac{\cosh x}{x^2} \right), \tag{3.188}$$

respectively, which can verified easily by using the series representation of the function $\mathscr{I}_p$ and of the hyperbolic cosine and hyperbolic sine functions, respectively. Now, choosing in (3.181) the value $p = -1/2$, in view of (3.186) we obtain the following sharp Redheffer-type inequalities (see L. Zhu and J. Sun [243, Theorem 5] for $r = \pi/2$)

$$\left(\frac{\pi^2 + 4x^2}{\pi^2 - 4x^2} \right)^{\alpha_{-1/2}} \leq \cosh x \leq \left(\frac{\pi^2 + 4x^2}{\pi^2 - 4x^2} \right)^{\beta_{-1/2}} \quad \text{for all } x \in \left[-\frac{\pi}{2}, \frac{\pi}{2} \right],$$

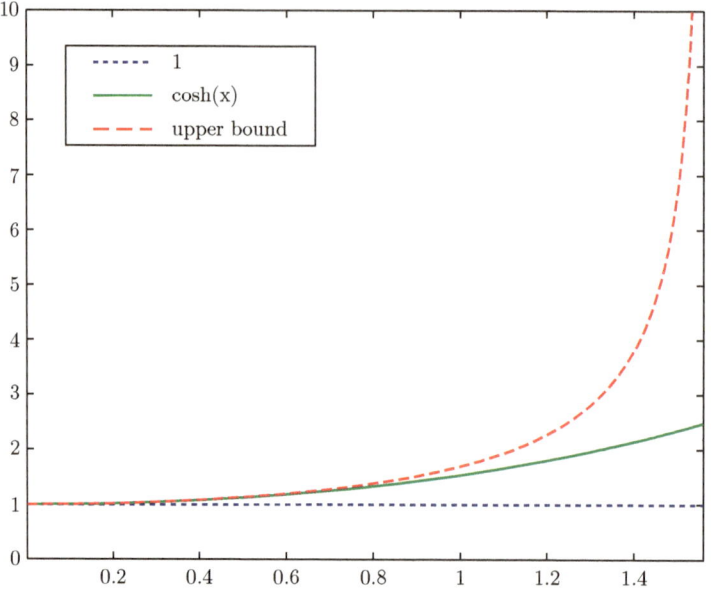

Fig. 3.6 The graph of the functions $1, \cosh x$ and $\left(\dfrac{\pi^2 + 4x^2}{\pi^2 - 4x^2}\right)^{\pi^2/16}$ on $(0, \pi/2)$

with the best possible constants $\alpha_{-1/2} = 0$ and $\beta_{-1/2} = \pi^2/16$ (see Fig. 3.6). Similarly, taking $p = 1/2$ in (3.181), in view of (3.187), we reobtain the following sharp inequalities (see L. Zhu and J. Sun [243, Theorem 4] for $r = \pi$)

$$\left(\frac{\pi^2 + x^2}{\pi^2 - x^2}\right)^{\alpha_{1/2}} \leq \frac{\sinh x}{x} \leq \left(\frac{\pi^2 + x^2}{\pi^2 - x^2}\right)^{\beta_{1/2}} \quad \text{for all } x \in [-\pi, \pi],$$

with the best possible constants $\alpha_{1/2} = 0$ and $\beta_{1/2} = \pi^2/12$ (see Fig. 3.7).

Finally, if we take $p = -1/2$ in (3.182), then in view of (3.186) and (3.187) we get the following sharp Redheffer-type inequalities (see L. Zhu and J. Sun [243, Theorem 6] for $r = \pi/2$)

$$\left(\frac{\pi^2 - 4x^2}{\pi^2 + 4x^2}\right)^{\delta_{-1/2}} \leq \frac{\tanh x}{x} \leq \left(\frac{\pi^2 - 4x^2}{\pi^2 + 4x^2}\right)^{\alpha_{-1/2}} \quad \text{for all } x \in \left[-\frac{\pi}{2}, \frac{\pi}{2}\right],$$

with the best possible constants $\alpha_{-1/2} = 0$ and $\delta_{-1/2} = \pi^2/24$ (see Fig. 3.8).

Observe that combining (3.170) with the right hand side of (3.181) we easily obtain that $\mathcal{J}_p(x)\mathcal{I}_p(x) \leq 1$ for all $|x| < j_{p,1}$ and $p \geq -7/8$. Moreover, combining (3.171) with the left hand side of (3.182) we obtain that $\mathcal{J}_p(x)\mathcal{I}_p(x) \leq \mathcal{J}_{p+1}(x)\mathcal{I}_{p+1}(x)$ for all $|x| < j_{p,1}$ and $p \geq -7/8$. The next result shows that the above properties hold true for all $p > -1$.

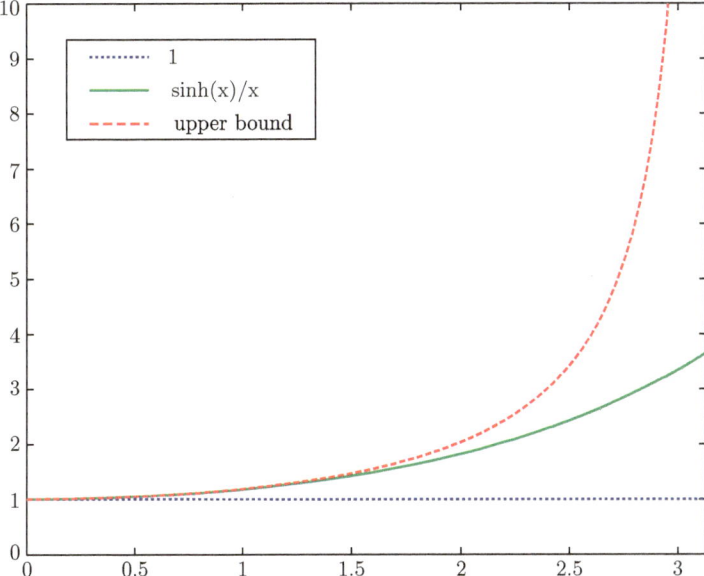

Fig. 3.7 The graph of the functions 1, $\dfrac{\sinh x}{x}$ and $\left(\dfrac{\pi^2+x^2}{\pi^2-x^2}\right)^{\pi^2/12}$ on $(0,\pi)$

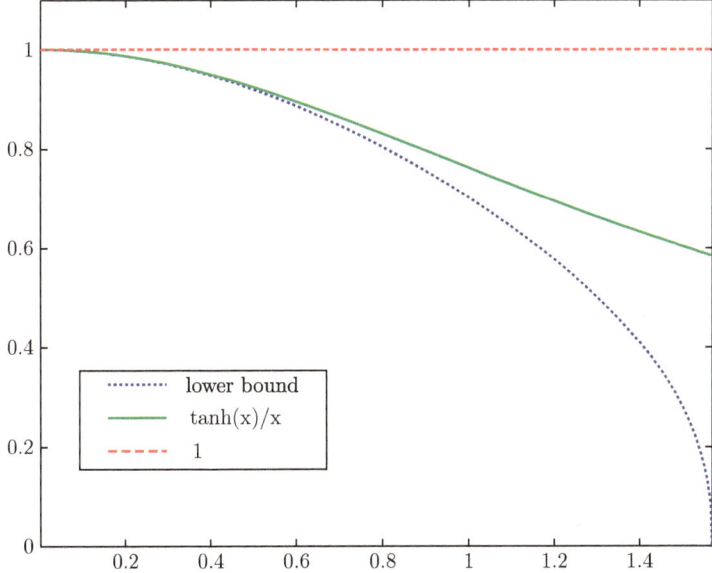

Fig. 3.8 The graph of the functions $\left(\dfrac{\pi^2-4x^2}{\pi^2+4x^2}\right)^{\pi^2/24}$, $\dfrac{\tanh x}{x}$ and 1 on $(0,\pi/2)$

Theorem 3.42. (Á. Baricz and S. Wu [63]) *The following assertions are true:*

(a) *The function* $x \mapsto \mathscr{J}_p(x)\mathscr{I}_p(x)$ *is increasing on* $(-j_{p,1}, 0]$ *and decreasing on* $[0, j_{p,1})$ *for all* $p > -1$.
(b) *The function* $p \mapsto \mathscr{J}_p(x)\mathscr{I}_p(x)$ *is increasing on* $(-1, \infty)$ *for all* $|x| < j_{p,1}$ *fixed.*
(c) *The following inequalities hold*

$$0 < \mathscr{J}_p(x)\mathscr{I}_p(x) \leq \mathscr{J}_{p+1}(x)\mathscr{I}_{p+1}(x) \leq 1 \qquad (3.189)$$

for all $|x| < j_{p,1}$ *and* $p > -1$.

Proof. (a) By using the differentiation formulas (3.173) and (3.183) we obtain

$$\frac{d}{dx}\left[\mathscr{J}_p(x)\mathscr{I}_p(x)\right] = \frac{x}{2(p+1)}\left[\mathscr{I}_{p+1}(x)\mathscr{J}_p(x) - \mathscr{I}_p(x)\mathscr{J}_{p+1}(x)\right].$$

Since the function $x \mapsto \mathscr{J}_p(x)\mathscr{I}_p(x)$ is even, it is enough to show that the above expression is negative for all $[0, j_{p,1})$ and all $p > -1$. For this recall that (see Á. Baricz [51]) the function $p \mapsto \mathscr{J}_p(x)$ is increasing on $(-1, \infty)$ for each fixed $|x| < j_{p,1}$, while the function $p \mapsto \mathscr{I}_p(x)$ is decreasing on $(-1, \infty)$ for all $x \in \mathbb{R}$ fixed. These properties in particular imply that for all $|x| < j_{p,1}$ and $p > -1$ we have

$$\mathscr{I}_{p+1}(x)/\mathscr{I}_p(x) \leq 1 \leq \mathscr{J}_{p+1}(x)/\mathscr{J}_p(x),$$

and thus the proof of this part is complete. Another proof can be obtained if we consider the factorizations (3.184) and (3.185). Namely, in view of these formulas, it is enough to show that

$$\prod_{n\geq 1}\left(1 + \frac{x^2}{j_{p+1,n}^2}\right)\left(1 - \frac{x^2}{j_{p,n}^2}\right) \leq \prod_{n\geq 1}\left(1 - \frac{x^2}{j_{p+1,n}^2}\right)\left(1 + \frac{x^2}{j_{p,n}^2}\right),$$

which clearly holds because each terms in the above products are positive and

$$\left(1 + \frac{x^2}{j_{p+1,n}^2}\right)\left(1 - \frac{x^2}{j_{p,n}^2}\right) \leq \left(1 - \frac{x^2}{j_{p+1,n}^2}\right)\left(1 + \frac{x^2}{j_{p,n}^2}\right)$$

holds for all $p > -1$, $n \in \{1, 2, 3, \dots\}$ and $x \in [0, j_{p,1})$. Here we have used that

$$1/j_{p+1,n}^2 - 1/j_{p,n}^2 \leq 1/j_{p,n}^2 - 1/j_{p+1,n}^2,$$

that is, $j_{p,n} < j_{p+1,n}$ holds for all $p > -1$ and $n \in \{1, 2, 3, \dots\}$.
(b) Recall that the function $p \mapsto j_{p,n}$ is increasing on $(-1, \infty)$ for each $n \in \{1, 2, 3, \dots\}$. From this we deduce that the function $p \mapsto \log(1 - x^4/j_{p,n}^4)$ is

increasing too on $(-1,\infty)$ for each $n \in \{1,2,3,\dots\}$ and $|x| < j_{p,1}$ fixed. Conse-
quently, by using the infinite product formulas (3.184) and (3.185), the function

$$p \mapsto \log\left[\mathscr{J}_p(x)\mathscr{I}_p(x)\right] = \sum_{n\geq 1}\log\left(1 - \frac{x^4}{j_{p,n}^4}\right)$$

is increasing on $(-1,\infty)$ for each $|x| < j_{p,1}$ fixed.
(c) This follows from part (a) and (b).   □

Remark 3.38. We note that if we choose $p \in \{-1/2, 1/2, 3/2\}$ in part (a) of
Theorem 3.42, then in view of (3.177), (3.178), (3.179), (3.186), (3.187) and (3.188)
we obtain the following inequalities:

$$0 < (\cos x)(\cosh x) \leq 1 \quad \text{for all } |x| < \pi/2,$$

$$0 < \left(\frac{\sin x}{x}\right)\left(\frac{\sinh x}{x}\right) \leq 1 \quad \text{for all } |x| < \pi,$$

$$0 < 9\left(\frac{\sin x}{x^3} - \frac{\cos x}{x^2}\right)\left(\frac{\cosh x}{x^2} - \frac{\sinh x}{x^3}\right) \leq 1 \quad \text{for all } |x| < j_{3/2,1},$$

where $j_{3/2,1} = 4.493409$ in view of (3.179) is in fact the first positive zero of the
equation $\tan x = x$. We note that the first two inequalities presented above were communicated to the author by M. Vuorinen. Thanks are due to him for this information.
Finally, observe that using (3.189) for all $|x| < \pi/2$ the following chain of inequali-
ties holds

$$0 < (\cos x)(\cosh x) \leq \left(\frac{\sin x}{x}\right)\left(\frac{\sinh x}{x}\right)$$

$$\leq 9\left(\frac{\sin x}{x^3} - \frac{\cos x}{x^2}\right)\left(\frac{\cosh x}{x^2} - \frac{\sinh x}{x^3}\right) \leq 1.$$

3.8.3 A Lower Bound for the Gamma Function

In this section we use the idea of C.P. Chen et al. [81] to deduce a new lower bound
for the Γ function.

Theorem 3.43. (Á. Baricz [47]) *If $x \in (0,1]$, then*

$$\Gamma(x) \geq \frac{1-x}{1+x} \cdot \frac{e^{(1-\gamma)x}}{x}, \tag{3.190}$$

where γ is the Euler-Mascheroni constant.

Proof. From the Euler infinite product formula (see M. Abramowitz and I.A. Stegun [1, p. 255])

$$\frac{1}{xe^{\gamma x}\Gamma(x)} = \prod_{n\geq1}\left(1+\frac{x}{n}\right)e^{-\frac{x}{n}},$$

we have

$$\frac{e^{(1-\gamma)x}}{x\Gamma(x)} = \frac{1+x}{1-x}\left[(1-x)\lim_{n\to\infty}S_n\right], \tag{3.191}$$

where

$$S_n = \prod_{k=2}^{n}\left(1+\frac{x}{k}\right)e^{-\frac{x}{k}} \quad \text{for all } n \in \{2,3,\dots\}.$$

In order to prove (3.190) it is enough to show that

$$(1-x)S_n < \left(1-\frac{x}{n}\right) \quad \text{for all } n \in \{2,3,\dots\}, \tag{3.192}$$

hence from this we get $\lim_{n\to\infty}(1-x)S_n \leq 1$, and consequently the inequality (3.190) follows from (3.191).

To prove (3.192) we use mathematical induction. For $n = 2$ we easily get for $x \in (0,1)$ that

$$(1-x)S_2 < \left(1-\frac{x}{2}\right)$$

or equivalently

$$e^{x/2} > \frac{(1-x)\left(1+\frac{x}{2}\right)}{1-\frac{x}{2}},$$

which clearly holds because

$$e^{x/2} - \frac{(1-x)\left(1+\frac{x}{2}\right)}{1-\frac{x}{2}} = \sum_{k\geq1}\left(x+\frac{1}{k!}\right)\frac{x^k}{2^k} > 0.$$

Now suppose that (3.192) holds for some $m \geq 2$. Then from (3.191) and (3.192) we obtain that

$$(1-x)S_{m+1} - \left(1-\frac{x}{m+1}\right)$$
$$= (1-x)S_m\left(1+\frac{x}{m+1}\right)e^{-\frac{x}{m+1}} - \left(1-\frac{x}{m+1}\right)$$
$$< \left(1-\frac{x}{m}\right)\left(1+\frac{x}{m+1}\right)e^{-\frac{x}{m+1}} - \left(1-\frac{x}{m+1}\right)$$

and this is negative if and only if

$$e^{\frac{x}{m+1}} - \frac{\left(1-\frac{x}{m}\right)\left(1+\frac{x}{m+1}\right)}{\left(1-\frac{x}{m+1}\right)} = \sum_{k\geq 1}\left(\frac{x}{m} + \frac{1}{m} - 1 + \frac{1}{k!}\right)\frac{x^k}{(m+1)^k} > 0.$$

$\square$

Remark 3.39. Numerical experiments in Maple6 show that the lower bound in (3.190) is far from being the best possible one. For example, when $x = 0.5$, we have $\Gamma(0.5) = \sqrt{\pi} = 1.772453851\ldots$, while the right-hand side of (3.190) is $0.8235978\ldots$. Similarly for $x = 0.25$ we have $\Gamma(0.25) = 3.6256099082\ldots$, while the right-hand side of (3.190) is $2.667561665\ldots$. In fact, graphics in Maple6 suggest that the function $\varphi : (-1,\infty) \to \mathbb{R}$, defined by

$$\varphi(x) = \Gamma(x) - \frac{1-x}{1+x} \cdot \frac{e^{(1-\gamma)x}}{x},$$

is convex and satisfies the inequality $\varphi(x) \geq 1$ for all $x \in (-1,0] \cup [1,\infty)$. Moreover, $\varphi(x) \in (0.9483, 1]$ for all $x \in [0,1]$ (see Fig. 3.9).

Remark 3.40. During the course of writing this monograph J. Sándor communicated to us the following elegant improvement of (3.190): by using the Euler infinite product formula directly we have for all $x > 0$

$$\Gamma(x) = \frac{e^{-\gamma x}}{x}\prod_{n\geq 1}\frac{e^{x/n}}{1+x/n} = \frac{e^{(1-\gamma)x}}{x(1+x)}\prod_{n\geq 2}\frac{e^{x/n}}{1+x/n} \geq \frac{e^{(1-\gamma)x}}{x(1+x)}, \tag{3.193}$$

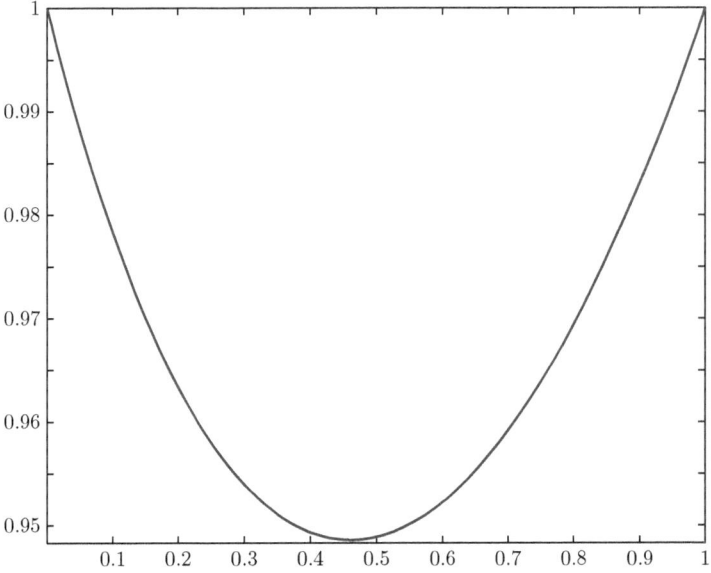

Fig. 3.9 The graph of the function φ on $[0,1]$

where we used the inequality $e^t \geq 1 + t$, which holds for all $t \geq 0$. Sándor's refinement (3.193) not only improve (3.190), but even extend the range of validity. Thanks are due to him for the above information.

It is also worth to mention here that another improvement of (3.190) has been found very recently by P. Ivády [123]. Namely, in [123] it is proved that for all $x \in (0,1)$ we have

$$\Gamma(x) > \frac{x^2 + 1}{x(x+1)}.$$

Finally, observe that these two refinements are not directly comparable to each other on the whole interval $(0,1)$. More precisely, computer experiments show that there is an $x^* \in (0.4, 0.5)$ such that on $(0, x^*)$ Sándor's lower bound is tighter than of Ivády's and on $(x^*, 1)$ the situation is reversed.

Appendix A

A.1 Conjectures

1. (Á. Baricz [45]) The inequality (3.23) holds for all $a, b > 0$ and for all $r, s \in (0, 1)$.
2. (Á. Baricz and E. Neuman [59]) The function $x \in \mathbb{R} \mapsto \mathscr{I}_p(x) \in [1, \infty)$ is strictly log-convex for all $p > -1$, while the function $x \in (0, \infty) \mapsto I_p(x) \in (0, \infty)$ is strictly log-concave for all $p > 0$.
3. (Á. Baricz [54]) The function $x \in \mathbb{R} \mapsto \mathscr{I}_p^{(2k)}(x) \in (0, \infty)$ is strictly log-convex for all $k \in \mathbb{N}$ and $p > -1$, while the function $x \in (0, \infty) \mapsto \mathscr{I}_p^{(2k+1)}(x) \in (0, \infty)$ is strictly log-concave for all $k \in \mathbb{N}$ and $p > -1$.
4. (Á. Baricz [54]) The inequality (3.97) holds for all $a, b \geq 0$ and $p > -1$.
5. (Á. Baricz and S.Wu [63]) The inequality (3.169) holds for all $p \in (-1, p_0)$ and $|x| < j_{p,1}$, while (3.170) and (3.171) hold for all $p > -1$ and $|x| < j_{p,1}$.

A.2 Open Problems

1. (Á. Baricz [43]) Find the explicit range of b, c, p for which u_p is univalent in $\mathbb{D}$.
2. (S. András and Á. Baricz [24]) Is it true that if $p > -1$ increases, then the image region $\mathscr{I}_p(\mathbb{D})$ decreases, more precisely if $p > q > -1$, then $\mathscr{I}_p(\mathbb{D}) \subset \mathscr{I}_q(\mathbb{D})$? This problem is related to Corollary 2.14 which provides a partial affirmative answer. Namely, it is true that if $p > q \geq -1/4$, then $\mathscr{I}_p(\mathbb{D}) \subset \mathscr{I}_q(\mathbb{D})$.
3. (Á. Baricz [41]) For which $b, p, c \in \mathbb{R}$ does $\sigma(r_1) + \sigma(r_2) \leq 2\sigma(\sqrt{r_1 r_2})$ hold for all $r_1, r_2 \in (0, 1)$, where $\sigma(r) = u_p(1 - r^2)/u_p(r^2)$?
4. (Á. Baricz [41]) Suppose that the power series $f(r) = \sum_{n \geq 0} A_n r^n$ is convergent for all $r \in (0, 1)$, where $A_n > 0$ for all $n \geq 0$. Let us consider the function $m_f : (0, 1) \to (0, \infty)$, defined by $m_f(r) = f(1 - r^2)/f(r^2)$. Find conditions concerning the coefficients A_n, which guarantee that $m_f(r_1) + m_f(r_2) \leq 2m_f(\sqrt{r_1 r_2})$ for all $r_1, r_2 \in (0, 1)$.

A.3   Matlab Programs for Graphs

1. **Fig. 1.1**
```
x=0:0.1:10;
y1=besselj(-1/2,x);
y2=besselj(0,x);
y3=besselj(1/2,x);
plot(x,y1,'b.');hold on;
plot(x,y2,'g');
plot(x,y3,'r--');
```
2. **Fig. 1.2**
```
x=0:0.1:4;
y1=besseli(-1/2,x);
y2=besseli(0,x);
y3=besseli(1/2,x);
plot(x,y1,'b.');hold on;
plot(x,y2,'g');
plot(x,y3,'r--');
```
3. **Fig. 2.1**
```
clear;
fi=0:0.001:2*pi;
z=cos(fi)+i*sin(fi);
gz=cos(z);
plot(z,'g'); hold on;
plot(gz,'b');
axis equal;
axis([-1.1 1.6 -1.1 1.1]);
```
4. **Fig. 2.2**
```
clear;
fi=0:0.001:2*pi;
z=cos(fi)+i*sin(fi);
fz=sin(z)./z;
plot(fz,'r');hold on;
plot(z,'g'); hold on;
axis equal;
axis([-1.1 1.6 -1.1 1.1]);
```
5. **Fig. 2.3**
```
clear;
fi=0:0.001:2*pi;
z=cos(fi)+i*sin(fi);
fz=sin(z)./z;
gz=cos(z);
plot(fz,'r');hold on;
plot(z,'g'); hold on;
```

```
plot(gz,'b');
axis equal;
axis([-1.1 1.6 -1.1 1.1]);
```

6. **Fig. 2.4**

```
clear;
fi=0:0.001:2*pi;
z=cos(fi)+i*sin(fi);
fz=sin(z)./z;
gz=cos(z);
hz=3*((sin(z))./(z.^3)-(cos(z))./(z.^2));
plot(fz,'r');hold on;
plot(z,'g'); hold on;
plot(gz,'b');
plot(hz,'y');
axis equal;
axis([-1.1 1.6 -1.1 1.1]);
```

7. **Fig. 3.1**

```
clear;
plot(0,1,'o');hold on; %point A
plot(pi/2,2/pi,'o'); %point B
x=0:0.05:2;
y1=2/pi-4/(pi^2)*(x-pi/2);
y2=sin(x)./x;
y3=1+2/pi*(2/pi-1)*x;
plot(x,y1,'r.'); hold on;
plot(x,y2,'g');
plot(x,y3,'b-.');
```

8. **Fig. 3.2**

```
clear;
x=-1.5:0.1:1.5;
y1=2/9*(2+5/2*cos(sqrt(3/5)*x));
y2=sin(x)./x;
y3=cos(x/sqrt(3));
plot(x,y1,'b.-'); hold on;
plot(x,y2,'r.');
plot(x,y3,'g');
```

9. **Fig. 3.3**

```
clear;
x=0:0.05:pi/2;
y1=cos(x);
y2=(pi^2-4*x.^2)./(pi^2+4*x.^2);
y3=((pi^2-4*x.^2)./(pi^2+4*x.^2)).^(pi^2/16);
plot(x,y1,'g'); hold on;
plot(x,y2,'b.');
plot(x,y3,'r-.');
```

10. **Fig. 3.4**

```
clear;
x=0:0.01:pi;
y1=sin(x)./x;
y2=(pi^2-x.^2)./(pi^2+x.^2);
y3=((pi^2-x.^2)./(pi^2+x.^2)).^(pi^2/12);
p1=plot(x,y1,'g'); hold on;
p2=plot(x,y2,'b.');
p3=plot(x,y3,'r-.');
set(p2,'Markersize',1);
set(p3,'Markersize',1);
```

11. **Fig. 3.5**

```
clear;
x=0:0.01:1.5;
y1=tan(x)./x;
y2=(pi^2+4*x.^2)./(pi^2-4*x.^2);
y3=((pi^2+4*x.^2)./(pi^2-4*x.^2)).^(pi^2/24);
p1=plot(x,y1,'g'); hold on;
p2=plot(x,y2,'r.-');
p3=plot(x,y3,'b.');
set(p2,'Markersize',4);
set(p3,'Markersize',4);
```

12. **Fig. 3.6**

```
clear;
x=0:0.01:1.5;
y1=cosh(x);
y2=ones(1,length(x));
y3=((pi^2+4*x.^2)./(pi^2-4*x.^2)).^(pi^2/16);
p1=plot(x,y1,'g'); hold on;
p2=plot(x,y2,'b.');
p3=plot(x,y3,'r-.');
set(p2,'Markersize',4);
set(p3,'Markersize',4);
```

13. **Fig. 3.7**

```
clear;
x=0:0.01:3;
y1=sinh(x)./x;
y2=ones(1,length(x));
y3=((pi^2+x.^2)./(pi^2-x.^2)).^(pi^2/12);
p1=plot(x,y1,'g'); hold on;
p2=plot(x,y2,'b.');
p3=plot(x,y3,'r-.');
set(p2,'Markersize',4);
set(p3,'Markersize',4);
```

14. **Fig. 3.8**
```
clear;
x=0:0.01:pi/2;
y1=tanh(x)./x;
y2=ones(1,length(x));
y3=((pi^2-4*x.^2)./(pi^2+4*x.^2)).^(pi^2/24);
p1=plot(x,y1,'g'); hold on;
p2=plot(x,y2,'b.');
p3=plot(x,y3,'r-.');
set(p2,'Markersize',4);
set(p3,'Markersize',4);
ylim([0 1.1]);
```
15. **Fig. 3.9**
```
clear;
x=0:0.01:1;
g=0.577215665;
y1=gamma(x)-(1-x)./(1+x).*(exp((1-g)*x)./(x));
p1=plot(x,y1,'b');
set(p1,'Markersize',4);
```

References

1. M. ABRAMOWITZ and I.A. STEGUN (Eds.): Handbook of Mathematical Functions with Formulas, Graphs, and Mathematical Tables. Dover Publications, New York, 1965.
2. A.P. ACHARYA: Univalence Criteria for Analytic Functions and Applications to Hypergeometric Functions. Ph.D. thesis, Universität Würzburg, Würzburg, 1997.
3. J. ACZÉL: Lectures on Functional Equations and Their Applications. Academic Press, New York, 1966.
4. L.V. AHLFORS: Classical Analysis: Present and Future. A. Baernstein II, D. Drasin, P.L. Duren, and A. Marden (Eds.): Proc. Congr. on the Occasion of the Solution of the Bieberbach Conjecture. Math. Surveys Monographs 21, Amer. Math. Soc., Providence, RI, (1986), 1–6.
5. J.W. ALEXANDER: Functions which map the interior of the unit circle upon simple regions. Ann. of Math. 17 (1915), 12–29.
6. R.M. ALI and V. SINGH: Convexity and starlikeness of functions defined by a class of integral operators. Complex Variables Theory Appl. 26(4) (1995), 299–309.
7. G. ALMKVIST and B.C. BERNDT: Gauss, Landen, Ramanujan, the arithmetic-geometric mean, ellipses, π, and the Ladies Diary. Amer. Math. Monthly 95 (1988), 585–608.
8. H. ALZER and S. KOUMANDOS: Sub- and superadditive properties of Fejér's sine polynomial. Bull. London Math. Soc. 38(2) (2006), 261–268.
9. H. ALZER and S.L. QIU: Monotonicity theorems and inequalities for the complete elliptic integrals. J. Comput. Appl. Math. 172(2) (2004), 289–312.
10. D.E. AMOS: Computation of modified Bessel functions and their ratios. Math. Comp. 28 (1974), 239–251.
11. G.D. ANDERSON, R.W. BARNARD, K.C. RICHARDS, M.K. VAMANAMURTHY and M. VUORINEN: Inequalities for zero-balanced hypergeometric functions. Trans. Amer. Math. Soc. 347 (1995), 1713–1723.
12. G.D. ANDERSON, S.L. QIU, M.K. VAMANAMURTHY and M. VUORINEN: Generalized elliptic integrals and modular equations. Pacific J. Math. 192 (2000), 1–37.
13. G.D. ANDERSON, S.L. QIU and M. VUORINEN: Modular equations and distortion functions. Ramanujan J. 18(2) (2009), 147–169.
14. G.D. ANDERSON, T. SUGAWA, M.K. VAMANAMURTHY and M. VUORINEN: Hypergeometric functions and hyperbolic metric. Comput. Methods Funct. Theory 9(1) (2009), 269–284.
15. G.D. ANDERSON, M.K. VAMANAMURTHY and M. VUORINEN: Functional inequalities for complete elliptic integrals and their ratios. SIAM J. Math. Anal. 21(2) (1990), 536–549.
16. G.D. ANDERSON, M.K. VAMANAMURTHY and M. VUORINEN: Functional inequalities for hypergeometric functions and complete elliptic integrals. SIAM J. Math. Anal. 23(2) (1992), 512–524.
17. G.D. ANDERSON, M.K. VAMANAMURTHY and M. VUORINEN: Hypergeometric functions and elliptic integrals. In H.M. Srivastava, S. Owa (Eds.): Current Topics in Analytic Function Theory. World Scientific, Singapore/London, (1992), 48–85.

18. G.D. ANDERSON, M.K. VAMANAMURTHY and M. VUORINEN: Inequalities for quasicon-formal mappings in space. Pacific J. Math. 160(1) (1993), 1–18.

19. G.D. ANDERSON, M.K. VAMANAMURTHY and M. VUORINEN: Conformal Invariants, Inequalities, and Quasiconformal Maps. Wiley, New York, 1997.

20. G.D. ANDERSON, M.K. VAMANAMURTHY and M. VUORINEN: Topics in special functions. Papers on analysis, 5–26, Rep. Univ. Jyvskyl Dep. Math. Stat., 83, Univ. Jyvskyl, Jyvskyl, 2001.

21. G.D. ANDERSON, M.K. VAMANAMURTHY and M. VUORINEN: Generalized convexity and inequalities. J. Math. Anal. Appl. 335(2) (2007), 1294–1308.

22. G.D. ANDERSON, M.K. VAMANAMURTHY and M. VUORINEN: Topics in special functions. II. Conform. Geom. Dyn. 11 (2007), 250–270.

23. S. ANDRÁS and Á. BARICZ: Properties of the probability density function of the non-central chi-squared distribution. J. Math. Anal. Appl. 346(2) (2008), 395–402.

24. S. ANDRÁS and Á. BARICZ: Monotonicity property of generalized and normalized Bessel functions of complex order. Complex Var. Elliptic Equ. 54(7) (2009), 689–696.

25. S. ANDRÁS and Á. BARICZ: Bounds for complete elliptic integrals of the first kind. Expo. Math. (2010), doi:10.1016/j.exmath.2009.12.005.

26. G.E. ANDREWS, R. ASKEY and R. ROY: Special Functions. Cambridge University Press, 1999.

27. T.M. APOSTOL: Calculus. Vol. I: One-variable calculus, with an introduction to linear algebra. Blaisdell Publishing Co. Ginn and Co., Waltham, Mass.-Toronto, Ont.-London, 1967.

28. R. ASKEY: Grünbaum's inequality for Bessel functions. J. Math. Anal. Appl. 41 (1973), 122–124.

29. K.E. ATKINSON: An Introduction to Numerical Analysis. 2nd ed. Wiley, New York, 1989.

30. G. AUMANN: Konvexe Funktionen und Induktion bei Ungleichungen zwischen Mittelverten, Bayer. Akad. Wiss. Math.-Natur. Kl. Abh., Math. Ann. 109 (1933), 405–413.

31. F.G. AVHADIEV and L.A. AKSENTEV: Fundamental results on sufficient conditions for the univalence of analytic functions. Uspehi Mat. Nauk 30 (1975), 3–60. English translation: Russian Math. Surveys 30 (1975), 1–63.

32. R. BALASUBRAMANIAN, S. NAIK, S. PONNUSAMY and M. VUORINEN: Elliott's identity and hypergeometric functions. J. Math. Anal. Appl. 271(1) (2002), 232–256.

33. R. BALASUBRAMANIAN, S. PONNUSAMY and D.J. PRABHAKARAN: Duality techniques for certain integral transforms to be starlike. J. Math. Anal. Appl. 293 (2004), 355–373.

34. R. BALASUBRAMANIAN, S. PONNUSAMY and D.J. PRABHAKARAN: Convexity of integral transforms and function spaces. Integral Transforms Spec. Funct. 18 (2007), 1–14.

35. R. BALASUBRAMANIAN, S. PONNUSAMY and D.J. PRABHAKARAN: On extremal prob-lems related to integral transforms of a class of analytic functions. J. Math. Anal. Appl. 336(1) (2007), 542–555.

36. R. BALASUBRAMANIAN, S. PONNUSAMY and M. VUORINEN: Functional inequalities for the quotients of hypergeometric functions. J. Math. Anal. Appl. 218 (1998), 256–268.

37. R. BALASUBRAMANIAN, S. PONNUSAMY and M. VUORINEN: On hypergeometric func-tions and function spaces. J. Comput. Appl. Math. 139(2) (2002), 299–322.

38. Á. BARICZ: Applications of the admissible functions method for some differential equations. Pure Math. Appl. 13(4) (2002), 433–440.

39. Á. BARICZ: Landen-type inequality for Bessel functions. Comput. Method Funct. Theory 5(2) (2005), 373–379.

40. Á. BARICZ: Univalent functions in simply connected domains. Libertas Math. 25 (2005), 97–103.

41. Á. BARICZ: Functional inequalities involving special functions. J. Math. Anal. Appl. 319(2) (2006), 450–459.

42. Á. BARICZ: Geometric properties of generalized Bessel functions of complex order. Mathe-matica 48(71)(1) (2006), 13–18.

43. Á. BARICZ: Bessel transforms and Hardy space of generalized Bessel functions. Mathematica 48(71)(2) (2006), 127–136.

44. Á. BARICZ: Grünbaum-type inequalities for special functions. J. Inequal. Pure Appl. Math. 7(5) (2006), Article 175, 8 pp. (electronic).

45. Á. BARICZ: Functional inequalities involving special functions II. J. Math. Anal. Appl. 327(2) (2007), 1202–1213.

46. Á. BARICZ: Functional inequalities for Galué's generalized modified Bessel functions. J. Math. Inequal. 1(2) (2007), 183–193.

47. Á. BARICZ: Redheffer type inequality for Bessel functions. J. Inequal. Pure Appl. Math. 8(1) (2007), Article 11, 6 pp. (electronic).

48. Á. BARICZ: Convexity of the zero-balanced Gaussian hypergeometric functions with respect to Hölder means. J. Inequal. Pure Appl. Math. 8(2) (2007), Article 40, 9 pp. (electronic).

49. Á. BARICZ: Some inequalities involving generalized Bessel functions. Math. Inequal. Appl. 10(4) (2007), 827–842.

50. Á. BARICZ: Turán type inequalities for generalized complete elliptic integrals. Math. Z. 256(4) (2007), 895–911.

51. Á. BARICZ: Functional inequalities involving Bessel and modified Bessel functions of the first kind. Expo. Math. 26(3) (2008), 279–293.

52. Á. BARICZ: Geometric properties of generalized Bessel functions. Publ. Math. Debrecen 73 (2008), 155–178.

53. Á. BARICZ: Jordan-type inequalities for generalized Bessel functions. J. Inequal. Pure Appl. Math. 9(2) (2008), Article 39, 6 pp. (electronic).

54. Á. BARICZ: Generalized Bessel Functions of the First Kind. Ph.D. thesis, Babeş-Bolyai University, Cluj-Napoca, 2008.

55. Á. BARICZ: Turán type Inequalities for some Special Functions. Ph.D. thesis, University of Debrecen, Debrecen, 2008.

56. Á. BARICZ: Tight bounds for the generalized Marcum Q−function. J. Math. Anal. Appl. 360(1) (2009), 265–277.

57. Á. BARICZ: Geometrically concave univariate distributions. J. Math. Anal. Appl. 363(1) (2010), 182–196.

58. Á. BARICZ and E. NEUMAN: Inequalities involving generalized Bessel functions. J. Inequal. Pure Appl. Math. 6(4) (2005), Article 126, 9 pp. (electronic).

59. Á. BARICZ and E. NEUMAN: Inequalities involving modified Bessel functions of the first kind II. J. Math. Anal. Appl. 332(1) (2007), 256–271.

60. Á. BARICZ and J. SÁNDOR: Extensions of the generalized Wilker inequality to Bessel functions. J. Math. Inequal. 2(3) (2008), 397–406.

61. Á. BARICZ and Y. SUN: New bounds for the generalized Marcum Q−function. IEEE Trans. Inform. Theory 55(7) (2009), 3091–3100.

62. Á. BARICZ and S. WU: Sharp Jordan-type inequalities for Bessel functions. Publ. Math. Debrecen 74 (2009), 107–126.

63. Á. BARICZ and S. WU: Sharp exponential Redheffer-type inequalities for Bessel functions. Publ. Math. Debrecen 74 (2009), 257–278.

64. Á. BARICZ and L. ZHU: Extension of Oppenheim's problem to Bessel functions. J. Inequal. Appl. 2007 (2007), Art. 82038.

65. R.W. BARNARD, S. NAIK and S. PONNUSAMY: Univalency of weighted integral transforms of certain functions. J. Comput. Appl. Math. 193(2) (2006), 638–651.

66. J. BECKER: Löwnersche Differentialgleichung und quasikonform fortsetzbare schlichte Funktionen. J. Reine Angew. Math. 255 (1972), 23–43.

67. J. BECKER and C. POMMERENKE: Schlichtheitskriterien und Jordangebiete. J. Reine Angew. Math. 354 (1984), 74–94.

68. B.C. BERNDT: Ramanujan's Notebooks, Part I, Springer-Verlag, New York, 1985.

69. B.C. BERNDT: Ramanujan's Notebooks, Part II, Springer-Verlag, New York, 1987.

70. B.C. BERNDT: Ramanujan's Notebooks, Part III, Springer-Verlag, New York, 1991.

71. B.C. BERNDT: Ramanujan's Notebooks, Part IV, Springer-Verlag, New York, 1994.

72. B.C. BERNDT: Ramanujan's Notebooks, Part V, Springer-Verlag, New York, 1998.

73. B.C. BERNDT, S. BHARGAVA and F.G. GARVAN: Ramanujan's theories of elliptic functions to alternative bases. Trans. Amer. Math. Soc. 347 (1995), 4163-4244.

74. M. BIERNACKI and J. KRZYŻ: On the monotonity of certain functionals in the theory of analytic functions. Ann. Univ. Mariae Curie-Skłodowska, Sect. A. 9 (1955), 135–147.

75. D. BORWEIN, J. BORWEIN, G. FEE and R. GIRGENSOHN: Refined convexity and special cases of the Blascke-Santalo inequality. Math. Inequal. Appl. 5 (2002), 631–638.

76. R. BOUCEKKINE and J.R. RUIZ-TAMARIT: Special functions for the study of economic dynamics: the case of the Lucas-Uzawa model. J. Math. Econom. 44(1) (2008), 33–54.

77. L. DE BRANGES: A proof of the Bieberbach conjecture. Acta Math. 154 (1985), 137–152.

78. R.K. BROWN: Univalence of Bessel functions. Proc. Amer. Math. Soc. 11(2) (1960), 278–283.

79. R.K. BROWN: Univalent solutions of $W'' + pW = 0$. Canad. J. Math. 14 (1962), 69–78.

80. B.C. CARLSON and D.B. SHAFFER: Starlike and prestarlike hypergeometric functions. SIAM J. Math. Anal. 15 (1984), 737–745.

81. C.P. CHEN, J.W. ZHAO and F. QI: Three inequalities involving hyperbolically trigonometric functions. Octogon Math. Mag. 12(2) (2004), 592–596. RGMIA Res. Rep. Coll. 6(3) (2003), Article 4, 6 pp. (electronic).

82. J.H. CHOI, Y.C. KIM, S. PONNUSAMY and T. SUGAWA: Norm estimates for the Alexander transforms of convex functions of order alpha. J. Math. Anal. Appl. 303(2) (2005), 661–668.

83. J.H. CHOI, Y.C. KIM and M. SAIGO: Geometric properties of convolution operators defined by Gaussian hypergeometric functions. Integral Transforms Spec. Funct. 13(2) (2002), 117–130.

84. J.H. CHOI, Y.C. KIM and H.M. SRIVASTAVA: Convex and starlike generalized hypergeometric functions associated with the Hardy space. Complex Variables Theory Appl. 31(4) (1996), 345–355.

85. Y. CHU, G. WANG, X. ZHANG and S.L. QIU: Generalized convexity and inequalities involving special functions. J. Math. Anal. Appl. 336(2) (2007), 768–776.

86. Z. DARÓCZY: On a class of means of two variables. Publ. Math. Debrecen 55 (1999), 177–197.

87. A. DEAÑO, A. GIL and J. SEGURA: New inequalities from classical Sturm theorems. J. Approx. Theory 131 (2004), 208–230.

88. L. DEBNATH and C.J. ZHAO: New strengthened Jordan's inequality and its applications. Appl. Math. Lett. 16 (2003), 557–560.

89. P.L. DUREN: Theory of $\mathscr{H}^p$ Spaces. Academic Press, New York/London, 1970.

90. P.L. DUREN: Univalent Functions. Grundlehren Math. Wiss. 259, Springer, New York, 1983.

91. J. DUTKA: The early history of the hypergeometric function. Arch. Hist. Exact Sci. 31 (1984), 15–34.

92. P.J. EENIGENBURG: A class of starlike mappings of the unit disk. Compositio Math. 24 (1972), 235–238.

93. P.J. EENIGENBURG and F.R. KEOGH: The Hardy class of some univalent functions and their derivatives. Michigan Math J. 17 (1970), 335–346.

94. Á. ELBERT and A. LAFORGIA: Monotonicity properties of the zeros of Bessel functions. SIAM J. Math. Anal. 17(6) (1986), 1483-1488.

95. Á. ELBERT and P.D. SIAFARIKAS: On the square of the first zero of the Bessel function $J_v(z)$. Canad. Math. Bull. 42(1) (1999), 56–67.

96. E.B. ELLIOTT: A formula including Legendre's $EK' + KE' - KK' = \frac{1}{2}\pi$. Messenger Math. 33 (1904), 31–40.

97. R.J. EVANS: Ramanujan's second notebook: asymptotic expansions for hypergeometric series and related functions. G.E. Andrews, R.A. Askey, B.C. Berndt, R.G. Ramanathan, R.A. Rankin (Eds.): Proc. of the Centenary Conference Univ. of Illinois at Urbana-Champaign. Academic Press, Boston, (1988), 537–560.

98. L. FEJÉR: Untersuchungen über Potenzreihen mit mehrfach monotoner Koeffizientenfolge. Acta Litterarum ac Scientiarum 8 (1936), 89–115.

99. R. FOURNIER and S. RUSCHEWEYH: On two extremal problems related to univalent functions. Rocky Mountain J. Math. 24(2) (1994), 529–538.

100. C. GIORDANO, A. LAFORGIA and J. PEČĂRIĆ: Supplement to known inequalities for some special functions. J. Math. Anal. Appl. 200 (1996), 34–41.

101. A.W. GOODMAN: Univalent Functions. Vol. I. Mariner Publishing Co., Tampa, Florida, 1983.
102. A.W. GOODMAN: Univalent Functions. Vol. II. Mariner Publishing Co., Tampa, Florida, 1983.
103. D. GRONAU and J. MATKOWSKI: Geometrical convexity and generalizations of the Bohr-Mollerup theorem on the gamma function. Math. Pannon. 4 (1993), 153–160.
104. D. GRONAU and J. MATKOWSKI: Geometrically convex solutions of certain difference equations and generalized Bohr-Mollerup type theorems. Results. Math. 26 (1994), 290–297.
105. F.A. GRÜNBAUM: A property of Legendre polynomials. Proc. Nat. Acad. Sci. 67 (1970), 959–960.
106. F.A. GRÜNBAUM: Linearization for the Boltzmann equation. Trans. Amer. Math Soc. 165 (1972), 425–449.
107. F.A. GRÜNBAUM: On the existence of a "wave operator" for the Boltzmann equation. Bull. Amer. Math. Soc. 78(5) (1972), 759–762.
108. F.A. GRÜNBAUM: A new kind of inequality for Bessel functions. J. Math. Anal. Appl. 41 (1973), 115–121.
109. B.N. GUO and F. QI: Some bounds for the complete elliptic integrals of the first and second kinds. http://arxiv.org/PS_cache/arxiv/pdf/0905/0905.2787v1.pdf.
110. D. HAMMARWALL: Resource Allocation in Multi–Antenna Communication Systems with Limited Feedback. Doctoral Thesis in Telecommunications, Stockholm, Sweden, 2007.
111. P.A. HANSEN: Ermittelung der absoluten Strungen in Ellipsen von beliebiger Excentricitt und Neigung, I. Schriften der Sternwarte Seeberg. Gotha, 1843.
112. G.H. HARDY, J.E. LITTLEWOOD and G. PÓLYA: Inequalities. Cambridge Mathematical Library, 1952.
113. V. HEIKKALA, H. LINDÉN, M.K. VAMANAMURTHY and M. VUORINEN: Generalized elliptic integrals and the Legendre $\mathscr{M}$-function. J. Math. Anal. Appl. 338(1) (2008), 223–243.
114. V. HEIKKALA, M.K. VAMANAMURTHY and M. VUORINEN: Generalized elliptic integrals. Comput. Methods Funct. Theory 9(1) (2009), 75–109.
115. E. HILLE: Remarks on a paper by Zeev Nehari. Bull. Amer. Math. Soc. 55 (1949), 552–553.
116. E.K. IFANTIS and P.D. SIAFARIKAS: A differential equation for the zeros of Bessel functions. Applicable Anal. 20 (1985), 269–281.
117. E.K. IFANTIS and P.D. SIAFARIKAS: Ordering relations between the zeros of miscellaneous Bessel functions. Applicable Anal. 23 (1986), 85–110.
118. E.K. IFANTIS and P.D. SIAFARIKAS: Inequalities involving Bessel and modified Bessel functions. J. Math. Anal. Appl. 147(1) (1990), 214–227.
119. M.E.H. ISMAIL: Remarks on a Paper by Giordano, Laforgia, and Pečarić. J. Math. Anal. Appl. 211 (1997), 621–625.
120. M.E.H. ISMAIL and M.E. MULDOON: On the variation with respect to a parameter of zeros of Bessel and $q-$Bessel functions. J. Math. Anal. Appl. 135(1) (1988), 187–207.
121. M.E.H. ISMAIL and M.E. MULDOON: Bounds for the small real and purely imaginary zeros of Bessel and related functions. Methods Appl. Anal. 2(1) (1995), 1–21.
122. A.P. IUSKEVICI: History of Mathematics in 16th and 17th Centuries. Moskva, 1961.
123. P. IVÁDY: A note on a gamma function inequality. J. Math. Inequal. 3(2) (2009), 227–236.
124. I.S. JACK: Functions starlike and convex of order α. J. London Math. Soc. 3(2) (1971), 469–474.
125. C.M. JOSHI and S.K. BISSU: Some inequalities of Bessel and modified Bessel functions. J. Austral. Math. Soc. (Series A) 50 (1991), 333–342.
126. C.M. JOSHI and S.K. BISSU: Inequalities for some special functions. J. Comput. Appl. Math. 69 (1996), 251–259.
127. J. KAMPÉ DE FÉRIET: La Fonction Hypergéométrique. Gauthier-Villars, Paris, 1937.
128. S. KANAS and J. STANKIEWICZ: Univalence of confluent hypergeometric functions. Ann. Univ. Marie-Curie Skłodowska 52(7) (1998), 51–56.
129. W. KAPLAN: Close to convex schlicht functions. Michigan Math. J. 2(1) (1952), 169–185.
130. E.A. KARATSUBA and M. VUORINEN: On hypergeometric functions and generalizations of Legendre's relation. J. Math. Anal. Appl. 260(2) (2001) 623–640.

131. H. KAZI and E. NEUMAN: Inequalities and bounds for elliptic integrals. J. Approx. Theory 146(2) (2007), 212–226.

132. H. KAZI and E. NEUMAN: Inequalities and bounds for elliptic integrals. II. Special functions and orthogonal polynomials, 127–138, Contemp. Math., 471, Amer. Math. Soc., Providence, RI, 2008.

133. Y.C. KIM and S. PONNUSAMY: Sufficiency of Gaussian hypergeometric functions to be uniformly convex. Internat. J. Math. Math. Sci. 22(4) (1999), 765–773.

134. Y.C. KIM and F. RØNNING: Integral transforms of certain subclasses of analytic functions. J. Math. Anal. Appl. 258 (2001), 466–489.

135. N. KISHORE: The Rayleigh function. Proc. Amer. Math. Soc. 14 (1963), 527–533.

136. H. KOBER: Approximation by integral functions in the complex domain. Trans. Amer. Math. Soc. 56(22) (1944), 7–31.

137. S. KOUMANDOS: Some inequalities for the sine integral. J. Inequal. Pure Appl. Math. 6(1) (2005), Article 25, 5 pp. (electronic).

138. E. KREYSZIG and J. TODD: The radius of univalence of Bessel functions. Illinois J. Math. 4 (1960), 143–149.

139. R. KÜSTNER: Mapping properties of hypergeometric functions and convolutions of starlike or convex functions of order α. Comput. Methods Funct. Theory 2(2) (2002), 597–610.

140. R. KÜSTNER: On the order of starlikeness of the shifted Gauss hypergeometric function. J. Math. Anal. Appl. 334(2) (2007), 1363–1385.

141. L. LORCH and M.E. MULDOON: An inequality for concave functions with applications to Bessel and trigonometric functions. Facta. Univ. Ser. Math. Inform. 2 (1987), 29–34.

142. T.H. MACGREGOR: The radius of convexity for starlike functions of order $1/2$. Proc. Amer. Math. Soc. 14 (1963), 71–76.

143. A. MAHAJAN: A Bessel function inequality. Univ. Beograd Publ. Elektrotehn. Fak. Ser. Mat. Fiz. 634 (1979), 70–71.

144. J. MATKOWSKI: Invariant and complementary quasi-arithmetic means. Aequationes Math. 57 (1999), 87–107.

145. J. MATKOWSKI: Convex functions with respect to a mean and a characterization of quasi-arithmetic means. Real Anal. Exchange 29 (2003/2004), 229–246.

146. J. MATKOWSKI and J. RÄTZ: Convexity of power functions with respect to symmetric homogeneous means. Internat. Ser. Numer. Math. 123 (1997), 231–247.

147. J. MATKOWSKI and J. RÄTZ: Convexity with respect to an arbitrary mean. Internat. Ser. Numer. Math. 123 (1997), 249–258.

148. A. MCD. MERCER: Grünbaum's inequality for Bessel functions and its extensions. SIAM J. Math. Anal. 6(6) (1975), 1021–1023.

149. A. MCD. MERCER: Integral representations and inequalities for Bessel functions. SIAM J. Math. Anal. 8(3) (1977), 486–490.

150. E.P. MERKES, M.S. ROBERTSON and W.T. SCOTT: On products of starlike functions. Proc. Amer. Math. Soc. 13 (1962), 960–964.

151. E.P. MERKES and W.T. SCOTT: Starlike hypergeometric functions. Proc. Amer. Math. Soc. 12 (1961), 885–888.

152. S.S. MILLER and P.T. MOCANU: Differential subordinations and univalent functions. Michigan Math. J. 28 (1981), 157–171.

153. S.S. MILLER and P.T. MOCANU: Differential subordinations and inequalities in the complex plane. J. Differential Equations 67 (1987), 199–211.

154. S.S. MILLER and P.T. MOCANU: Univalence of Gaussian and confluent hypergeometric functions. Proc. Amer. Math. Soc. 110 (1990), 333–342.

155. S.S. MILLER and P.T. MOCANU: Differential Subordinations. Theory and Applications. M. Dekker Inc., New York - Basel, 2000.

156. D.S. MITRINOVIĆ: Analytic Inequalities. Springer-Verlag, Berlin, 1970.

157. D.S. MITRINOVIĆ and I. LACKOVIĆ: Hermite and convexity. Aequat. Math. 28 (1985), 229–232.

158. Z. NEHARI: The Schwarzian derivative and schlicht functions. Bull. Amer. Math. Soc. 55 (1949), 545–551.

159. E. NEUMAN: Inequalities involving modified Bessel functions of the first kind. J. Math. Anal. Appl. 171 (1992), 532–536.
160. E. NEUMAN: Bounds for symmetric elliptic integrals. J. Approx. Theory 122(2) (2003), 249–259.
161. E. NEUMAN: Inequalities involving Bessel functions of the first kind. J. Inequal. Pure Appl. Math. 5(4) (2004), Article 94, 4 pp. (electronic).
162. I.R. NEZHMETDINOV and S. PONNUSAMY: New coefficient conditions for the starlikeness of analytic functions and their applications. Houston J. Math. 31(2) (2005), 587–604.
163. C.P. NICULESCU: Convexity according to the geometric mean. Math. Inequal. Appl. 3(2) (2000), 155–167.
164. D.W. NIU: Generalizations of Jordan's Inequality and Applications. M.Sc. Thesis, Henan Polytechnic University, 2007.
165. D.W. NIU, Z.H. HUO, J. CAO and F. QI: A general refinement of Jordan's inequality and a refinement of L. Yang's inequality. Integral Transforms Spec. Funct. 19(3) (2008), 157–164.
166. K. NOSHIRO: On the theory of schlicht functions. J. Fac. Sci. Hokkaido Univ. 2 (1935), 129–155.
167. S. OZAKI: On the theory of multivalent functions. Sci. Rep. Tokyo Bunrika Daigaku A 40(2) (1935), 167–188.
168. S. OWA and H.M. SRIVASTAVA: Univalent and starlike generalized hypergeometric functions. Canad. J. Math. 39(5) (1987), 1057–1077.
169. N.N. PASCU, D. RĂDUCANU, M.N. PASCU and R.N. PASCU: On convex function in an elliptical domain. Studia Univ. Babeş-Bolyai Math. 46(2) (2001), 97–100.
170. M. PETROVIĆ: Sur une fonctionnelle. Publ. Math. Univ. Belgrade 1 (1932), 149–156.
171. R. PIESSENS: A series expansion for the first positive zero of Bessel function. Math. Comp. 42 (1984), 195–197.
172. B. PINCHUK: On starlike and convex functions of order α. Duke Math. J. 35 (1968), 721–734.
173. C. POMMERENKE: Univalent Functions. Vandenhoeck & Ruprecht, Gttingen, 1975.
174. C. POMMERENKE: Boundary Behaviour of Conformal Maps. Springer-Verlag, Berlin, 1992.
175. S. PONNUSAMY: The Hardy space of hypergeometric functions. Complex Variables Theory Appl. 29 (1996), 83–96.
176. S. PONNUSAMY: Univalence of Alexander transform under new mapping properties. Complex Variables Theory Appl. 30(1) (1996), 55–68.
177. S. PONNUSAMY: Close-to-convexity properties of Gaussian hypergeometric functions. J. Comput. Appl. Math. 88 (1997), 327–337.
178. S. PONNUSAMY: Hypergeometric transforms of functions with derivative in a half plane. J. Comput. Appl. Math. 96 (1998), 35–49.
179. S. PONNUSAMY: Starlikeness properties for convolutions involving hypergeometric series. Ann. Univ. Marie Curie-Skłodowska, Sect. A 52 (1998), 1–16.
180. S. PONNUSAMY: Inclusion theorems for convolution product of second order polylogarithms and functions with the derivative in a halfplane. Rocky Mountain J. Math. 28(2) (1998), 695–733.
181. S. PONNUSAMY and F. RØNNING: Duality for Hadamard products applied to certain integral transforms. Complex Variables Theory Appl. 32(3) (1997), 263–287.
182. S. PONNUSAMY and F. RØNNING: Starlikeness properties for convolutions involving hypergeometric series. Ann. Univ. Mariae Curie-Skłodowska Sect. A 52(1) (1998), 141–155.
183. S. PONNUSAMY and F. RØNNING: Geometric properties for convolutions of hypergeometric functions and functions with the derivative in a halfplane. Integral Transforms Special Funct. 8 (1999), 121–138.
184. S. PONNUSAMY and F. RØNNING: Integral transforms of functions with the derivative in a halfplane. Israel J. Math. 114 (1999), 177–188.
185. S. PONNUSAMY and S. SABAPATHY: Geometric properties of generalized hypergeometric functions. Ramanujan J. 1(2) (1997), 187–210.
186. S. PONNUSAMY and M. VUORINEN: Asymptotic expansions and inequalities for hypergeometric functions. Mathematika 44 (1997), 43–64.

187. S. PONNUSAMY and M. VUORINEN: Univalence and convexity properties for confluent hypergeometric functions. Complex Variables Theory Appl. 36 (1998), 73–97.

188. S. PONNUSAMY and M. VUORINEN: Univalence and convexity properties of Gaussian hypergeometric functions. Rocky Mountain J. Math. 31 (2001), 327–353.

189. F. QI: Jordan's inequality: Refinements, generalizations, applications and related problems. RGMIA Res. Rep. Coll. 9(3) (2006), Article 12, 16 pp. (electronic). Bùděngshi Yánjiū Tōngxùn (Communications in Studies on Inequalities) 13(3) (2006), 243–259.

190. F. QI, L.H. CUI and S.L. XU: Some inequalities constructed by Tchebysheff's integral inequality. Math. Inequal. Appl. 2(4) (1999), 517–528.

191. F. QI and Q.D. HAO: Refinements and sharpenings of Jordan's and Kober's inequality. Mathematics and Informatics Quarterly 8(3) (1998), 116–120.

192. F. QI and D.W. NIU: Refinements, generalizations and applications of Jordan's inequality and related problems. RGMIA Res. Rep. Coll. 11(2) (2008), Article 9, 38 pp. (electronic).

193. F. QI, D.W. NIU and B.N. GUO: Refinements, generalizations, and applications of Jordan's inequality and related problems. J. Inequal. Appl. 2009 (2009), Art. 271923.

194. S.L. QIU and M. VUORINEN: Landen inequalities for hypergeometric functions. Nagoya Math. J. 154 (1999), 31–56.

195. S.L. QIU and M. VUORINEN: Duplication inequalities for the ratios of hypergeometric functions. Forum Math. 12(1) (2000), 109–133.

196. S.L. QIU and M. VUORINEN: Special functions in geometric function theory. R. Kühnau (Ed.): Handbook of complex analysis: geometric function theory. Vol. 2, Elsevier Sci. B. V., Amsterdam, (2005), 621–659.

197. E.D. RAINVILLE: Special Functions. Chelsea Publishing Company, New York, 1960.

198. R. REDHEFFER: Problem 5642. Amer. Math. Monthly 76 (1969), 422.

199. A.W. ROBERTS and D.E. VARBERG: Convex Functions, Academic Press, New York, 1973.

200. M.S. ROBERTSON: On the theory of univalent functions. Ann. of Math. 37 (1936), 374–408.

201. S. RUSCHEWEYH and V. SINGH: On the order of starlikeness of hypergeometric functions. J. Math. Anal. Appl. 113(1) (1986), 1–11.

202. J. SÁNDOR: On the concavity of $\sin x / x$. Octogon Math. Mag. 13(1) (2005), 404.

203. J. SÁNDOR: Selected Chapters of Geometry, Analysis and Number Theory: Classical Topics in New Perspectives. Lambert Academic Publishing, 2009.

204. J. SÁNDOR: A note on certain Jordan type inequalities. RGMIA Res. Rep. Coll. 10(1) (2007), Article 1, 11 pp. (electronic).

205. J. SÁNDOR and M. BENCZE: On Huygens' trigonometric inequality. RGMIA Res. Rep. Coll. 8(3) (2005), Article 14, 4 pp. (electronic).

206. A. SCHILD: On starlike functions of order α. Amer. J. Math. 87 (1965), 65–70.

207. O.X. SCHLÖMILCH: Über die Besselschen Functionen. Zeitschrift fr Mathematik und Physik 2 (1857), 137–165.

208. V. SELINGER: Geometric properties of normalized Bessel functions. Pure Math. Appl. 6 (1995), 273–277.

209. T.N. SHANMUGAM: Hypergeometric functions in the geometric function theory. Appl. Math. Comput. 187(1) (2007), 433–444.

210. H. SILVERMAN: Univalent functions with negative coefficients. Proc. Amer. Math. Soc. 51 (1975), 109–116.

211. H. SILVERMAN: Subclasses of starlike functions. Rev. Roumaine Math. Pures Appl. 23 (1978), 1093–1099.

212. H. SILVERMAN: Starlike and convexity properties for hypergeometric functions. J. Math. Anal. Appl. 172 (1993), 574–581.

213. H.C. SIMPSON and S.J. SPECTOR: Some monotonicity results for ratios of modified Bessel functions. Quart. Appl. Math. 42 (1984), 95–98.

214. H.C. SIMPSON and S.J. SPECTOR: On barelling for a special material in finite elasticity. Quart. Appl. Math. 42 (1984), 99–105.

215. H.M. SRIVASTAVA, G. MURUGUSUNDARAMOORTHY and S. SIVASUBRAMANIAN: Hypergeometric functions in the parabolic starlike and uniformly convex domains. Integral Transforms Spec. Funct. 18 (2007), 511–520.

216. K.R. STROMBERG: Introduction to Classical Real Analysis, Wadsworth, Belmont, 1981.
217. A. SWAMINATHAN: Certain sufficiency conditions on Gaussian hypergeometric functions. J. Inequal. Pure Appl. Math. 5(4) (2004), Article 83, 10 pp. (electronic).
218. A. SWAMINATHAN: Hypergeometric functions in the parabolic domain. Tamsui Oxf. J. Math. Sci. 20(1) (2004), 1–16.
219. A. SWAMINATHAN: Sufficiency for hypergeometric transforms to be associated with conic regions. Math. Comput. Modelling 44 (2006), 276–286.
220. A. SWAMINATHAN: Inclusion theorems of convolution operators associated with normalized hypergeometric functions. J. Comput. Appl. Math. 197 (2006), 15–28.
221. A. SWAMINATHAN: Convexity of the incomplete beta functions. Integral Transforms Spec. Funct. 18 (2007), 521–528.
222. G. SZEGŐ: Orthogonal Polynomials. 4th ed. American Mathematical Society, Providence, RI, 1975.
223. N.M. TEMME: Special Functions. An Introduction to the Classical Functions of Mathematical Physics. John Wiley & Sons, Inc., New York, 1996.
224. V.K. THIRUVENKATACHAR and T.S. NANJUNDIAH: Inequalities concerning Bessel functions and orthogonal polynomials. Proc. Indian Nat. Acad. Part A 33 (1951), 373–384.
225. T. TRIF: Convexity of the gamma function with respect to Hölder means. In: Inequality theory and applications. Nova Sci. Publ. Hauppauge, New York, 3 (2003), 189–195.
226. S.E. WARSCHAWSKI: On the higher derivatives at the boundary in conformal mapping. Trans. Amer. Math. Soc. 38 (1935), 310–340.
227. G.N. WATSON: A Treatise on the Theory of Bessel Functions. Cambridge University Press, Cambridge, 1944.
228. E.T. WHITTAKER and G.N. WATSON: A Course of Modern Analysis. 4th ed. Cambridge University Press, 1958.
229. H.S. WILF: The radius of univalence of certain entire functions. Illinois J. Math. 6 (1962), 242–244.
230. S. WU and L. DEBNATH: A new generalized and sharp version of Jordan's inequality and its applications to the improvement of the Yang Le inequality. Appl. Math. Lett. 19 (2006), 1378–1384.
231. S. WU and L. DEBNATH: A new generalized and sharp version of Jordan's inequality and its applications to the improvement of the Yang Le inequality. II. Appl. Math. Lett. 20(5) (2007), 532–538.
232. S. WU and L. DEBNATH: Jordan-type inequalities for differentiable functions and their applications. Appl. Math. Lett. 21(8) (2008), 803–809.
233. S. WU and L. DEBNATH: Generalizations of a parametrized Jordan-type inequality, Janous's inequality and Tsintsifas's inequality. Appl. Math. Lett. 22(1) (2009), 130–135.
234. S. WU and H.M. SRIVASTAVA: A further refinement of a Jordan type inequality and its application. Appl. Math. Comput. 197(2) (2008), 914–923.
235. S. WU, H.M. SRIVASTAVA and L. DEBNATH: Some refined families of Jordan-type inequalities and their applications. Integral Transforms Spec. Funct. 19 (2008), 183–193.
236. X. ZHANG, G. WANG and Y. CHU: Extensions and sharpenings of Jordan's and Kober's inequalities. J. Inequal. Pure Appl. Math. 7(2) (2006), Article 63, 3 pp. (electronic).
237. X. ZHANG, G. WANG and Y. CHU: Convexity with respect to Hölder mean involving zero-balanced hypergeometric functions. J. Math. Anal. Appl. 353(1) (2009), 256–259.
238. L. ZHU: A solution of a problem of Oppeheim. Math. Inequal. Appl. 10(1) (2007), 57–61.
239. L. ZHU: Sharpening of Jordan's inequalities and its applications. Math. Inequal. Appl. 9(1) (2006), 103–106.
240. L. ZHU: A general form of Jordan's inequality and its applications. Math. Inequal. Appl. 11(4) (2008), 655–665.
241. L. ZHU: A general refinement of Jordan-type inequality. Comput. Math. Appl. 55(11) (2008), 2498–2505.
242. L. ZHU: General forms of Jordan and Yang Le inequality. Appl. Math. Lett. 22(2) (2009), 236–241

243. L. ZHU and J. SUN: Six new Redheffer-type inequalities for circular and hyperbolic functions. Comput. Math. Appl. 56(2) (2008), 522–529.
244. L. ZHU: Jordan type inequalities involving the Bessel and modified Bessel functions. Comput. Math. Appl. 59(2) (2010), 724–736.
245. L. ZHU: A general form of Jordan-type double inequality for the generalized and normalized Bessel functions. Appl. Math. Comput. 215(11) (2010), 3802–3810.
246. A. ZYGMUND: Trigonometric Series. 3rd ed. Cambridge University Press, 2002.

Index

Lecture Notes in Mathematics

For information about earlier volumes
please contact your bookseller or Springer
LNM Online archive: springerlink.com

Vol. 1852: A.S. Kechris, B.D. Miller, Topics in Orbit Equivalence (2004)

Vol. 1853: Ch. Favre, M. Jonsson, The Valuative Tree (2004)

Vol. 1854: O. Saeki, Topology of Singular Fibers of Differential Maps (2004)

Vol. 1855: G. Da Prato, P.C. Kunstmann, I. Lasiecka, A. Lunardi, R. Schnaubelt, L. Weis, Functional Analytic Methods for Evolution Equations. Editors: M. Iannelli, R. Nagel, S. Piazzera (2004)

Vol. 1856: K. Back, T.R. Bielecki, C. Hipp, S. Peng, W. Schachermayer, Stochastic Methods in Finance, Bressanone/Brixen, Italy, 2003. Editors: M. Fritelli, W. Runggaldier (2004)

Vol. 1857: M. Émery, M. Ledoux, M. Yor (Eds.), Séminaire de Probabilités XXXVIII (2005)

Vol. 1858: A.S. Cherny, H.-J. Engelbert, Singular Stochastic Differential Equations (2005)

Vol. 1859: E. Letellier, Fourier Transforms of Invariant Functions on Finite Reductive Lie Algebras (2005)

Vol. 1860: A. Borisyuk, G.B. Ermentrout, A. Friedman, D. Terman, Tutorials in Mathematical Biosciences I. Mathematical Neurosciences (2005)

Vol. 1861: G. Benettin, J. Henrard, S. Kuksin, Hamiltonian Dynamics – Theory and Applications, Cetraro, Italy, 1999. Editor: A. Giorgilli (2005)

Vol. 1862: B. Helffer, F. Nier, Hypoelliptic Estimates and Spectral Theory for Fokker-Planck Operators and Witten Laplacians (2005)

Vol. 1863: H. Führ, Abstract Harmonic Analysis of Continuous Wavelet Transforms (2005)

Vol. 1864: K. Efstathiou, Metamorphoses of Hamiltonian Systems with Symmetries (2005)

Vol. 1865: D. Applebaum, B.V. R. Bhat, J. Kustermans, J. M. Lindsay, Quantum Independent Increment Processes I. From Classical Probability to Quantum Stochastic Calculus. Editors: M. Schürmann, U. Franz (2005)

Vol. 1866: O.E. Barndorff-Nielsen, U. Franz, R. Gohm, B. Kümmerer, S. Thorbjønsen, Quantum Independent Increment Processes II. Structure of Quantum Lévy Processes, Classical Probability, and Physics. Editors: M. Schürmann, U. Franz, (2005)

Vol. 1867: J. Sneyd (Ed.), Tutorials in Mathematical Biosciences II. Mathematical Modeling of Calcium Dynamics and Signal Transduction. (2005)

Vol. 1868: J. Jorgenson, S. Lang, $Pos_n(R)$ and Eisenstein Series. (2005)

Vol. 1869: A. Dembo, T. Funaki, Lectures on Probability Theory and Statistics. Ecole d'Eté de Probabilités de Saint-Flour XXXIII-2003. Editor: J. Picard (2005)

Vol. 1870: V.I. Gurariy, W. Lusky, Geometry of Müntz Spaces and Related Questions. (2005)

Vol. 1871: P. Constantin, G. Gallavotti, A.V. Kazhikhov, Y. Meyer, S. Ukai, Mathematical Foundation of Turbulent Viscous Flows, Martina Franca, Italy, 2003. Editors: M. Cannone, T. Miyakawa (2006)

Vol. 1872: A. Friedman (Ed.), Tutorials in Mathematical Biosciences III. Cell Cycle, Proliferation, and Cancer (2006)

Vol. 1873: R. Mansuy, M. Yor, Random Times and Enlargements of Filtrations in a Brownian Setting (2006)

Vol. 1874: M. Yor, M. Émery (Eds.), In Memoriam Paul-André Meyer - Séminaire de Probabilités XXXIX (2006)

Vol. 1875: J. Pitman, Combinatorial Stochastic Processes. Ecole d'Eté de Probabilités de Saint-Flour XXXII-2002. Editor: J. Picard (2006)

Vol. 1876: H. Herrlich, Axiom of Choice (2006)

Vol. 1877: J. Steuding, Value Distributions of L-Functions (2007)

Vol. 1878: R. Cerf, The Wulff Crystal in Ising and Percolation Models, Ecole d'Eté de Probabilités de Saint-Flour XXXIV-2004. Editor: Jean Picard (2006)

Vol. 1879: G. Slade, The Lace Expansion and its Applications, Ecole d'Eté de Probabilités de Saint-Flour XXXIV-2004. Editor: Jean Picard (2006)

Vol. 1880: S. Attal, A. Joye, C.-A. Pillet, Open Quantum Systems I, The Hamiltonian Approach (2006)

Vol. 1881: S. Attal, A. Joye, C.-A. Pillet, Open Quantum Systems II, The Markovian Approach (2006)

Vol. 1882: S. Attal, A. Joye, C.-A. Pillet, Open Quantum Systems III, Recent Developments (2006)

Vol. 1883: W. Van Assche, F. Marcellàn (Eds.), Orthogonal Polynomials and Special Functions, Computation and Application (2006)

Vol. 1884: N. Hayashi, E.I. Kaikina, P.I. Naumkin, I.A. Shishmarev, Asymptotics for Dissipative Nonlinear Equations (2006)

Vol. 1885: A. Telcs, The Art of Random Walks (2006)

Vol. 1886: S. Takamura, Splitting Deformations of Degenerations of Complex Curves (2006)

Vol. 1887: K. Habermann, L. Habermann, Introduction to Symplectic Dirac Operators (2006)

Vol. 1888: J. van der Hoeven, Transseries and Real Differential Algebra (2006)

Vol. 1889: G. Osipenko, Dynamical Systems, Graphs, and Algorithms (2006)

Vol. 1890: M. Bunge, J. Funk, Singular Coverings of Toposes (2006)

Vol. 1891: J.B. Friedlander, D.R. Heath-Brown, H. Iwaniec, J. Kaczorowski, Analytic Number Theory, Cetraro, Italy, 2002. Editors: A. Perelli, C. Viola (2006)

Vol. 1892: A. Baddeley, I. Bárány, R. Schneider, W. Weil, Stochastic Geometry, Martina Franca, Italy, 2004. Editor: W. Weil (2007)

Vol. 1893: H. Hanßmann, Local and Semi-Local Bifurcations in Hamiltonian Dynamical Systems, Results and Examples (2007)

Vol. 1894: C.W. Groetsch, Stable Approximate Evaluation of Unbounded Operators (2007)

Vol. 1895: L. Molnár, Selected Preserver Problems on Algebraic Structures of Linear Operators and on Function Spaces (2007)

Vol. 1896: P. Massart, Concentration Inequalities and Model Selection, Ecole d'Été de Probabilités de Saint-Flour XXXIII-2003. Editor: J. Picard (2007)

Vol. 1897: R. Doney, Fluctuation Theory for Lévy Processes, Ecole d'Été de Probabilités de Saint-Flour XXXV-2005. Editor: J. Picard (2007)

Vol. 1898: H.R. Beyer, Beyond Partial Differential Equations, On linear and Quasi-Linear Abstract Hyperbolic Evolution Equations (2007)

Vol. 1899: Séminaire de Probabilités XL. Editors: C. Donati-Martin, M. Émery, A. Rouault, C. Stricker (2007)

Vol. 1900: E. Bolthausen, A. Bovier (Eds.), Spin Glasses (2007)

Vol. 1901: O. Wittenberg, Intersections de deux quadriques et pinceaux de courbes de genre 1, Intersections of Two Quadrics and Pencils of Curves of Genus 1 (2007)

Vol. 1902: A. Isaev, Lectures on the Automorphism Groups of Kobayashi-Hyperbolic Manifolds (2007)

Recent Reprints and New Editions

Additional technical instructions, if necessary, are available on request from: lnm@springer.com.

4. Careful preparation of the manuscripts will help keep production time short besides ensuring satisfactory appearance of the finished book in print and online. After acceptance of the manuscript authors will be asked to prepare the final LaTeX source files and also the corresponding dvi-, pdf- or zipped ps-file. The LaTeX source files are essential for producing the full-text online version of the book (see http://www.springerlink.com/openurl.asp?genre=journal&issn=0075-8434 for the existing online volumes of LNM).

The actual production of a Lecture Notes volume takes approximately 12 weeks.

5. Authors receive a total of 50 free copies of their volume, but no royalties. They are entitled to a discount of 33.3% on the price of Springer books purchased for their personal use, if ordering directly from Springer.

6. Commitment to publish is made by letter of intent rather than by signing a formal contract. Springer-Verlag secures the copyright for each volume. Authors are free to reuse material contained in their LNM volumes in later publications: a brief written (or e-mail) request for formal permission is sufficient.

Addresses:
Professor J.-M. Morel, CMLA,
École Normale Supérieure de Cachan,
61 Avenue du Président Wilson, 94235 Cachan Cedex, France
E-mail: Jean-Michel.Morel@cmla.ens-cachan.fr

Professor F. Takens, Mathematisch Instituut,
Rijksuniversiteit Groningen, Postbus 800,
9700 AV Groningen, The Netherlands
E-mail: F.Takens@rug.nl

Professor B. Teissier, Institut Mathématique de Jussieu,
UMR 7586 du CNRS, Équipe "Géométrie et Dynamique",
175 rue du Chevaleret,
75013 Paris, France
E-mail: teissier@math.jussieu.fr

For the "Mathematical Biosciences Subseries" of LNM:

Professor P.K. Maini, Center for Mathematical Biology,
Mathematical Institute, 24-29 St Giles,
Oxford OX1 3LP, UK
E-mail: maini@maths.ox.ac.uk

Springer, Mathematics Editorial, Tiergartenstr. 17,
69121 Heidelberg, Germany,
Tel.: +49 (6221) 487-259
Fax: +49 (6221) 4876-8259
E-mail: lnm@springer.com